Design and Optimization of Biogas Energy Systems

Design and Optimization of Biogas Energy Systems

Prashant Baredar

Vikas Khare

Savita Nema

ELSEVIER

ACADEMIC PRESS

An imprint of Elsevier

Academic Press is an imprint of Elsevier
125 London Wall, London EC2Y 5AS, United Kingdom
525 B Street, Suite 1650, San Diego, CA 92101, United States
50 Hampshire Street, 5th Floor, Cambridge, MA 02139, United States
The Boulevard, Langford Lane, Kidlington, Oxford OX5 1GB, United Kingdom

British Library Cataloguing-in-Publication Data
A catalogue record for this book is available from the British Library

Library of Congress Cataloging-in-Publication Data
A catalog record for this book is available from the Library of Congress

ISBN: 978-0-12-822718-3

For Information on all Academic Press publications
visit our website at https://www.elsevier.com/books-and-journals

Publisher: Brian Romer
Acquisitions Editor: Peter Adamson
Editorial Project Manager: Andrae Akeh
Production Project Manager: Sojan P. Pazhayattil
Cover Designer:

Typeset by MPS Limited, Chennai, India

Contents

Chapter 1

Biogas energy system

Chapter outline

Objectives

- To provide knowledge about the importance of biogas energy systems in the field of electricity generation
- To provide knowledge about the worldwide evaluation of biogas technology
- To provide knowledge about the dissemination of biogas systems
- To provide knowledge about the pros and cons of the biogas energy system

1.1 Introduction

Biogas energy systems are one of the prominent energy sources of renewable energy (RE) systems, and they are utilized for electricity generation, where biogas energy is the energy generated or produced by living or once-living organisms. The energy from these organisms can be burned to create heat and converted into electricity. Biogas is a versatile energy source that can be used for the production of heat, power, transport fuels, and biomaterials, apart from making a significant contribution to climate change mitigation. Currently, biogas-driven combined heat and power, cofiring, and combustion plants provide reliable, efficient, and clean power that generates a lot of electrical energy for fulfilling the load demand. The feedstock for biogas energy plants can include residues from agriculture, forestry, wood processing and food processing industries, municipal solid wastes, industrial wastes, and biogas produced from degraded and marginal lands. The terms biogas energy, bioenergy, and biofuel cover any energy products derived from a plant or an

Design and Optimization of Biogas Energy Systems. DOI: https://doi.org/10.1016/B978-0-12-822718-3.00001-0

animal or an organic material. The increasing interest in biogas energy and biofuel has been the result of the following associated benefits:

- potential to reduce greenhouse gas (GHG) emissions;
- energy security benefits;
- substitution for diminishing global oil supplies;
- potential impacts on waste management strategy;
- capacity to convert a wide variety of wastes into clean energy; and
- technological advancement in thermal and biochemical processes for waste-to-energy transformation;

Biogas can play the pivotal role in the production of carbon-neutral fuels of high quality as well as providing feedstocks for various industries. This is a unique property of biogas compared with other REs, which makes biogas a prime alternative to the use of fossil fuels. The performance of biogas-based systems for heat and power generation has already been proved in many situations on commercial as well as domestic scales.

In addition, biogas wastes can also yield liquid fuels, such as cellulosic ethanol, which can be used to replace petroleum-based fuels. Cellulosic ethanol can be produced from grasses, wood chips, and agricultural residues by biochemical route using heat, pressure, chemicals, and enzymes unlocking the sugars in cellulosic biogas. Algal biogas is also emerging as a good source of energy, because it can serve as a natural source of oil, which conventional refineries can transform into jet fuel or diesel fuel. Biogas energy systems have the potential to address many environmental issues, especially global warming and GHG emissions, and foster sustainable development among poor communities. Biogas fuel sources are readily available in the rural and urban areas of all countries. Biogas-based industries can provide appreciable employment opportunities and promote biogas regrowth through sustainable land management practices. The negative aspects of traditional biogas utilization in developing countries can be mitigated by the promotion of modern waste-to-energy technologies that provide solid, liquid, and gaseous fuels as well as electricity for house appliances, commercial buildings, and different types of industries. Biogas wastes can be transformed into clean and efficient energy by biochemical as well as thermochemical technologies. Fig. 1.1 shows the types of biogases, which include different forms of biogas such as virgin wood, energy crops, agricultural waste, and waste from other materials. Virgin wood is also classified into tree surgery residues, forestry residues, and fuel wood, as shown in Fig. 1.2. Figs. 1.3–1.5 also show the types of energy crops, agricultural wastes, and other wastes, respectively.

Rather than thermochemical technologies, the most common technique for producing both heat and electrical energies from biogas wastes is direct combustion. In direct combustion, the thermal efficiencies as high as 80%–90% can be achieved by advanced gasification technology with greatly reduced atmospheric emissions. Combined heat and power (CHP) systems,

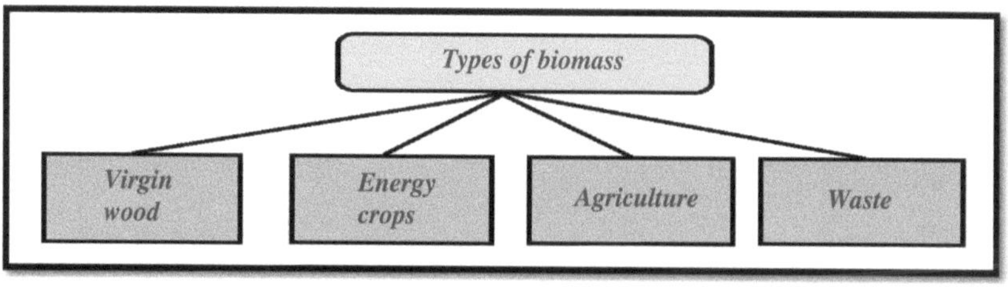

FIGURE 1.1 Different forms of biogas.

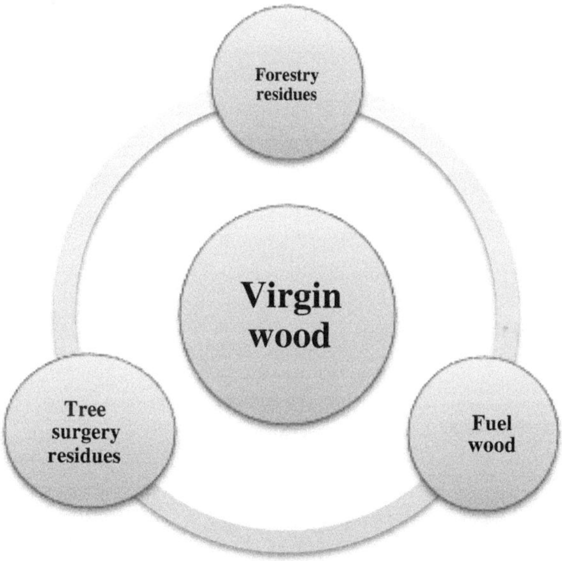

FIGURE 1.2 Types of virgin wood.

ranging from small-scale technology to large grid-connected facilities, provide significantly higher efficiencies than systems that only generate electricity. Biochemical processes, like anaerobic digestion and sanitary landfills, can also produce clean energy in the form of biogas and producer gas, which can be converted into power and heat using a gas engine. A biogas system also uses plant or animal material for energy generation, for heat production, or in various industrial processes as the raw material for a range of products. It can be purposely grown energy crops, wood or forest residues, waste from food crops (wheat straw and bagasse), horticulture (yard waste), food processing (corn cobs), animal farming (manure, rich in nitrogen and phosphorus), or human waste from sewage plants. Burning plant-derived biogas releases CO_2, but it has still been classified as an RE source in the European Union and United Nation legal frameworks, because photosynthesis cycles the CO_2 back into new crops. In some cases, this recycling of CO_2 from plants to the

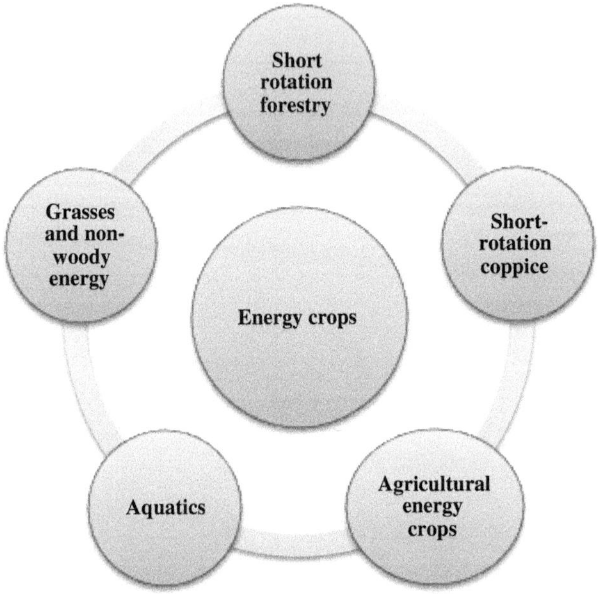

FIGURE 1.3 Types of energy crops.

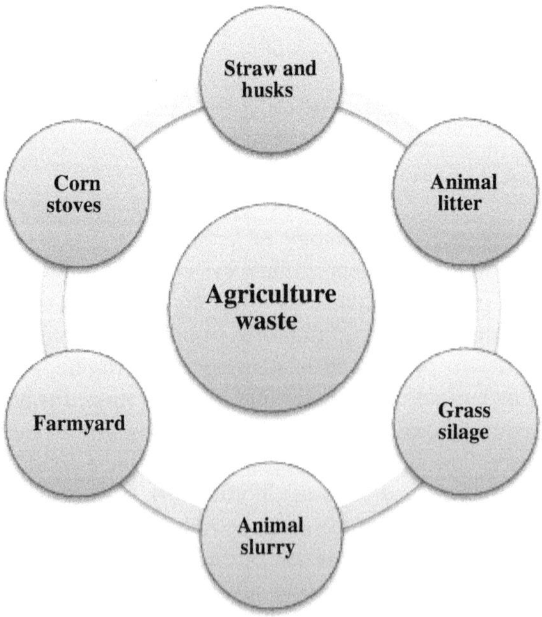

FIGURE 1.4 Types of agricultural wastes.

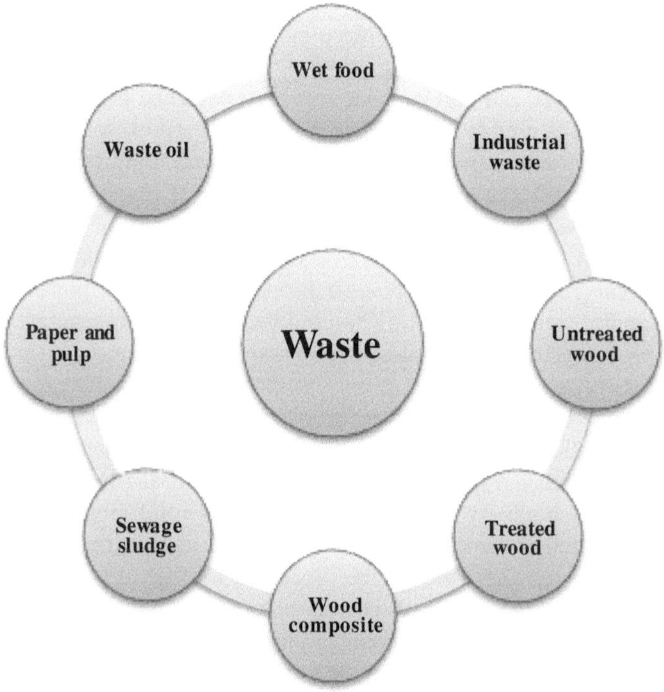

FIGURE 1.5 Types of wastes.

atmosphere and back into plants can even be CO_2 negative, as a relatively large portion of the CO_2 is moved into the soil during each cycle. Cofiring using biogas has increased in coal power plants, because it makes it possible to release less CO_2 without the cost associated with building new infrastructure. Cofiring is not without issues however; often an upgrade of the biogas has been the beneficiary.

There are four principal ways that are mentioned below, in which organic materials and crops can be used as a biogas or biofuel as part of an RE resource.

1. *Thermal combustion of biogas*: The burning of solid biogas materials, also called incineration, in the air is the most common use of biogas fuels today. The bioenergy released from this process is used directly for space and water heating, as well as for cooking and washing. Domestic thermal combustion of solid biogas materials for cooking and space heating may be an attractive alternative to conventional fossil fuels, especially where the biofuel is available at economic prices or, particularly, in isolated or rural areas, where the biofuel may be available for gathering by the consumer locally. Small-scale use, such as for home cooking and burning in open fireplaces, is usually very inefficient as most of the bioenergy

goes up the chimney as wasted heat. Highly efficient cooking stoves, home heating stoves, and fireplace systems have been developed especially for biogas energy combustion and are now widely available. Larger furnaces and boilers have also been designed for burning various types of solid biogas materials such as waste wood, wood chips, logs, and sawdust and nutshells either with or without high moisture content. The larger units can be very efficient over a wide range of particle sizes and compositions, nearly matching the performance of traditional oil- or gas-fired furnaces.

2. *Electrical generation uses biogas*: Biopower or biogas power uses the heat and/or steam produced by burning the feedstock to generate electricity. Most electricity generating stations and conventional power stations use fossil fuels in the form of coal as their primary fuel source. Current predictions indicate that the utilization of steam coal for power production worldwide will increase substantially over the next few decades, so the market demand for a high-efficiency, clean, coal-fired power generation plant is high. By premixing the coal with solid biogas feedstock, a new type of fuel can be produced by burning in existing coal-fired boilers. The mixed fuel can still be processed through the same coal handling, milling, and firing systems as before—the advantage now is that the cofiring of solid biogas reduces the reliance of generating plants on fossil fuels only, reducing its waste ash content and harmful sulfur and CO_2 carbon emissions. Both dedicated biogas and biogas cofiring power plants are used in the production of electricity, with large-scale biogas cofiring being one of the most efficient and cost-effective approaches to generating electricity from renewable sources. The main advantage of biogas cofiring power generation plants is that biogas is much cheaper than coal, so cofiring is cheaper than burning coal alone. Nearly all current power generation from biogas energy uses steam turbines. The cofiring of biogas and coal to generate the required amount of steam to drive the turbines produces gases that can be captured, cleaned, and used as another bioenergy product. Biogas cofiring for electricity generation can also provide other industries, such as forestry, wood products, pulp and paper, agriculture, and food processing, with a way to sell their large quantities of combustible biogas residues instead of sending them to landfills. The cost of processed biogas fuels can be low when large amounts of wood and agricultural wastes are available.
3. *Gasification of biogas*: A subsection of biogas is biogas gasification, which is a naturally produced gas generated from biological sources such as animal manure, rotting waste, and algae. The gasification of biogas into a fuel, which can then be used directly or transported by road or pipeline to the final consumer for heating or power generation, is another useful form of bioenergy. Biogas is the biological anaerobic (bacterial decomposition without oxygen) digestion of biogas materials, producing

a 60/40 gaseous mixture of methane (CH_4) and carbon dioxide (CO_2). If you have ever seen bubbles rising from a swampy, marshy area or algae filled pond, that is naturally produced methane. The methane is separated from the carbon dioxide using conventional technology to supply gas to a natural gas system or other consumer. Like the gas in the orange liquid petroleum gas canisters, biogas can be burnt in conventional ovens, stoves, and boilers for cooking, heating, or lighting the home. It can also be used to power internal combustion engines to drive a car or generate electricity. Biogas energy offers many advantages over a conventional natural gas. Biogas powered electricity plants can be built quickly, simply, and for much less money per kilowatt than coal, oil, or nuclear power plants. Unlike fossil fuels, _"biogas" is a renewable resource. The methane produced by the bacterial decomposition of organic matter is still going to be naturally produced in landfills and farmlands whether it is used as a gas or not. Methane is also an important GHG and is a major contributor to the global warming problem, so burning biogas provides an excellent source of energy that is helpful to the environment. Finally, the residue left over by the burning of biogas, called activated sludge, can be dried and used on the land as fertilizer.

4. *Liquid conversion of biogas*: Another exciting alternative energy is the production of biogas fuels called biofuel. Biofuel is a liquid fuel made from biogas, usually from plant matter. There are many types of biofuels with some common ones including methanol and ethanol, as well as synthetic gasoline, biodiesel, and aviation fuels. Ethanol (ethyl alcohol) biofuel, which is harvested from corn, sugarcane, and soya beans, is now the most common biofuel worldwide and is generally considered to be an _"RE resource," because it is primarily the result of the conversion of the Sun's energy into usable energy.

1.2 Why biogas energy?

Biogas technology may have the potential to short-circuit the _"energy transition" from biogas to _"modern" fuels. Biogas technology is a particularly useful system in the worldwide rural economy and can fulfill several end-uses. The gas is useful as a fuel substitute for firewood, dung, agricultural residues, petrol, diesel, and electricity, depending on the nature of the task, and local supply conditions and constraints, thus supplying energy for cooking and lighting. Biogas systems also provide a residue organic waste after anaerobic digestion that has superior nutrient qualities over the usual organic fertilizer, cattle dung, as it is in the form of ammonia. Anaerobic digesters also function as a waste disposal system, particularly for human waste, and can, therefore, prevent potential sources of environmental contamination and the spread of pathogens. Small-scale industries are also made possible, from the sale of surplus gas to the provision of power of a rural-based industry;

therefore biogas may also provide the user with income generating opportunities. The gas can also be used to power engines, in a dual fuel mix with petrol and diesel, and can aid in pumped irrigation systems.

Apart from the direct benefits gleaned from biogas systems, there is another, perhaps less tangible benefit associated with this renewable technology. By providing an alternative source of fuel, biogas can replace the traditional biogas-based fuels, notably wood. Introduced on a significant scale, biogas may reduce the dependence on wood from forests and create a vacuum in the market, at least for firewood (whether this might reduce pressure on forests, however, is contestable). Promoted by Khadi and Village Industries Commission (KVIC) and other bodies as _"eliminating drudgery of women," a regular supply of energy piped to the home reduces, if not removes, the daily task of fuel wood gathering, which can, in areas of scarcity, be the single most time-consuming task of a woman's day taking more than 3 hours in some areas. Freeing up energy and time for a woman in such circumstances often allows for other activities, some of which may be income generating. A clean and particulate-free source of energy also reduces the likelihood of chronic diseases that are associated with the indoor combustion of biogas-based fuels, such as respiratory infections, ailments of the lungs, bronchitis, asthma, lung cancer, and increased severity of coronary artery disease. Benefits can also be scaled up, when the potential environmental impacts are also taken into account; significant reductions in emissions associated with the combustion of biofuels, such as sulfur dioxide (SO_2), nitrogen dioxide (NO_2), carbon monoxide (CO), total suspended particles, and polyaromatic hydrocarbons, are possible with the large-scale introduction of biogas technology. The use of biogas systems in an agrarian community can increase agricultural productivity. All the agricultural residues and dung generated within the community are available for anaerobic digestion, whereas previously, a portion would be composted daily for fuel. Therefore more is returned to the land. Moreover, as mentioned earlier, the slurry that is returned after methanogenesis is superior in terms of its nutrient content; the process of methane production serves to narrow the carbon/nitrogen ratio, while a fraction of the organic nitrogen is mineralized to ammonium (NH_4^+) and nitrate (NO_3^-), the form that is immediately available to plants. According to the many researchers in 1991, the resulting slurry has double the short-term fertilizer effect of dung, while the long-term fertilizer effects are cut by half. However, in the tropics, the short-term effects are the most critical, as even the slow degrading manure fraction is quickly degraded, due to rapid biological activity. An increase in land fertility, then, can result in an increase in agricultural production. The knock on the benefits may include improved subsistence, increased local food security, or income generation of a higher output. Biogas systems, then, offer an integrated system that lends itself to a rural setting; the plants can be maintained with a variety of organic residues from humans, animals, crops, and domestic

TABLE 1.1 System perspective: impact of various parameters on the competitiveness of biogas.

Parameter	Effect on
Positive	
Higher fossil-fuel prices	Biogas utilization
Higher CO_2 quota price	Biogas utilization
Higher national biogas targets	Biogas investments
Lower biogas price	Biogas utilization and investments
Specific subsidy scheme	Biogas utilization and investments
Negative	
Lower wind and solar cost	Biogas investments
Availability of cheap storage	Biogas investments
More stringent standards and policies on pollutant emissions	Biogas investments
Lower power demands	Biogas investments

food waste. Many researchers successfully maintained a biogas plant of 25 L capacity, fed with market waste, in Pune, western India, and suggest such a system to be a viable option for solid waste disposal in the areas of rapid urbanization. Although this essay is more concerned with biogas in rural areas, the example does, nonetheless, demonstrate the potential of biogas technology and its multifunctional and flexible applications. For biogas systems to be truly viable and workable in rural areas, the technology demands to be preferably generated from within the community. As will be seen later, this may not always be possible logistically among other reasons. If not actually produced by the community it is to serve, then the technology must be amenable and possible to manage and modify the individuals within the community, preferably the plant owner, and the reliance on _"outside" assistance is kept to a minimum. Without this basic requirement being fulfilled, biogas technology will not be a truly viable option for meeting rural area's energy demands. With this in mind, the government agencies involved in designing biogas plants have attempted to create plants that could be maintained locally. Although the designs have evolved over the last 40 years since their inception, which will be outlined later, the microbial processes around which they are built, methanogenesis, remain the same. Table 1.1 shows the impact of various parameters on the competitiveness of biogas.

1.3 History of biogas system

The first person to observe that decaying vegetation produced a combustible gas was Alessandro Volta. In 1776 Volta has noticed that more bubbles are coming out when he disturbed the bottom sediment containing more plant material. In 1806 William Henry showed that Voltas gas was identical to methane gas. Humphrey Devy in the early 1800s noticed that methane was present in farmyard manure piles. In 1868 Bechamp demonstrated that manure was formed from carbon compounds by the action of microorganisms. Tappeiner, in 1882–84, showed conclusively that methane was of microbiological origin. The first plant for the production of methane was set up at the Ackworth leper asylum, Mathunga, India, in 1900. Another plant was set up in Indonesia in 1914, which used strawboard wastes as the source of gas. A prototype of a biogas plant was first developed on an experimental basis in Germany during World War II when there was a shortage of fuel. The interest in biogas reached a peak at the beginning of World War II. In 1940 French scientists working in North Africa developed the technology of production of biogas from agricultural wastes and it was reported that about 1000 biogas plants were in operation in France and French North Africa by 1950. Germany was forced to develop new energy resources during the war. Reportedly, 90,000 vehicles were converted to operate on methane gas in order to save petroleum fuel. For every 1000 kg of rice that is milled, 182 kg of husk is produced. Chinese engineers have developed a way to convert this abundant by-product into gas that can be used as a substitute for diesel oil. A pilot plant built in China uses up to 7 tonnes of husk a day and runs a 140 kW generator to provide power for a local factory. It is claimed that the cost of the electricity produced is 60% lower than that generated by diesel.

In addition, this operation disposes of a problematic waste product in a virtually pollution-free manner unlike the usual technique of burning the husk. Nonconventional energy sources like cattle waste and agricultural residue attracted the attention of the developed and underdeveloped countries due to the steep rise in the fuel prices during the 1970s. In India, 80% of the population lives in villages and burns nearly 50% of cattle waste (dung) for cooking. The government of India has launched a program for the construction of community biogas plants in the villages to supply clean fuel and save the rich source of crop nutrients from burning.

In India, the development of biogas plants has passed through three stages: the experimental stage (1937–50), the pilot plant stage (1950–63), and the field stage (since 1964), as shown in Fig. 1.6.

1.4 Worldwide evaluation of biogas technology

The availability of biogas is commonly described in terms of a hierarchy of potentials. In the order of decreasing size, these are theoretical, technical,

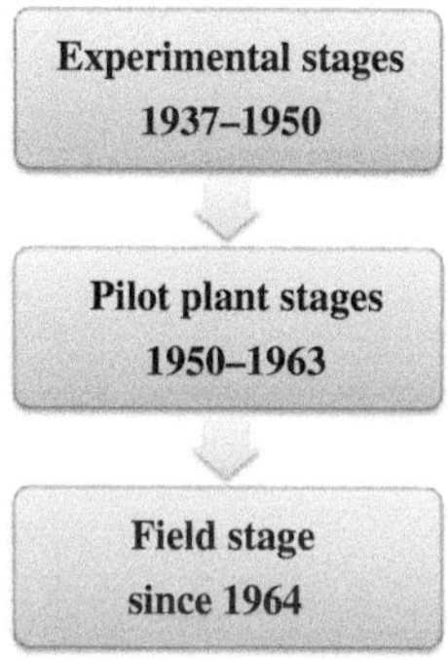

FIGURE 1.6 Three stages of biogas plant.

economic, and realistic. A theoretical potential estimate, for example, might be made by assuming that all net primary productivity not needed for food could be available for bioenergy purposes. This assumption would lead to a very large and abstract number, because it would ignore all competing land uses and socioeconomic constraints. At the other end of the spectrum, an economic potential would constrain the usable quantity of biogas to the amount that could be produced at a specific price. This would lead to a smaller number, but one that was necessarily more subjective. Adding additional constraints reduce the size of a biogas potential estimate. Therefore in order to compare studies on a similar basis, it is important that the definitions are aligned. The majority of studies considered here estimate technical potentials, but there is considerable disagreement between the definitions. Bioenergy accounts for roughly 9% of world total primary energy supply today. Over half of this relates to the traditional use of biogas in developing countries for cooking and heating using inefficient open fires or simple cookstoves with an impact on the health and the environments. A new market analysis from Germany-based ecoprog state biogas subsidization schemes has recently experienced positive amendments in Europe, while Asian countries are currently reducing this kind of support for the first time. The research shows that, in 2018, the number of biogas power plants (BMPPs) increased again by about 300 facilities. Today, there are about 3800 BMPPs with an installed capacity of around 60 GW.

Subsidies for REs are the most important factor driving the BMPP market, especially in Europe. The markets in South and North America as well as in many Asian countries are rather stimulated by fuel availability; however, RE subsidies are an important factor for the development of new capacities in these countries as well. Poland organized BMPP auctions for the first time in 2018, after the introduction had been awaited for many years. However, according to the report, these auctions showed very limited success that only one project was approved for subsidies. This is because only few

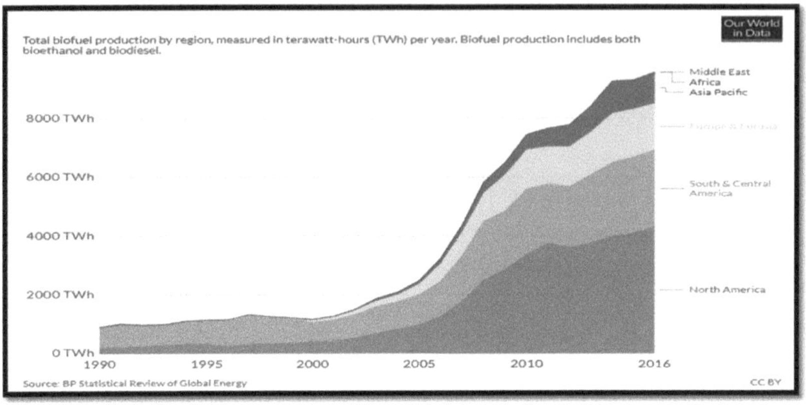

FIGURE 1.7 Biofuel production by regions in terawatt-hours per year.

project developers participated, possibly due to a wait-and-see attitude by many investors. In late 2018, Finland also introduced an auctioning system that could benefit electricity generation from biogas. Ireland passed an auctioning scheme, which should increase the establishment of REs (including biogas) until 2025.

Outside of Europe, the number of countries cutting biogas subsidies increased for the first time in 2018. Thailand drastically reduced the feed-in tariff for biogas electricity and Japan lowered the subsidization for biogas projects with capacities of over 10 MWel and introduced a cap of 200 MWel per year for additional constructions. We expect the construction of about 1900 additional BMPPs with an installed capacity of around 25 GWel. About 50% of this growth will be realized in Asia, especially in the two lead markets China and India. In addition, North and South America will remain attractive markets for electricity generated from solid biogas and, particularly, their lead markets Brazil, Canada, and the United States, the report states. Attractive subsidization terms remain in place in China and India, the countries with the strongest growth potentials. In 2018 India additionally introduced a nationwide support scheme for building biogas CHP plants (based on grants for plant construction). From a global perspective, biogas electricity subsidization continues to promote the market development for the construction of BMPPs. Until 2027, the worldwide market for BMPPs is expected to remain on its dynamic development path. Figs. 1.7 and 1.8 show biofuel production by regions in terawatt hour per year from 1990 to 2016. Table 1.2 shows the forecasting of bioenergy potential in 2050. Fig. 1.9 shows the total primary energy supply globally of all renewables until 2018. Fig. 1.10 shows the distribution of biogas waste and material. Table 1.3 shows fuel price assumptions in the reference and biogas scenario in Germany. Table 1.4 shows total primary energy from biogas globally. Table 1.5 shows the total biogas energy supply in continents until 2018.

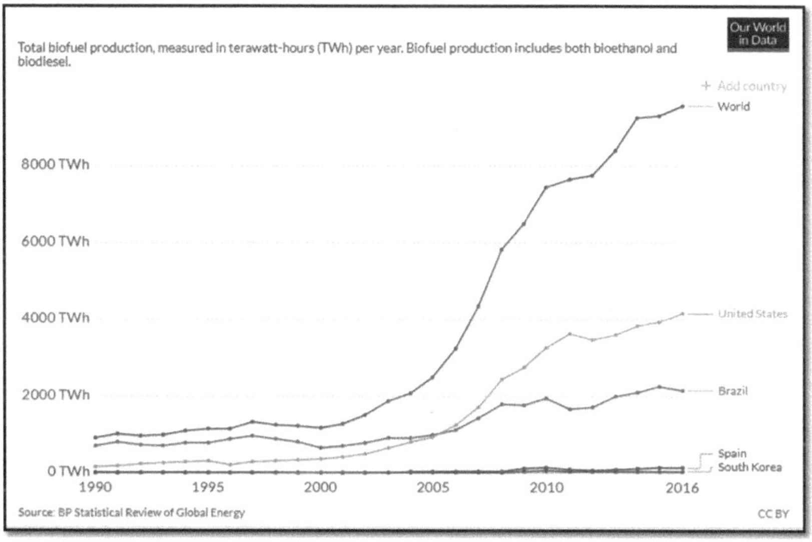

FIGURE 1.8 Graphical representation of biofuel production by regions in terawatt-hours per year.

Table 1.6 shows the top 10 countries of biogas supply until 2018 (EJ). Table 1.7 shows electricity generation from biogas TWh. Table 1.8 shows the derived heat generation from the renewable globally. Table 1.9 shows liquid biofuel production globally.

1.5 Dissemination of biogas system

Since 1960, biogas systems have been implemented in India, but it was in 1981 with the beginning of the sixth 5-year plan and the formation of the National Project for Biogas Development (NPBD), the drive to step-up dissemination was taken, perhaps also reflecting the alarm of fuel wood shortages at the time. Currently, there are thought to be about 2.5 million biogas plants installed around the country, though the potential of large-scale implementation of biogas technology remains unrealized. According to Ministry of non-conventional energy source (MNES), in 1991, the use of electricity for cooking, which includes biogas, only accounted for about 2% and 3% for rural and urban areas, respectively, and sharply demonstrates the continued minority status of this alternative fuel. The Tata Research Institute, New Delhi, estimates that 12 million biogas systems in total could be installed over the subcontinent, while GATE, an alternative energy Non governmental organization (NGO) based in Germany, estimates the total potential number of plants that could usefully be employed to be 30 million household sizes and nearly 600,000 community-sized plants, one for each

TABLE 1.2 Technical bioenergy potential from forestry residues in 2050.

Area	Low estimate (EJ/year)	High estimate (EJ/year)	Arithmetic mean (EJ/year)
North America	6	12	9
Western Europe	4	7	6
Pacific OECD	1	2	2
Central and Eastern Europe	1	2	2
Former Soviet Union	2	4	3
China	2	3	3
South Asia	0	0	0
Pacific Asia	0	1	1
Middle East and North Africa	0	0	0
Latin America and Caribbean	2	4	3
Sub Saharan Africa	0	1	1
Grand total	19	35	27

Anttila, P., Karjalainen, T., Asikainen, A., 2009. Global Potential of Modern Fuelwood. Finnish Forest Research Institute. *OECD*, Organisation for Economic Co-operation and Development.

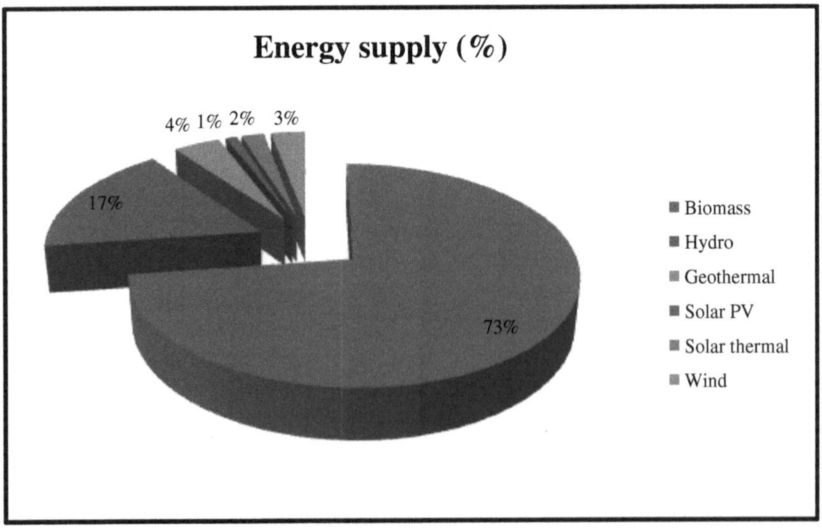

FIGURE 1.9 Total primary energy supply globally of all renewables until 2018.

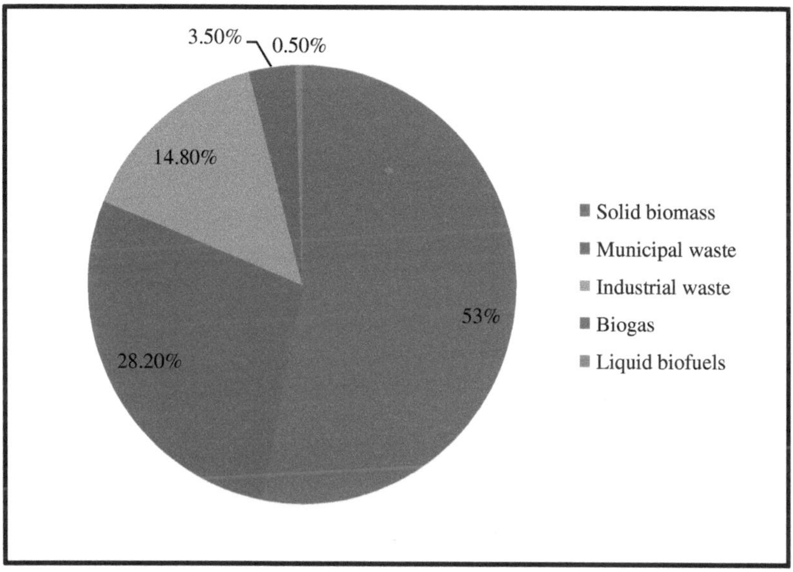

FIGURE 1.10 Distribution of waste and biogas material.

TABLE 1.3 Fuel price assumptions in the reference and biogas scenario in Germany.

($/GJ)		2017	2020	2030	2040
	Coal	3.3	3	2.6	2.5
	Natural gas	6.5	8	8	8.6
	Fuel oil	8.2	9.5	12.5	11.3
Reference	Wood chips	7.7	7.8	8.3	8.4
	Wood pellets	8.8	10.5	10.6	10.5
Biogas +	Wood chips	7.7	3.9	4.1	4.2
	Wood pellets	8.8	5.3	5.3	5.3

village. Nonetheless, there is still enormous potential for biogas technology, and the government continues in its drive for more widespread implementation. However, for biogas to be considered as a viable source of fuel depends not only on an effective dissemination program and extension but also on the success of existing plants in the field. Although literature could not be found regarding the success rate of the 2.5 million biogas plants installed to date, for example, how many are fully operational, which may be indicative of a lack of consequent monitoring, it would be instructive to examine the

TABLE 1.4 Total primary energy from biogas globally.

Year	Energy supply (EJ)
2000	43.0
2005	47.5
2010	54.2
2014	59.2
2018	65.1

TABLE 1.5 Total biogas energy supply in continents until 2018.

Continents	Energy supply (EJ)
Africa	16.2
America	12.7
Asia	28
Europe	7.91
Oceania	0.35

TABLE 1.6 Top 10 countries of biogas supply until 2018 (EJ).

Country	Biogas energy supply (EJ)
China	11.1
India	10.7
Unites States	5.32
Brazil	4.41
Nigeria	5.01
Indonesia	3.50
Canada	0.85
Ethiopia	2.30
Pakistan	1.67
Congo DR	1.30

TABLE 1.7 Electricity generation from biogas (TWh).

Year	Total	Municipal waste	Industrial waste	Solid biogas	Biogas	Liquid biofuels
2000	164	34.2	15.2	101	13.0	0.00
2005	227	46.1	11.7	147	21.0	1.97
2010	372	60.5	20.6	238	46.1	5.06
2015	528	71.3	22.7	345	82.4	7.61

TABLE 1.8 Derived heat generation from renewable globally (TJ).

Year	Total	Municipal waste	Industrial waste	Solid biogas	Biogas	Liquid biofuels
2000	414,081	125,141	74,975	208,995	4931	39
2005	530,237	152,549	82,630	284,745	6615	3698
2010	781,020	206,212	126,337	426,477	12,296	9698
2015	940,492	265,300	138,958	498,795	32,948	4491

WBA Worldbioenergy.org 2018.

TABLE 1.9 Liquid biofuel production globally.

	Total	Bioethanol	Biodiesel	Other biofuels
2000	15.9	12.2	0.78	2.97
2005	34.1	24.5	3.42	6.16
2010	94.4	60.5	18.9	15.0
2015	125	82.0	28.9	14.6
2016	132	85.6	32.6	13.6
2017	143			

implementation of biogas systems in rural India to determine how the technology has been received on the ground. Implementation of biogas technology is overseen centrally by MNES, but actual dissemination is devolved to the individual state governments, public corporations, such as KVIC, the National Dairy Development Board, and NGOs. Although there will be differences between the states, the general approach to disseminate biogas technology is based on a system of subsidies and concessions to encourage the uptake. Subsidies are granted on plants up to 10 m^3 (a large family-sized system) and usually for the models recognized by the government though there may be regional differences. Allowances are paid toward investment costs, for every user and for every biogas plant that is installed, in what may be interpreted as a measure of intent to promote biogas technology and perhaps the most critical instrument in determining initial uptake. The extent of the allowance is dependent on the size of plant, socioeconomic status of the user, and geographical region, according to the rules worked out by central government. India has been divided into three areas according to altitude; the mountainous northeast is where the highest allowances are paid, perhaps reflecting the commonly held notion that tribal communities are depleting forests. Mountains or high-altitude areas in other states from the second category and the remaining states make up the last category. Here, socioeconomic status largely determines the size of the allowance, with priorities for the scheduled caste and tribe, and smallholders. The landless and marginal farmers are entitled to higher allowances than farmers not in the aforementioned groups who have more than 5 ha. Other allowances exist for bodies to establish and maintain an organizational infrastructure, subject to reaching certain targets, of which a percentage must be allocated in the provision of follow-up services and monitoring.

Subsidies certainly appear to have encouraged uptake, and participation seems to be high among target groups, such as margins and smallholders. This can be demonstrated in the size and type of the digester opted for a particular purpose. In India, Orissa, on the east coast, is one of the poorest states in India and characterized by smallholders of approximately 1.6 ha, less than the average of other states, and agriculture is the principal industry in Orissa. Therefore it is not surprising that of all the digesters, the most popular is the smallest capacity fixed-dome Deenbandhu model, at 6 m^3, which accounts for 84% of all plants installed. Similarly, in Sangli, Maharashtra, western India, where there are 345,000 biogas digesters, more than any other state, the same Deenbandhu model accounts for 85% of all systems constructed.

However, many researchers note that advancement does not guarantee a successful operation of biomass power plant. By studying installed systems in Maharashtra, western India, the workers note a correlation between the decreasing land size and the nonfunctioning plants. Similarly, Moulik (1982) maintains that the early biogas plants installed at a great percentage, perhaps as many as 70%, are inoperative. Moulik explains that in the enthusiasm to promote biogas technology, many _"marginal" farmers and landless were hastily provided with plants, as full subsidies were given, and NGOs and other organizations had targets to reach. However, many were to remain inoperative, due to a variety of reasons, but critically, due to an inability to fulfill the requirements necessary for operating the plant. Moulik states that, however well intentioned, the biogas program cannot cater to the needs of the poorest and marginalized, as these groups fail in the technical requirements to maintain a viable plant. More specifically, for even the smallest sized plant, three to four cattle are needed to provide the necessary quantity of dung. Less than this, the plants are not economically or operationally viable. Moreover, considerable constraints may also exist in the provision of space and water that are likewise necessary for a biogas plant. According to Moulik, the smallest 3 m^3 family-sized plant requires about 27 m^2 of land, when an area of the plant and a compost pit for the slurry is taken into account, which in many circumstances may not be available. The characteristic clustering of houses in a village between the networks of narrow lanes may render land enough around the homestead to accommodate a biogas plant as the exception, rather than the rule. Even if surplus land is available, issues of land tenure and ownership may prohibit the construction of a plant.

Water scarcity, or difficulty in obtaining water, for example, from a distant source, may also impose further constraints on the viability of biogas technology in a rural environment. To function properly, a biogas plant requires feeding a mixture of cow dung and water in the ratio of 1:1 or 4:5, thus imposing a significantly higher daily water demand over domestic needs. If there is difficulty in obtaining water, particularly resonant for low-caste groups in a village environment, who may not have the same resource access rights as others, or general scarcity, then the maintenance of a biogas plant may not be possible. Given the above, Moulik estimates that perhaps only 10%–15% of the rural population fulfills the technical requirements. Despite a well-intentioned attempt to cater for the energy needs of rural India, particularly the poor, as defined by _"scheduled caste" and _"scheduled tribe," the biogas program seemingly cannot meet these needs, through insurmountable constraints associated with their very marginality, ironically. In this sense, then, the biogas program may be an unrealizable notion and the Gandhian aspirations of Swadeshi, little more than a bucolic dream. However, it may be instructive to consider briefly a case study to understand how biogas technology has been received in targeted areas. In 1980 the NPBD was active in promoting biogas in low-caste and tribal areas of Udaipur, Rajasthan, and northwestern India. One of the researcher conducted a survey in eight villages of mixed caste and tribe, in an attempt to assess the impact and effectiveness of NPBD in these areas. One hundred and fourteen samples of families who had installed biogas plants under the NPBD program up to 1985, notably the cheaper fixed-dome Janata, were considered. The data revealed some interesting findings; of the 114 beneficiaries, 107 were registered as _"landless" or _"marginal," though the survey discovered that the plant owners were mostly the wives or sons of landowners who owned between 6 and 20 acres of land. These family members had been encouraged to apply to make use of the higher rate of subsidies available for marginal and landless groups. Only 10 were found to be scheduled caste or tribe with poor landholdings.

Curiously, Nag et al. interpret the results as a success for the NPBD and describe the scheme as a _"peoples" program. That participation among farmers is high is a positive sign of the potential role of biogas in an agricultural community; however, the program does not appear to be delivering to the rural poor, as defined by the scheduled caste and tribe, which may be indicative of the inherent incompatibility of the technology with regard to marginalized groups. Uptake of biogas technology among scheduled caste and adivasi (tribal) groups is then found to vary across the subcontinent, though even where participation is high, the technology may not be truly viable. Biogas, however, does appear to be taken up more successfully by the wealthiest sectors of the agricultural community. As Nag et al. (1990) note, over 30% of the families with biogas plants sampled were found to be engaged in more than one service or business, which is usually an indication of entrepreneurship and solvency. Further, according to Ritchie and Nesmith (1991), biogas technology appears to be associated with status and wealth and was observed more commonly in top income groups in a study in West Bengal, eastern India. (This association with wealth may well be a hindrance to the wider dissemination of biogas technology among groups who may view themselves as perhaps not fully entitled to it.)

As household-sized plants may be generally nonviable to many scheduled caste and adivasi groups, community-sized plants might be more appropriate. Larger sized plants, servicing a cluster of houses or indeed a whole village, may overcome the seemingly insurmountable problems apparent regarding individual plants and the rural poor, as discussed earlier. However, Lichtman (1983) states that the government subsidy system has discriminated against the provision of community-sized plants, by subsidizing up to 6 m^3 plants only (and later up to 10 m^3). Thus wealthier farmers have been able to apply for grants and loans to construct household-sized systems, while larger plants that may benefit the wider community have been ineligible for support. In this way, the government subsidy program may be interpreted as discriminating against the poorer sections of the community, while supporting the wealthier farmers.

However, where community plants have been constructed, many problems have been encountered. Singh (1988) randomly sampled half the beneficiaries of seven community biogas plants in Punjab, northern India, after the first year of operation and discovered considerable technical, economic, and social problems. Singh found that all the plants were being routinely underfed with dung, by 30%–50%, as shown in Table 1.10. In one case, the entire daily dung load needed to be brought from the nearest city. Although, in theory, there were enough cattle to provide the required amounts of dung, competing demands with nonbeneficiaries were evident, who collected dung for fuel, in the absence of crop residues. Gas production was also found to fall to 30% of its rated production in the winter months, due to greater direct use of dung, for fuel.

TABLE 1.10 Daily dung requirements and dung fed (quintal = 100 kg) (Singh, 1986).

Name of village	Dung requirement	Dung fed			Difference
		From village	From outside	Total	
Mehdoodan	30	14	–	4	16
Pehar Kalan	30	16	4	0	10
Ablowal	30	10	10	0	10
Passiana	30	12	10	2	8
Hambran	95	45	–	5	50
Pandori	30	12	3	5	15
Chabewal	30	–	30	0	–

At the time of writing the paper, Singh noted two plants to be nonoperational, principally due to problems associated with the availability of labor. Labor shortages were attributed to economic factors, such as low pay compared with agricultural labor. Social factors were also evident in the nonavailability of labor, particularly the stigma associated with working with dung, considered as a low-cost task, and usually performed by women. However, in this instance, the volume of dung involved in the daily maintenance of the community plants, 3000 kg, was considered beyond the physical strength of women laborers, given its dispersed nature and distance of some of the sources. Laborers were found to complain about the logistical difficulties in collecting dung from diffused sources, weighing, and recording it to the satisfaction of the donor and for the community records of dung input, etc. Four of the community latrines were also not functioning due to labor shortage. Supervision problems were also identified by Singh, principally relating to low pay, which resulted in an ad hoc arrangement and a high turnover of supervisors. Sometimes closure of the plants occurred as a consequence.

Singh describes the experience of scheduled castes and tribes, the targeted beneficiaries of the community biogas system. It was found that dung was having to be purchased in substantial quantities to feed some of the plants, up to 1000 kg in several, while in one, the entire 3000 kg daily need was having to be imported (see Table 1.3). While dung purchasing costs were high and increasing, returns on the sale of slurry were considerably smaller than estimated, between 15% and 30% of the expected revenue. Consequently, an increase in the gas charges was necessary to cover costs, and prices were raised from □30 to □50 per month. The increased prices could not be borne by many of the scheduled caste and adivasi communities, and many disconnected themselves from the supply. In one village, Mehdoodan, 24 of the 29 scheduled caste and tribe connections to the biogas supply were duly removed.

Community biogas plants, then, appear to be logistically difficult to coordinate and, certainly in the Punjab, similarly failing the sections of the community most in need of a reliable source of energy. Other workers have reported community biogas plants failing for reasons such as political feuds (Lichtman, 1983) and due to variable climatic conditions that resulted in the forced sale of cattle (Lichtman, 1983). However, there have also been reports of community biogas plants successfully maintained by collective management efforts. The eventual success of a community biogas system in Pura, southern India, after several years of problems, and a change in the end use of gas. The program was implemented with the help of The Centre for Application of Science and Technology to Rural Areas (ASTRA), which considered Pura, a village of 430, with 240 cattle, suitable for a community biogas plant. ASTRA calculated that manure from the village could fuel a biogas plant sufficient to provide for all cooking needs and generate surplus gas for lighting and pumping drinking water. The plant became operative in 1982, but serious logistical problems became apparent, as gas would run out before the cooking of the second daily meal. Conflicts ensued between the villagers regarding contributions and share of benefits, and the project stopped in 1984. Interestingly, when ASTRA attempted to revive the project and suggested that the gas could be used solely for generating electricity for lighting, it was discovered to ASTRA's surprise that the villagers' top priority was actually the provision of safe drinking water. ASTRA duly acted according to the village needs, rather than working to their own assumptions, and by all accounts, the program is now a success. The standard of living has been raised, and management is possible by the tangible benefits enjoyed by the whole village. At the time of writing the paper, Hall et al. report that the success of the program has encouraged residents to consider building a wood gasifier to bolster their energy supply.

1.6 Factor hindering biogas system

Biogas is produced through anaerobic digestion of animal excreta and other agricultural wastes. This energy can be harnessed successfully to meet the existing as well as future energy requirements of the rural areas. Biogas is the fourth largest source of energy in the world, supplying about 13% (55 EJ/year), which is equivalent to 25 million barrels of primary energy (Mittal, 1997). The main raw material used for biogas production is cattle dung. India has the largest population of livestock of over 300 millions, which produce about 980 million tonnes of dung. If the entire quantity of available cattle dung is used for biomethanation, it could generate nearly 195 billion kWh of energy annually (Govil and Gaur, 2000). Most of the biogas plants installed all over the world are based on cattle dung and operate at 10% total solids (TS) concentration. To achieve this, equal quantity of water is to be added with dung. The large quantity of water required for the operation of a biogas plant is an important constraint in the propagation of the biogas technology, particularly in water scarce regions of the country. The solid-state anaerobic digestion process could be used as an effective tool for solving the above problem. By using solid-state digestion technology, loading rate could be increased to the optimized level, the output rate of gas per unit digester volume could be increased, water requirement could be minimized, and the digested sludge handling problem could be eliminated to a greater extent and can be easily handled and directly transported to the field. Shyam (2001) reported that the modified biogas plants for solid-state digestion of cattle dung required very little or no water for mixing with the cattle dung and generated 50% higher gas than common biogas plants. He also reported that solid-state digestion makes handling of input slurry and digested slurry much easier than the conventional plants. Anonmyous (2002) reported that average biogas production in the modified Janata biogas plant was 204.30 L per kg of dry matter at 16% TS, and at 10% solid content, it was 176.54 L per kg of dry matter, which showed a 15.72% increase in gas production. Several researchers have done considerable research work on the various aspects of solid-state anaerobic digestion, and few designs to modify Janata biogas plant have been evolved in various regions of the country. However, no feasibility survey work on solid-state digestion of cattle dung has been carried out in Raichur region. Hence, a study was undertaken to evaluate the performance of cattle dung in solid state in fixed-dome type biogas plant in a water scarce region of Raichur district. It would be worth briefly considering the problems associated with the alternative technology, in terms of technical/operational, economic, and cultural aspects, which may potentially hinder its spread. Finally, the government's overall approach in disseminating biogas technology will be considered.

Technically, problems have arisen from installing too large a capacity plant, either by accident or by design. A too large plant was found to lead to underfeeding and eventual failure of the plants to produce gas. Underfeeding was also found to occur due to the undercollection of dung, estimated typically at 30%–40% of the required capacity and principally due to cattle being worked in the field, which would also lead to a reduction in gas production. In the areas of climatic instability, the occurrence of drought may reduce dung availability, by forcing the sale of cattle, or even death of cattle. In some areas, the plant may not be technically feasible all year round due to low winter temperatures that inhibit methanogenesis (Sudhakar and Gusain, 1991). Sometimes the plants are faulty in their construction or have developmental problems that lead to the nonfunctioning of the plant due to shoddy construction (more relevant to the fixed-dome models and then the floating dome, which comes precast). Chand and Murthy (1988) analyzed factors in the nonfunctioning of plants in Maharashtra, western India. The workers discovered that specially trained masons in the biogas plant construction were overlooked often, due to their higher cost, in favor of cheaper trainees or those with no training at all and often encouraged local by the government agencies to meet ambitious targets. Chand and Murthy identified 50% of 1670 plants in the study as incapable of ever being made functional.

Economically, biogas systems have been shown to be cost-effective. Lichtman (1983) modeled different energy use scenarios of village-sized plants in Pura. The analysis was site-specific and localized in its approach. Lichtman found that in 78% of the situations modeled, the village showed a net gain. This percentage is likely to decrease in the consideration of smaller household-sized systems. Lichtman concedes, however, that it is more profitable to maintain a community-sized system as a public utility and fertilizer plant than as a source of cooking gas, subject to the viable provision of an alternative energy source for cooking, such as woodlots, and for fodder. Biogas production could perhaps be linked to small-scale industries. Despite the positive cost–benefit of biogas technology, the _"macroenvironment," may discriminate against the uptake of biogas. The macroenvironment that determines the price structures of conventional fuels most likely acts as a disincentive to adopt renewable technologies, generally. Subsidized conventional fuels, such as electricity, along with free connection to the grid for farmers, will continue to make nonrenewable technology the cheapest option, unless subsidies for biogas can be brought into line or prices of conventional fuels rose. The system of grants and loans may hinder the correct choice of plant for different users, such as the ineligibility of community-sized systems due to their size. While finally, another point in prohibiting uptake may be the perceived unnecessary switch from the existing free source of energy, such as wood and crop residues.

Cultural practices may also hinder general uptake due to the reluctance to adopt different behaviors, particularly regarding the use of latrines in biogas systems. Traditional cooking practices may also need to be altered. Researcher reports that a common complaint about the use of gas burners for cooking is that the staple bread *chapati* cannot be properly roasted and the cooking of *dal* (pulses) may be increased. Further, women are not necessarily the decision makers in a household, and the men of the household may not consider benefits, which mainly accrue to women, to be of significant urgency. Some of the problems discussed above may be overcome through effective selection processes of the technology, and proper extension and support services. By all accounts, the government does not seem to be effectively organized to achieve such a goal, and a high number of nonoperative biogas plants are likely to continue. Criticisms of NPBD have been widely articulated, from the lax selection process, to the arbitrary fixing of regional targets, which are then pursued. Chand and Murthy (1988) discovered in the study of biogas uptake in Maharashtra that in a sample of 1670 plants, 1086 beneficiaries were found not to qualify under the feasibility criteria. Further, when complications have arisen in the functioning of plants, a common complaint articulated is that there is a lack of available technical support (Sudhakar and Gusain, 1991). In this way, plants may be allowed to fall into disrepair when their functioning may depend on adequate maintenance skills, which should be available in every village. There is a danger that biogas may come to be thought of as a useless and inappropriate initiative, a folly imposed from policy makers and NGOs.

Compared with the biogas program in China, where 7 million households and community biogas systems have been successfully installed, India has a long way to go to realize the benefits of biogas technology. China, through the creation of effective institutions and by placing an emphasis on training and education, has achieved widespread dissemination of biogas technology (Daxiong et al., 1990), though the social organization may particularly facilitate the spread of new, community-focused technologies. Workers stress the need for microplanning (Lichtman, 1983), so that genuinely appropriate biogas technology is made available to rural communities. Other workers also propose coordinated management information systems as part of biogas development in order for problems to be identified and remedial measures undertaken (Chand and Natarajan, 1987; Chand and Murthy, 1988).

1.7 Pros and cons of biogas system

1.7.1 Advantages

Bioenergy systems offer significant possibilities for reducing GHG emissions due to their immense potential to replace fossil fuels in energy production. Biogas reduces emissions and enhances carbon sequestration, since short-rotation crops or forests established on abandoned agricultural land accumulate carbon in the soil. Bioenergy usually provides an irreversible mitigating effect by reducing carbon dioxide at the source, but it may emit more carbon per unit of energy than fossil fuels unless biogas fuels are produced unsustainably. Biogas can play a major role in reducing the reliance on fossil fuels by making use of thermochemical conversion technologies. In addition, the increased utilization of biogas-based fuels will be instrumental in safeguarding the environment, generation of new job opportunities, sustainable development, and health improvements in rural areas. When compared with wind and solar energies, BMPPs are able to provide crucial, reliable baseload generation. Biogas plants provide fuel diversity, which protects communities from volatile fossil fuels. Since biogas energy uses domestically produced fuels, biogas power greatly reduces our dependence on foreign energy sources and increases national energy security. A large amount of energy is expended in the cultivation and processing of crops like sugarcane, coconut, and rice, which can be met by utilizing energy-rich residues for electricity production. The integration of biogas-fueled gasifiers in coal-fired power stations would be advantageous in terms of improved flexibility in response to fluctuations in biogas availability and lower investment costs. The growth of the bioenergy industry can also be achieved by laying more stress on green power marketing. The development of efficient biogas handling technology, improvement of agroforestry systems, and establishment of small- and large-scale biogas-based power plants can play a major role in rural development. Biogas energy could also aid in modernizing the agricultural economy.

1.7.1.1 The leading standard reference in the industry for electricity generation from solid biogas

The updated ninth edition includes the following:

- Specific data on around 3700 units in nearly 3400 BMPPs with a capacity of over 55 GWel.
- A description and evaluation of more than 1080 projects worldwide, including essential data such as on who is responsible for the project as well as project status, planned commissioning date, and fuel type.
- A worldwide market development forecast 2018–27 by country, including assessment of new constructions, shutdowns, and investment volumes based on over 600 cost examples.
- An analysis of the existing plants by country, for example, age, feedstock, capacities, and competition.
- An analysis of biogas electricity generation subsidization (feed-in tariffs, quota systems, and auctions) by country (for the world's most important markets).
- The overview of market factors, fuels, treatment technologies, as well as investment and operational costs and revenues (with exemplary calculations).
- The description and market shares of all important operators and technology providers.

1.7.1.2 Environmental impact

On combustion, the carbon from biogas is released into the atmosphere as carbon dioxide. After a few months, or years, or decades, the CO_2 has been absorbed back by growing plants or trees. However, the carbon storage capacity of forests may be reduced overall if destructive forestry techniques are employed. All biogas crops sequester carbon. For example, soil organic carbon has been observed to be greater below switch grass crops than under-cultivated cropland, especially at depths below 30 cm (12 in.). For Miscanthus × giganteus, one of the researcher found accumulation rates ranging from 0.42 to 3.8 tonnes per hectare per year, with a mean accumulation rate of 1.84 tonnes (0.74 tonnes per acre per year), or 20% of total harvested carbon per year. The grass sequesters carbon in its continually increasing root biogas, together with carbon input from fallen leaves. Typically, perennial crops sequester significantly more carbon than annual crops due to greater nonharvested living biogas (roots and residues), both living and dead, built up over years, and less soil disruption in cultivation. The simple proposal that biogas is carbon-neutral put forward in the early 1990s has been superseded by the more nuanced proposal that for a particular bioenergy project to be carbon-neutral, the total carbon sequestered by a bioenergy crop's root system must compensate for all the emissions from the related, aboveground bioenergy project. This includes any emissions caused by direct or indirect land use change. Many first-generation bioenergy projects are not carbon-neutral given these demands. Some have even higher total GHG emissions than some fossil-based alternatives. Transport fuels might be worse than solid fuels in this regard.

Some are carbon-neutral or even negative, though, especially perennial crops. The amount of carbon sequestrated and the amount of GHG emitted will determine if the total GHG life cycle cost of a bioenergy project is positive, neutral, or negative. A carbon negative life cycle is possible if the total below-ground carbon accumulation more than compensates for the aboveground total life cycle GHG emissions. The graphic on the right displays two CO_2 negative Miscanthus × giganteus production pathways, represented in gram CO_2 equivalents per megajoule. The yellow diamonds represent mean values. Successful sequestration is dependent on planting sites, as the best soils for sequestration are those that are currently low in carbon. The varied results displayed in the graph highlight this fact. For the United Kingdom, successful sequestration is expected for arable land over most of England and Wales, with unsuccessful sequestration expected in parts of Scotland, due to already carbon-rich soils (existing woodland) plus lower yields. A soil already rich in carbon includes peatland and mature forest. Grassland can also be rich in carbon; however, One of the researcher argue that the most successful carbon sequestration in the United Kingdom takes place below improved grasslands. The bottom graphic displays the estimated yield necessary to compensate for the disturbance caused by planting plus life cycle GHG emissions for the related aboveground operation.

Forest-based biogas projects have received criticism for ineffective GHG mitigation from a number of environmental organizations, including Greenpeace and the Natural Resources Defense Council. Environmental groups also argue that it might take decades for the carbon released by burning biogas to be recaptured by new trees. Biogas burning produces air pollution in the form of carbon monoxide, volatile organic compounds, particulates, and other pollutants. In 2009 a Swedish study of the giant brown haze that periodically covers large areas in South Asia determined that two-thirds of it had been principally produced by residential cooking and agricultural burning and one-third by fossil-fuel burning. Use of wood biogas as an industrial fuel produces fewer particulates and other pollutants than the burning sun in wildfires or open-field fires.

1.7.1.3 Biogas is eco-friendly

Biogas is a renewable, as well as a clean, source of energy. Gas generated through biodigestion is nonpolluting; it actually reduces greenhouse emissions (i.e., reduces the greenhouse effect). No combustion takes place in the process, meaning there is zero emission of GHG to the atmosphere; therefore using gas from waste as a form of energy is actually a great way to combat global warming. Unsurprisingly, concern for the environment is a major reason why the use of biogas has become more widespread. Biogas plants significantly curb the greenhouse effect: the plants lower methane emissions by capturing this harmful gas and using it as fuel. Biogas generation helps cut reliance on the use of fossil fuels, such as oil and coal. Another biogas advantage is that, unlike other types of REs, the process is natural, not requiring energy for the generation process. In addition, the raw materials used in the production of biogas are renewable, as trees and crops will continue to grow. Manure, food scraps, and crop residue are raw materials that will always be available, which makes it a highly sustainable option.

1.7.1.4 Biogas generation reduces soil and water pollution

Overflowing landfills not only spread foul smells but also allow toxic liquids to drain into underground water sources. Consequently, yet another advantage of biogas is that biogas generation may improve water quality. Moreover, anaerobic digestion deactivates pathogens and parasites; thus it is also quite effective in reducing the incidence of waterborne diseases. Similarly, waste collection and management significantly improve in areas with biogas plants. This, in turn, leads to improvements in the environment, sanitation, and hygiene.

1.7.1.5 Biogas generation produces organic fertilizer

The by-product of the biogas generation process is enriched organic waste (digestate), which is a perfect supplement to, or substitute for, chemical fertilizers. The fertilizer discharge from the digester can accelerate plant growth and resilience to diseases, whereas commercial fertilizers contain chemicals that have toxic effects and can cause food poisoning, among other things.

1.7.1.6 It is a simple and low-cost technology that encourages a circular economy

The technology used to produce biogas is quite cheap. It is easy to set up and needs a little investment when on a small scale. Small biodigesters can be used right at home, utilizing kitchen waste and animal manure. A household system pays for itself after a while, and the materials used for generation are absolutely free. The gas manifested can be used directly for cooking and the generation of electricity. This is what allows the cost of biogas production to be relatively low. Farms can make use of biogas plants and waste products produced by their livestock every day. The waste products of one cow can provide enough energy to power a light bulb for an entire day. In large plants, biogas can also be compressed to achieve the quality of natural gas and utilized to power automobiles. Building such plants requires relatively low capital investment and creates green jobs. For instance, in India, 10 million jobs were created, mostly in rural areas, plants, and inorganic waste collection.

1.7.1.7 Healthy cooking alternative for developing areas

Biogas generators save women and children from the daunting task of firewood collection. As a result, more time is left over for cooking and cleaning. More importantly, cooking on a gas stove, instead of over an open fire, prevents the family from being exposed to smoke in the kitchen. This helps prevent deadly respiratory diseases. Sadly, 4.3 million people a year die prematurely from illness attributable to the household air pollution caused by the inefficient use of solid fuels for cooking.

1.7.2 Disadvantages

1.7.2.1 Few technological advancements

An unfortunate disadvantage of biogas today is that the systems used in the production of biogas are not efficient. There are no new technologies yet to simplify the process and make it abundant and low cost. This means large-scale production to supply for a large population is still not possible. Although the bigger plants available today are able to meet some energy needs, many governments are not willing to invest in the sector.

1.7.2.2 Contains impurities

After refinement and compression, biogas still contains impurities. If the generated biofuel was utilized to power automobiles, it can corrode the metal parts of the engine. This corrosion would lead to increased maintenance costs. The gaseous mix is much more suitable for kitchen stoves, water boilers, and lamps.

1.7.2.3 Effect of temperature on biogas production

Like other RE sources, biogas generation is also affected by the weather. The optimal temperature that bacteria need to digest waste is around 37°C. In cold climates, digesters require heat energy to maintain a constant biogas supply.

1.7.2.4 Less suitable for dense metropolitan areas

Another biogas disadvantage is that industrial biogas plants only make sense where raw materials are in plentiful supply (food waste and manure). For this reason, biogas generation is much more suitable for rural and suburban areas.

1.8 Exercise

1. What is the requirement of biogas energy system in the current scenario of electricity production?
2. What is the significance of the biogas energy system in the field of RE system?
3. Write a short note on the current scenario of biogas energy system.
4. Write a short note on the worldwide current scenario of the biogas energy system.
5. Write a short note on the current scenario of the biogas energy system to the Indian prospectus.
6. What is the meaning of dissemination of biogas energy system?
7. Explain the advantages and disadvantages of biogas energy system.
8. Why biogas energy system is preferred in rural areas?

References

Anonmyous, 2002. Training Manual on Dry Fermentation of Cattle Dung Through Modified Janata Biogas Plant. Department of Renewable Sources of Energy, MPUAT, Udaipur Centre, pp. 1–7.

Chand, A.D., Natarajan, U., 1987. Management information system for biogas development. Econ. Polit. Wkly. 22 (48), 153–160.

Chand, A.D., Murthy, N., 1988. 'District level management system for biogas programme: an analysis.'. Econ. Polit. Wkly. 23 (22), M80–84.

Daxiong, Q., Shuhua, G., Baofen, L., Gehua, W., 1990. Diffusion and innovation in the Chinese biogas program. World Dev. 18, 555–563.

Govil, G.P., Gaur, R.R., 2000. Development of conversion kits to promote the use of biogas in existing diesel engines for variable load rural applications. In: Proceedings of National Conference on Commercialization Aspects of Renewable Energy Sources, pp. 100–111.

Lichtman, R., 1983. Biogas Systems in India. Volunteers in Technical Assistance (VITA), Arlington, VA.

Mittal, K.M., 1997. Biogas System. New Age International (P) Ltd., p. 2.

Moulik, T.K., 1982. Biogas Energy in India. Academic Book Centre, Ahmedabad.

Nag, K., Boland, C., Rich, N.H., Keough, K.M.W., 1990. Design and construction of an epifluorescence microscopic surface balance for the study of lipid phase transition. Rev. Sci. Instrum, 61, 3425-3430

Ritchie, J.T., Nesmith, D.S., 1991. Temperature and crop development. In: Hanks, J, Ritchie, J. T. (Eds.). Modeling Plant and Soil Systems-Agronomy Monograph ASA-CSSA-SSSA, Madison, WI.

Shyam, M., 2001. Biogas generation made easy for water scarce regions. Kurukshetra 49 (11), 39–42.

Sudhakar, K., Gusain, P.P.S., 1991. Rural Energy Planning in Sikkim. Society for Development Alternatives, New Delhi.

Further reading

Bhatia, R., 1977. Economic appraisal of biogas units in India, frame work for social benefit cost analysis. Econ. Polit. Wkly. 12 (33–34), 1503–1518.

Moulik, T.K., Srivastave, U.K., Shingi, D.M., 1978. Biogas System in India, A Socio-Economic Evaluation. Centre for Management in Agriculture, Indian Institute of Management, Ahmedabad.

Nag, K., Harbottle, R.R., Panda, A.K., 2000. Molecular architecture of a self-assembled bio-interface: lung surfactant. J. Surf. Sci. Technol. 16, 157–170.

Ru Chen, C., 1983. Up-to-date status of anaerobic digestion technology in China. In: AD 83 Proceedings, pp. 415–428.

Singh, L., Maurya, M.S., Ramana, K.V., Alam, S.L., 1995. Production of biogas from night soil at psychrophilic temperature. Bioresour. Technol 53, 141–149.

Verma, A., Behera, B., 2003. Green Energy, Biogas Processing and Technology. Capital Publishing Company, New Delhi, p. 105.

Chapter 2

Optimum sizing and modeling of biogas energy system

Chapter Outline

Objectives

- To provide knowledge about prefeasibility assessment of bioenergy system.
- To provide knowledge about biogas production and different stages of biogas production.
- To provide knowledge about decomposition and digestible property of bioenergy system.
- To provide knowledge about the multifunctional biogas energy system.
- To provide knowledge about prefeasibility assessment of bioenergy system.

2.1 Prefeasibility analysis of biogas power plant

Prior to installation and operation, the prefeasibility study of biogas should be done. In biogas projects, an initial study is undertaken to determine whether it is worthwhile to continue to the feasibility study stage. A precise feasibility

Design and Optimization of Biogas Energy Systems. DOI: https://doi.org/10.1016/B978-0-12-822718-3.00002-2

study should provide a chronological background of the projects. In addition to climate condition of the application site, availability of input material for biogas power plant, potential of biogas energy sources, and load demand of application sites are included to find out the best location to develop a biogas power plant. Generally, feasibility precedes technical development and project implementation. It must, therefore, be conducted with a balanced approach to provide information upon which decisions can be based. Biogas can become a reliable and renewable local energy source to replace conventional fossil fuels in local industries and to reduce reliance on overloaded electricity grids. In this perspective, many medium-to-large agricultural, wood processing, or food processing industries in developing countries and emerging economies are well placed to benefit from the successful development of biogas to energy. Biogas resources are found almost everywhere and can become a reliable and renewable local energy source to replace fossil fuels. Energy produced from biogas can reduce reliance on an overloaded electricity grid and can replace expensive fuels used in local industries. Fig. 2.1 shows the parameter of prefeasibility analysis of biogas energy system.

The following conditions are satisfied during the prefeasibility assessment of biogas power plant.

- assessment of different technical options;
- approximate cost/benefits;
- permitting needs;
- market assessment for biogas feedstock and sale of producing energy;
- technical and financial evaluation of preferred option;
- assessment of environmental and social risks;

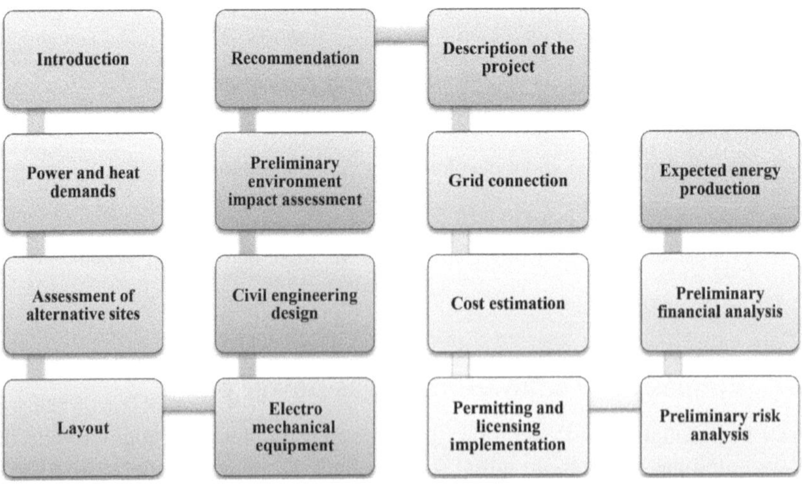

FIGURE 2.1 Typical content of a prefeasibility assessment.

- assessment of financing options; and
- initiation of permitting process.

The prefeasibility study is the first assessment of the potential project. It is a high-level review of the main aspects of the project, and the purpose is to decide if it is worth taking the project forward and investing further money and time.

The typical prefeasibility study may cover the following issues:

- description of the biogas fuel resource (amount, characteristics, price, transport, logistic, need for supplementary fuel, etc.);
- barriers for the project;
- potential technical concepts (several concepts may be identified and briefly assessed);
- calculation of expected energy production (electricity, steam, and heat);
- preliminary layout;
- preliminary assessment of energy sales (power purchase agreement (PPA), electricity price, heat price, steam price, etc.);
- preliminary assessment of alternative sites (access to site, size, connection to grid, sewer, etc.);
- preliminary assessment of alternative locations;
- preliminary assessment of environmental and social risks and impacts;
- preliminary assessment of construction costs (CAPEX) and operating costs (OPEX);
- preliminary financial analysis;
- preliminary risk assessment;
- preliminary assessment of necessary permitting and licensing; and
- planning and project implementation, including tentative time schedule.

If the outcome of the prefeasibility study is favorable, a detailed feasibility study will follow. This feasibility study consists of a significantly more detailed assessment of all aspects of the project. The purpose of the feasibility study is to explore the project in enough detail for the interested parties and stakeholders to make a commitment to proceed with its development. Financial institutions involved may require the preparation of a "bankable feasibility study." The bankable feasibility study may include an environmental and social impact assessment. A well-detailed technical description, rough layout, plant main data, etc. are needed in order to estimate the CAPEX/OPEX and to conduct, for example, a detailed environmental assessment, and consequently, a conceptual design study is necessary.

- *Conceptual design*:

The conceptual design typically comprises:

- Definition of fuel characteristics, such as composition and heating value.
- Description of applied technology.

- Evaluation of suitable technologies, including fuel handling, combustion system, boiler, ash handling and disposal, flue gas treatment technologies to meet applicable and relevant air emission standards, energy recovery system, etc.
- Assessment of potential plant location(s) following an evaluation of technical, environmental, and economic aspects, and local acceptability.
- Initial assessment of capital costs (CAPEX) and operational expenditures (OPEX).
- Assessment of potential use of steam and/or heat. Is it possible to use the heat for industrial purposes, perhaps as steam? Is there a market for district heating/cooling?
- Examination of the connections to the electrical grid, other external off-take customers, water and wastewater services, etc.

A preliminary business case, including cash flow for the project depreciation period, can be prepared based on the information collected above and on the budgetary figures for CAPEX and OPEX. International financial institutions normally require a full bankable feasibility study to be conducted before financing concepts can be finalized. The following items could be included in a bankable feasibility study, but the exact scope will be determined for each project, since different investors will have different demands for the study. The basis is the data determined under the conceptual design, but normally, some items will have to be investigated more thoroughly:

- the conceptual design and required investment;
- secured long-term supply of biogas (volume, heating value/properties, and price);
- financial and economic analysis, including cost–benefit calculations, calculations of net present value, and internal rate of return, and similar analyses;
- overview of current regulatory and policy framework relevant to the project;
- assessment of potential additional sources of financing, sensitivity analyses, and risk analyses important to financial institutions;
- assessment of potential risks to the financial viability of the project and suggestions of mitigation measures;
- environmental and social impact assessment, including identification of mitigation measures;
- organization studies of potential O&M service companies. Procurement plan and identification of potential equipment suppliers and contractors; and
- implementation plan, including time and financing schedule.

2.2 Decomposition of biogas

Biogas is a type of biofuel that is naturally produced from the decomposition of organic waste. When organic matter, such as food scraps and animal waste, breaks down in an anaerobic environment (an environment absent of oxygen), they release a blend of gases, primarily methane and carbon dioxide. Decomposition, or decay, is the breakdown of dead matter. The rate at which this happens depends on the number of decomposing microorganisms, the temperature, and water and oxygen availability. Because this decomposition happens in an anaerobic environment, the process of producing biogas is also known as anaerobic digestion, so that decomposition of biogas is also known as anaerobic digestion. The anaerobic digestion is a sustainable method to convert organic waste materials into renewable energy and recycled nutrients such as nitrogen and phosphorus. Decomposition occurs when bacteria and fungi break down dead matter without oxygen. This can happen naturally in some soils, particularly waterlogged soils, and in lakes and marshes. When people overwater their houseplants and flood the roots, anaerobic decay can occur, which can kill them. Controlled anaerobic, or oxygen-free, digestion of animal manure is a way to treat manure to prevent foul odor production while generating a usable energy product. Anaerobic digestion is a collection of processes by which microorganisms break down biodegradable material in the absence of oxygen. The process is used for industrial or domestic purposes to manage waste or to produce fuels. Much of the fermentation used industrially to produce food and drink products, as well as home fermentation, uses anaerobic digestion. Anaerobic digestion occurs naturally in some soils and in lake and oceanic basin sediments, where it is usually referred to as "anaerobic activity." This is the source of marsh gas methane as discovered by Alessandro Volta in 1776.

The digestion process begins with bacterial hydrolysis of the input materials. Insoluble organic polymers, such as carbohydrates, are broken down to soluble derivatives that become available for other bacteria. Acidogenic bacteria then convert the sugars and amino acids into carbon dioxide, hydrogen, ammonia, and organic acids. These bacteria convert these resulting organic acids into acetic acid, along with additional ammonia, hydrogen, and carbon dioxide. Finally, methanogens convert these products into methane and carbon dioxide. The methanogenic archaea populations play an indispensable role in anaerobic wastewater treatments.

Anaerobic digestion is used as part of the process to treat biodegradable waste and sewage sludge. As part of an integrated waste management system, anaerobic digestion reduces the emission of landfill gas into the atmosphere. Anaerobic digests can also be fed with purpose-grown energy crops, such as maize. Anaerobic digestion is widely used as a source of renewable energy. The process produces a biogas, consisting of methane, carbon dioxide, and traces of other "contaminant" gases. This biogas can be used directly as fuel, combined heat and power, in gas engines, or upgraded to natural gas-quality biomethane. The nutrient-rich digestate also produced can be used as fertilizer. With the reuse of waste as a resource and new technological approaches that have lowered capital costs, anaerobic digestion has in recent years received increased attention among governments in a number of countries, among these the United Kingdom, Germany, and Denmark.

Under the right conditions, liquid manure will break down into biogas and a low-odor effluent. Biogas can be burned to produce heat, electricity, or both, and the anaerobically digested manure can be stored and applied to fields with significantly less odor than stored, untreated liquid manure. Anaerobic digestion does not reduce the volume or nutrient value of manure. If dilution water is added to the system, the volume of material to handle is increased. The following test can help you determine if anaerobic digestion is a viable option for your farm. If most of the following statements describe your farm, anaerobic digestion may be compatible with your operation.

- Manure is currently handled as a liquid. Very little bedding or frozen manure is handled and the manure in the handling system is free from high levels of copper sulfate and antibiotics.
- Odor control is a major concern.
- There is space on the farm to expand the manure handling system with the possibility for gravity flow from a barn to an anaerobic digester or from a digester to manure storage.
- Someone on the farm has the interest, time, and technical skills to learn about the anaerobic digestion process, make repairs, and perform general maintenance on equipment.
- Resources are available to finance an anaerobic digestion system.
- Adhering to recommended safety practices is standard procedure on the farm.

2.2.1 Anaerobic digestion

Anaerobic digestion, or the decomposition of organic matter by bacteria in the absence of oxygen, occurs naturally in liquid manure systems. The lack of oxygen and abundance of organic matter in liquid manure provide the proper conditions for anaerobic bacteria to survive. Unfortunately, uncontrolled anaerobic decomposition can cause the foul odors sometimes associated with liquid manure storage and spreading. However, controlled anaerobic decomposition not only can reduce the odors in liquid manure systems but also can turn odorous compounds and organic matter into energy. The effluent remaining after controlling anaerobic decomposition, equal in volume to the influenced material, is liquefied, low in odor, and rich in nutrients. This digested material is biologically stable and will resist further breakdown and odor production when stored under normal conditions.

Anaerobic bacteria transform manure and other organic material into biogas and a liquefied effluent during the three stages of biogas production. In the liquefaction stage, liquefying bacteria convert insoluble, fibrous materials such as carbohydrates, fats, and proteins into soluble substances. However, some fibrous material cannot be liquefied and can accumulate in the digester or can pass through the digester intact. Water and other inorganic material also can accumulate in the digester or pass through the digester unchanged. Most of the liquefied, soluble compounds are converted into biogas by the acid- and methane-forming bacteria during steps 2 and 3 of biogas production. In the second stage of anaerobic digestion, acid-forming bacteria convert the soluble organic matter into volatile acids and the organic acids that can cause odor production from stored liquid manure. Finally, methane-forming bacteria convert those volatile acids into biogas, which is composed of about 60% of methane, 40% of carbon dioxide, and trace amounts of water vapor, hydrogen sulfide, and ammonia. Not all volatile acids and soluble organic compounds are converted into biogas; some become part of the effluent. Fig. 2.2 shows the steps of anaerobic digestion.

Methane-forming bacteria are more sensitive to their environment than acid-forming bacteria. Acid-forming bacteria can survive under a wide range of conditions, while methane-forming bacteria are more demanding (Fig. 2.3). Under the conditions typical of liquid manure storages, more acid-forming bacteria can survive than methane-forming bacteria. Therefore acids are formed and are not converted into biogas. This excess of volatile acids can result in a putrid odor. Fig. 2.3 shows the conditions for survival of acid- and methane-forming bacteria. In a controlled, optimum environment, methane-forming bacteria survive and convert most of the odor-producing volatile acids into biogas. Conditions that encourage activity of both acid- and methane-forming bacteria include:

- an oxygen-free environment;
- a relatively constant temperature of about 95°F;
- a pH between 6.6 and 7.6; and
- a consistent supply of organic matter to "feed" upon.

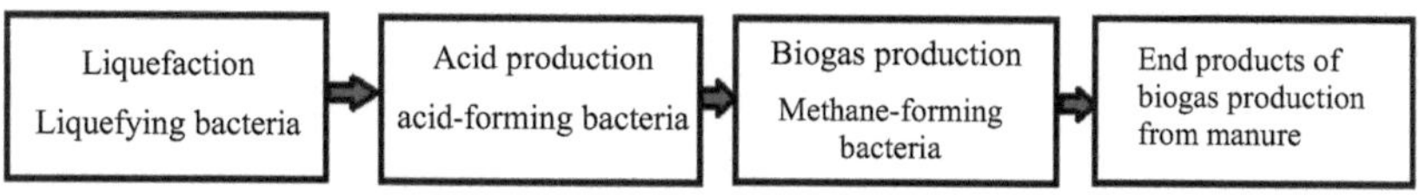

FIGURE 2.2 Steps of anaerobic digestion.

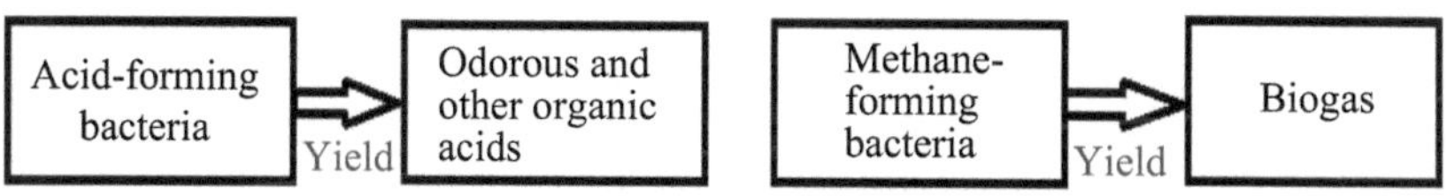

FIGURE 2.3 Conditions for survival of acid- and methane-forming bacteria.

For consistent operation of an anaerobic digester, the manure that "feeds" the bacteria should be:

- a flowable liquid, about 12% solids or less (for pump or flow requirements);
- not frozen;
- free from excess amounts of medication, feed additives, or chemical washes; and
- supplied fresh to the digester at least twice a day.

A uniform slurry of manure that does not separate easily, such as:

- dairy manure from scrape systems, which can include small amounts of fine, organic bedding such as sawdust, waste feed, milking center waste, or dilution water;
- swine manure from pull-plug or scrape systems; and
- poultry manure, diluted to about 10% solids with the grit settled out.

Acid-forming bacteria can survive (Fig. 2.3):

- with temperature fluctuations;
- in a wide range of pH conditions;
- with or without oxygen; and
- on a broad range of organic compounds as a food source.

Methane-forming bacteria can only survive:

- if the temperature is held relatively constant;
- in a narrow band of pH conditions;
- without oxygen; and
- on simple organic acids as a food source.

Anaerobic digestion is simply a continuation of the animal's digestive system, which is processed to turn manure into energy and effluent, just like an animal turns the feed into energy and manure.

2.2.2 Anaerobic digestion system

An anaerobic digestion system can provide an optimal environment for controlling anaerobic digestion. A typical system consists of liquid manure handling equipment, a heated anaerobic digester, gas utilization equipment, safety equipment, and effluent storage and handling systems. The anaerobic digestion system is an addition to the manure handling scheme, which is a step for manure processing between the barn and the storage facility. It does not replace any part of a typical manure handling system. Fig. 2.4 shows the schematic diagram of anaerobic digestion systems.

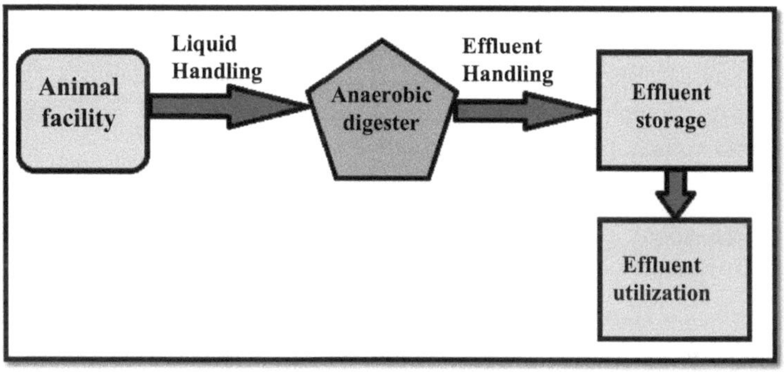

FIGURE 2.4 Schematic diagram of a typical anaerobic digestion system.

2.2.3 Liquid manure handling system

A liquid manure handling system (such as the system used to transport liquid manure from a barn to a storage facility) transports manure from the animal housing facility to the anaerobic digester and from the digester to the storage facility or spreader. When possible, the use of gravity flow is encouraged to reduce the energy consumption and complexity of the handling system. Bypass line routes manure around the digester when the manure is unsuitable for digestion or the digester is not operating.

2.3 Biogas production system

Biogas production is an eco-friendly strategy for energy production from biogas, and the residue can be used as a soil conditioner. Biogas is produced by the anaerobic biological breakdown of organic matter. It primarily consists of methane and carbon dioxide. Flammable methane is the main component of biogas (50%–85%), representing the main energy source. It can be used in boilers for heat generation. Upgraded biogas can be directly used in boilers. Biogas production from local agricultural waste using a laboratory-scale digester was evaluated by the many researchers. Various process variables affecting biogas production, like the nature of the feedstock and carbon-to-nitrogen (C/N) ratio, were evaluated. Among the different agroresidues screened, wheat stalk, soybean straw, and the black gram stalk were found suitable for biogas production. The production of biogas is done in the following way.

2.3.1 Waste collection

Municipal solid waste (MSW), often called garbage, is used to produce energy at waste-to-energy plants and at landfills in the United States. MSW contains

- biogas, or biogenic (plant or animal products), materials such as paper, cardboard, food waste, grass clippings, leaves, wood, and leather products;
- nonbiogas combustible materials such as plastics and other synthetic materials made from petroleum; and
- noncombustible materials such as glass and metals.

In 2018 about 262 million tons of MSW were generated in the United States, of which

- 52.5% were land-filled;
- 25.8% was recycled;
- 12.8% was burned with energy recovery; and
- 8.9% was composted.

Different wastes have been utilized for biogas production ranging from solids, semisolids, and liquids in the form of manure, wastes, and other residues obtained as by-products of industry, agricultural farms, disposal plants, etc. Biogas from these sources could be produced in various capacities with intent of meeting different energy demands (Schroder et al., 2008). However, production is strongly influenced by several factors including C/N ratio, temperature, pH, mineral composition, and presence of inhibitors (Esposito et al., 2012; Mata-Alvarez et al., 2000). The use of more than one substrate (codigestion) for biogas production has been tagged with some advantages including faster degradation rate, cost-effectiveness in terms of product formation, optimization of moisture, and nutrient contents, and reduction in concentration of inhibitory compounds (Divya et al., 2015; Luostarinen et al., 2009; Mata-Alvarez et al., 2000).

Utilization of municipal sludge from biogas production was reported by Kalloum et al. (2011), where the sludge was prepared to have 16 g/L of total solids (TS) with a total flora concentration of 1.67×10^6 germs/mL. It was subjected to anaerobic digestion for 33 days, and biogas production started from the seventh day and reached its maximum after the 26th day with about 280.31 N mL (45% methane content) based on a yield of 30 N mL of biogas/mg chemical oxygen demand (COD). This process resulted in the formation of digestate free of all tested pathogenic organisms with a reduction in sludge COD, BOD, and TS of 88%, 90%, and 81%, respectively. Connaughton et al. (2006) carried out a comparative study in two expanded granular sludge bed-anaerobic bioreactors at 15°C and 37°C using the brewery wastewater of 3136 ± 891 mg/L COD concentration. Following 194-day experiment, COD reduction was not significantly different between the two temperatures with a range of 85%–93%. Biogas production with a methane content of 50% was found when the organic loading rate (OLR) was fixed at 4.47 kg/(m^3 day) for 15°C with a liquid up-flow velocity of 5 m/h and at

37°C, and hydraulic loading rates of 1.33 kg/(m^3/day) were the optimum. Codigestion of cheese whey with dairy manure resulted in better biogas production. Kavacik and Topaloglu (2010) used two solid matter rates of 8% and 10% based on hydraulic retention time (HRT) of 5–20 days. Highest biogas production of 1.510 m^3 (m^3 day) with a methane content of about 60% was obtained from 8% total solid matter at HRT of 5 days and temperature of 34 °C. However, removal efficiencies of 49.5%, 49.4%, and 54% for TS, volatile solids (VS), and COD, respectively, were found to be optimum following the HRT of 10 days under the same conditions. Similarly, a mixture of equal ratio of cattle slurry and cheese whey was tested for biogas production. A methane yield of 343.43 L CH_4/kg volatile solid was achieved by using an OLR of 2.65 g volatile solid/L per day. Overall, total biogas production was found to be 621 L/kg volatile solid at an HRT of 42 days with 82% and 90% removal efficiencies of COD and BOD_5, respectively (Comino et al., 2012).

Cattle manure was supplemented with palm oil mill effluent (POME) for biogas production, and two bioreactors labeled R1 and R2 containing cattle manure in the absence and presence of POME. The digestion process was preceded for 5 days using batch mode of operation followed by semicontinuous operations using an HRT of 20 days. Higher biogas production was achieved in R2 with a methane content of 41% compared with 18% in R1. In the case of COD, R2 resulted in 10% higher reduction than R1 (Saidu et al., 2013). Enhanced biogas production was realized when olive mill effluent (OME) was mixed with laying hen litter (LHL) at a percentage dry matter of 10%. Biogas production was found to be several folds higher during the codigestion compared with when OME was used as monoculture. The COD conversion rates of 2.6-, 2.1-, and 1.94-folds were achieved for 3, 10, and 30 g/L. Thus an increase in the LHL concentration to 10% resulted in 90% increment in overall biogas production (Azbar et al., 2008).

Kafle et al. (2013) studied the potential of fish waste silage prepared by the addition of bread waste and brewery grain waste for biogas production. Following 96 days of digestion, maximum biogas production of 671–763 mL/g volatile solid with a methane recovery of 441–482 mL/g volatile solid was obtained. Thus fish waste silage digestion process was found to have significant HRTs and digestion periods of 21.0–23.8 days and 40.5–52.8 days, respectively. Five different feed mixes containing flushed dairy manure (FDM) and Turkey processing wastewater (TPW) were prepared by Ogejo and Li (2010) to determine the best substrate mixture for enhanced biogas production. Biogas production steadily increased from 0.072 to 0.8 m^3/kg volatile solid (methane content of 56%–70%) with an increasing concentration of TPW, and ratios of 1:1 and 1:2 FDM with TPW were found to produce biogas that can sufficiently generate electricity in a 50-kW generator for 5.5 and 9 hours, respectively.

2.3.2 Pretreatment of waste

2.3.2.1 Pretreatment methods

Utilization of organic wastes for biogas production using mono- and codigestion methods has been widely reported. Despite the multistage reactions associated with the production, hydrolysis as the first step is crucial and aids in increasing the overall yield. This is based on the fact that optimization of hydrolysis step results in decomposition of complex organic matter into high amounts of monomeric and oligomeric units that can easily be utilized under anaerobic condition for generation of biogas. The target of any pretreatment methods is to make the available nutrients accessible to most microbial species that speed up biogas utilization during anaerobic digestion process (Patil et al., 2016).

The pretreatment processes that can be applied in enhancing biogas production can be grouped into mechanical, thermal, chemical, and biological treatments, as described in the following sections. Mata-Alvarez et al. (2014) pointed out that vast majority of the studies associated with pretreatment of organic wastes for biogas production were devoted to mechanical, thermal, and chemical methods accounting for 33%, 24%, and 21%, respectively. The remaining percentage could be based on a combination of more than one method.

2.3.2.2 Mechanical pretreatment

Mechanical pretreatment aids in reducing the particles of the organic residues, without generating any products that may have inhibitory effect; thus this method is associated with increases in biogas production but its major drawback is that it is an energy requiring process. Based on this, advances in milling methods show that wet milling is more preferable to dry milling process due to its higher pulverization properties with minimal energy consumption (Fuerstenau and Abouzeid, 2002). Particle size reduction by household disposer and bead mill to enhance biogas production was studied by Izumi et al. (2010). Following the anaerobic digestion, methane yield was found to be 28% higher in a bead mill-treated waste at 1000 revolutions than those treated with household disposer. However, increasing the revolutions of bead mill resulted in further reduction of particle size, which leads to significant reduction in methane yield during the digestion process. Nah et al. (2000) reported the use of collision-plate at 300 kPa as a mechanical pretreatment for pilot scale anaerobic digestion of waste-activated sludge. The process resulted in solubilization of organic matter with five- to sevenfold increment in soluble Chemical Oxygen Demand (sCOD) and soluble total organic carbon, and solids retention time (SRT) of the digester was reduced to 6 days instead of 13 days. Biogas production of 790–850 L/kg VS was obtained, which corresponded to 30% VS removal efficiency.

High-pressure homogenization (HPH) has been reported to be used as a mechanical pretreatment method for anaerobic digestion of organic sludges with the ability to disrupt cells and sludge flocs, resulting in high sCOD and hydrolysis of macromolecules to their monomeric units. This method was not commonly applied for sludge pretreatment, as it was widely employed to stabilize food and dairy emulsions (Nah et al., 2000). Some of the advantages associated with the method include high disintegration potential, minimal operational costs, ease of operation, and handling with no chemical changes (Rai and Rao, 2009). The operation is generally dependent on shear stress as a result of pressure gradient build upon the sludge surfaces. Depending on the number of cycles, 50 and 40 MPa were found to be the optima for one and two cycles, respectively, and biogas production reached 3330 mL, which was 115% higher than that of unpretreated sludge. Thus HPH pretreatment led to 64% methane contents compared with 47% in unpretreated under the same conditions.

Ultrasonication has been found to be efficient in organic waste pretreatment prior to anaerobic digestion; the process involves releasing bioavailable nutrients through hydrolysis as a result of disruption of biosolid flocs and bacterial cells, which enhance nutrient solubilization as well as the overall process of anaerobic degradation (Elbeshbishy et al., 2011; Muller et al., 2009). Lehne et al. (2001) describe the complexity of sonication, as it involves a lot of processes including chemical reactions with radicals, shearing, pyrolysis, and combustion. The combined effects of these processes make ultrasonication efficient for pretreatment of organic sludges. Based on this, Pilli et al. (2011) developed an extensive review, where he describes ultrasonication as the most effective pretreatment method for sludge and the process efficiency is solely dependent on sludge characteristics. Thus Perez-Elvira et al. (2006) reported that pretreated organic wastes using ultrasonic methods resulted in an sCOD of six orders of magnitude compared with the unpretreated and this led to 10%–60% enhancement in biogas production during anaerobic digestion.

Using a frequency of 20 kHz for ultrasonication at different time on leachate samples showed higher organic matter solubilization based on increment in the ratio of sCOD to a total COD of 63% at 600 W/L following sonication for 45 minutes. Anaerobic digestion of the pretreated sample led to 40% higher biogas production than control with a methane production rate of 107 m^3 CH_4/day. This clearly demonstrated that low-frequency ultrasonication as a pretreatment step affects the overall performance of anaerobic digestion process (Oz and Yarimtepe, 2014).

In order to determine the effects of ultrasonic pretreatment on biogas production, Apul and Sanin (2010) studied three different sets of operational conditions in anaerobic digesters using both pretreated and unpretreated waste sludges. The parameters monitored were HRT and OLR, which were set at 15, 7.5, and 7.5 days (HRT) and 0.5, 1, and 0.5 kg VS/(m^3 day) (OLR) for setups 1, 2, and 3, respectively. Despite the fluctuations at early stages of operations, biogas production was apparent under steady-state conditions with 49%, 39%, and 56% higher production than control in setups 1, 2, and 3, respectively.

Specific energy of 500 kJ/kg TS was reported to be optimum for ultrasonic pretreatment of hog manure prior to anaerobic digestion. This was found to be sufficient for degradation of bound proteins as well as COD solubilization. Enhanced biogas production with 28% increment in methane content was observed. Based on this, low energy input was found to contribute significantly to the overall methane production rate (Elbeshbishy et al., 2011). Similarly, subjecting waste-activated sludge to ultrasonic pretreatment at 10,000 kJ/kg TS led to a better anaerobic digestion process. Biogas production was significantly enhanced by 172.56% compared with the control, which was apparent with high sCOD generated. A considerable increase up to 758 L/kg VS was obtained when relating the biogas production with the amount of volatile solid degraded.

Castrillon et al. (2011) carried out batch experiments to determine the effect of codigestion of cattle manure following its pretreatment using ultrasonication with glycerin for enhanced biogas production under temperature-controlled conditions in stirred tank reactors. At the initial stage, addition of 4% glycerin to the pretreated manure led to 400% increment in biogas production at 35°C. Further increment in biogas production up to 800% was obtained when the mixture of manure +4% glycerin was subjected to sonication at 20 kHz and 0.1 kW for 4 minutes. Highest biogas production was obtained at 55°C on a pretreated mixture of manure +6% glycerin, which yielded 348 L methane/kg COD utilized. Based on the available literature associated with mechanical pretreatment, ultrasonication has been the most widely reported for pretreatment of wastewater, sludge, and manure during anaerobic digestion processes for biogas production (Apul and Sanin, 2010; Elbeshbishy et al., 2011; Oz and Yarimtepe, 2014).

2.3.2.3 *Thermal pretreatment*

This method aids in hydrolyzing complex organic constituents of organic wastes and has been found to enhance anaerobic digestion. Li and Noike (1989) showed that pretreatment using this method results in high solubility of organic waste constituents, which promote the conversion of the hydrolyzed substances under anaerobic condition into biogas through volatile organic acid production. Pretreatment of several organic wastes including cow manure, pig manure, and municipal sewage sludge was studied by Qiao et al. (2011) using thermal method at 170°C for 1 hour. Increment in biogas production of 7.8%, 13.3%, and 67.8% was obtained for cow manure, pig manure, and municipal sewage sludge, respectively, compared with the control. Thus the commonly employed temperatures for thermal pretreatment were between 60°C and 180°C, and above the upper limit of this range, compounds that slow down the digestion process may be formed (Wilson and Novak, 2009). Methane production of swine waste using this pretreatment method at 170°C and 7 bar was reported to be 35% higher than the control (Gonzalez-Fernandez et al., 2008), while improvement of biogas production by 60%–70% was also reported during sludge digestion at 175°C by Haug et al. (1978).

Based on this, Bougrier et al. (2008) pointed out that thermal pretreatments can be grouped into two, where temperatures between 70°C and 121°C resulted in 20%–30% enhancement in biogas yield and up to 100% increment could be seen at 160°C–180°C.

Bougrier et al. (2006a) studied the effect of different thermal treatments of sewage sludge prior to digestion. The temperatures considered were 130°C, 150°C, and 170°C for 30 minutes. Remarkable results obtained for pretreated sludge at 150°C and 170°C showed 60% and 70% COD reduction, respectively, compared with 34% in the control. Thus considering 20°C rise in temperature (from 130°C to 150°C and to 170°C), methane yield improvement of 18 L CH_4/kg VS, which corresponds to 648 kJ/kg VS, was obtained. This indicated that biogas production was highest at 170°C, which could be related to higher sludge solubilization than other temperatures. Additionally, about twofold removal rates of TS and VS were seen in anaerobically digested pretreated sludges compared with the control and 170°C led to about 80% removal efficiency and biogas yield.

Sludge obtained from municipal wastewater treatment plant was subjected to pretreatment at 70°C in order to increase the amounts of utilizable organic matter in the form of volatile dissolved solids (VDS) and sCOD. The effects of pretreatment periods (9, 24, 48, and 72 hours) were considered, which resulted in significant enhancement of VDS from 1.5 g/L VDS in the untreated sludge to 11.9–13.9 g/L VDS in 9-, 24-, and 48-hour pretreated sludges. This shows an apparent increase in the ratio of soluble to total organic matter constituents from 5% to 50% after different periods of pretreatment at 70°C. Following the anaerobic digestion, difference in biogas yield was noticed after 10 days with 50% increment in 9-, 24-, and 48-hour pretreated sludges and the total biogas yield for the 37 day assay revealed 30% and 15% increments for 9- and 24- to 48-hour pretreatments, respectively. Overall pretreatment time of 72 hour does not show a better result, which could be linked to the presence of some inhibitory compounds associated with volatile fatty acids (VFAs; Ferrer et al., 2008).

Similarly, Mottet et al. (2009) carried out sludge pretreatment at various temperatures (110°C, 165°C, and 220°C) and maximum VDS and sCOD of 24% and 27% were obtained in 220°C pretreated sludge; however, biodegradability and methane yield were found to be lower than the control, which could be related to the production of recalcitrant compounds including Amadori and melanoidins. The best pretreatment was that of 165°C, where

VDS of 15% and sCOD of 18% were observed with significant increase in biodegradability under anaerobic process from 47% to 61%. Methane yield was found to be 30% higher in 165°C pretreated sludge (215 mL CH_4/g COD) than the control (165 mL CH_4/g COD).

In another development, freezing and thawing of sludge lead to cellular disruption accompanied by releasing constituent materials in the supernatant, as reported by Ormeci and Vesilind (2001). This motivated Montusiewicz et al. (2010) to study their effect on sewage sludge during anaerobic process. A remarkable reduction of 12%, 16.1%, and 16.9% in total COD, TS, and VDS, respectively, was observed, while sCOD was found to be two times higher in pretreated sludge by freezing and thawing than the control. Biogas production was enhanced by this method with 1.31 m^3/kg VS, and this value was 1.5 times higher than what was obtained in unpretreated sludge. Thus several reports have shown the effects of thermal pretreatment prior to anaerobic digestion for improving the biogas production.

Besides the commonly employed thermal pretreatment, microwave pretreatment has been found to be effective in organic waste stabilization as well as biogas production. This involves the use of electromagnetic radiation between the wavelength ranges 1 mm and 1 m, which is equivalent to 300-GHz to 300-MHz oscillation frequency and a frequency of 2450 MHz (12.24-cm wavelength) with an energy of 1.02×10^{-5} eV that could be sufficient for pretreating waste-activated sludge (Eskicioglu et al., 2007; Mudhoo and Sharma, 2011). This method has some advantages including fast heating and penetration, elimination of pathogens, ease of handling and control, efficient sludge dewaterability, and sludge reduction (Jones et al., 2002), and these make it more effective than conventional thermal technique.

The efficiency of microwave irradiation was explained based on two mechanisms: thermal effect, where temperature increase results in interaction of the electric field with dipolar molecules (water, proteins, fats, and other organic complexes), causing molecular rotation accompanied by internal pressure build up, which generate internal heating as well as cellular destruction. Alteration of polarized side chains of macromolecules caused by alternating electric field of microwaves resulting in disruption of intermolecular interactions which affect the secondary and tertiary interactions is considered as nonthermal effect (Appels et al., 2013; Eskicioglu et al., 2007; Solyom et al., 2011).

Eskicioglu et al. (2007) studied the effect of microwave irradiation on waste-activated sludge within a range of 50°C–96°C; no significant difference was observed in terms of sCOD, protein, and polysaccharide solubilization between the pretreated and control using conventional heating. An improved biogas production was observed in pretreated sludge at 96°C over the control, which indicated the contribution of nonthermal effect of microwave irradiation in anaerobic digestion process with $16 \pm 4\%$ higher yields than the control after 15 days of digestion.

Moreover, using continuous flow anaerobic sludge digesters, microwave pretreated sludge showed 3.6- and 3.2-fold increment in sCOD-to-total-COD ratios at 1.4 and 5.4 TS, respectively, with 17.0% improvement in biogas production compared with control following a 34-day digestion process (Eskicioglu et al., 2006). Similarly, Eskicioglu et al. (2009) irradiated waste-activated sludge at 50°C–175°C and observed that increasing trends existed between soluble to total COD and soluble solids to TS and temperature. The proportions of 24%, 28%, and 35% for soluble to total COD and 19%, 21%, and 32% for soluble solids to TS with a corresponding temperature of 120°C, 150°C, and 175°C, respectively, were recorded. Despite the initial inhibition of methane production in the first 9 days, sludge irradiated at 175°C showed better performance with 31% higher biogas production than the control after 18 days of digestion.

Thus organic matter solubilization and biogas production were used to study the effect of microwave absorbed energy in sludge samples; when the energy was 0.54 kJ/mL at 1000 W, higher solubilization effect was found, which yielded 7.1% higher methane content than the control. Further improvement in methane production of 15.4% was obtained when the energy was increased to 0.83 kJ/mL (Solyom et al., 2011). Park and Ahn (2011) reported the effect of microwave pretreatments on mixtures of primary and secondary sludges during anaerobic digestion, which resulted in 3.2-fold increment in sCOD-to-total COD ratio and VS removal of 41% with daily biogas production of 53% at a reduced HRT of 5. Thermal pretreatment using both conventional and microwave methods aids in efficient solubilization of sludge, which shortens the rate limiting steps (hydrolysis) during anaerobic digestion and overall results in higher biogas yield, as indicated in several studies reported herein.

2.3.2.4 Chemical pretreatment

This method is effective in the breakdown of organic constituents through the action of acids, alkali, and oxidants. Among the chemical-based methods, oxidation (ozonation and peroxidation) has been found useful in pretreatment resulting in sludge solubilization. A dose-dependent relationship exists between sludge solubilization and oxidant concentration up to a certain limit. Thus ozonation/peroxidation being oxidative process tends to show higher rate of sludge biodegradation with compromised biogas yield (Carrere et al., 2010). Acid and alkaline methods are mostly applied in combination with other methods for sludge solubilization.

Oxidation of both organic and inorganic compounds of sludge can be achieved in the presence of ozone; this leads to cellular disruption, flocs disintegration, and high COD solubilization, and under extreme condition, mineralization occurs. Depending on the intended applications, ozone concentration of 0.05 and 0.5 g O_3/g TS has been found to be adequate for the pretreatment of sludge (Kameswari et al., 2011; Tyagi and Lo, 2011). Bougrier et al. (2006b) found that biogas yield during digestion of ozonized sludge correlated with an increase in ozone concentration up to 0.15 g O_3/g TSS, followed by a sharp reduction.

Weemaes et al. (2000) studied the effect of ozone pretreatment prior to anaerobic digestion of sludge using 0.1 g O_3/g COD. About 38% of the organic constituents were oxidized and 29% were solubilized, which lead to changes in VSS compositions of the ozonized sludge. The digestion process resulted in 1.8 and 2.2 increments in methane yield and rate, respectively, compared with the control. Three reactors operated at SRT of 25 days were fed with different ozonized sludges pretreated at 2.65, 1.33, and 0.66 mg O_3/g VSS, and a control reactor was also used, which was fed with unpretreated sludge. Biogas production was found to be highest in ozonized sludge treated at 1.33 mg O_3/g VSS with more than 200% increment compared with the control. In addition, 33% enhancement in biogas production was found in 0.66 mg O_3/g VSS (Ak et al., 2013). Similarly, 20 mg O_3/g per TSS was found to be adequate for higher sludge solubilization with 28% and 17% improvement in daily biogas production in reactors operated under mesophilic and thermophilic conditions, respectively, when compared with the respective control (Carballa et al., 2007).

Furthermore, peroxidation involving the use of H_2O_2 activated by iron salts was reported to disintegrate sludge and rupture the cellular components that lead to increased concentration of sCOD. Three oxidative pretreatments were carried out using Fenton peroxidation, dimethyldioxirane (DMDO), and peroxymonosulfate (POMS) methods on activated sludge to determine their rate of solubilization as well as biogas production. The breakdown of organic matter was verified by monitoring the level of disolved solid (DS), COD, and Biological oxygen demand (BOD). The POMS and DMDO had higher solubilization effects than Fenton peroxidation. In addition, 2.5- and 2-fold increments in terms of biogas production were observed in sludge pretreated with DMDO and POMS, respectively. Comparatively, low biogas yield was recorded for peroxidation but the methane contents in all the three treatments were between 65% and 70% (Dewil et al., 2007).

Another oxidant with good potential is peracetic acid, which reacts with organic matter to form hydroxyl radicals that further react with other components. The process does not produce toxic by-products; instead, it dissociates into water and acetic acid, which can contribute its carbon skeleton for biogas production (Appels et al., 2011). Biogas enhancement of 21% was reported by Appels et al. (2011) with good sludge disintegration and organic matter solubilization compared with the control. Thus an oxidation process (ozonation, peroxidation, and peracetic acid) is capital-intensive with good sludge solubilization effect; however, higher ozonation results in the formation of soluble compounds, which may have deleterious effect when discharged into the environment, while low pH, proper handling, as well as specialized equipment associated with peroxidation based on its corrosiveness limit their applications. As such, the benefits as well as cost implication should be critically looked at before adopting any process on a large scale. Acid and alkali pretreatments have been used to solubilize sludges in an easy way with little or no energy demand and overall the process may result in pathogen-free digestate. However, extreme pH is required in both cases for efficient solubilization and this to some extent is a drawback to this method, as further treatment involving neutralization may be required. Devlin et al. (2011) showed that pretreatment of waste-activated sludge using HCl (pH 1–6) led to higher solubilization of macromolecules and COD; based on this, pH 2 was selected to be the best. Alkaline pretreatment was studied by Lin et al. (1997) where different concentrations of NaOH were used. Four anaerobic digestion setups were prepared containing 20 meq/L NaOH-pretreated sludge at 1% TSS, 40 meq/L NaOH-pretreated sludge at 1% TSS, 20 meq/L NaOH-pretreated sludge at 2% TSS, and a control where unpretreated sludge was used. Reactors fed with the pretreated sludges showed better performance with higher COD removal rate. Biogas production was found to be highest in 20 meq/L NaOH-pretreated sludge at 2% TSS with a yield of 163% in comparison with the control. Thus the alkali used in sludge solubilization has been ranked by Kim et al. (2003) as $NaOH > KOH > Mg(OH)_2 > Ca(OH)_2$. Lin et al. (2009) reported about 83% improvement in sCOD with high concentration of VFA in 8 g NaOH/100 g TS_{sludge}. Biogas production was found to be 183.5% higher with 0.32 m^3 CH_4/kg $VS_{removal}$ than the control.

Chemical pretreatment involving the use of alkali and acid is mostly used in combination with other treatment techniques as indicated in the following section.

2.3.2.5 Combined pretreatment

This involves the use of more than one pretreatment method where their combined effects result in better sludge solubilization, which in turn leads to higher biogas production. Several studies reported the combination of physical (mechanical and thermal) and chemical (acid, alkali, and ozone) methods, and the results seem to be attractive. Some of the combined pretreatments reported in the literature have been presented below and briefly explain the synergistic effects of the pretreatments in enhancing the biogas production.

Systematic combination of thermal and chemical methods was found to be appropriate for enhanced sludge disintegration and biogas production, as reported by Kim et al. (2003). Out of the four pretreatment methods tested on waste-activated sludge (WAS), thermochemical method at 121°C for 30 minutes and 7 g NaOH/L led to the maximum biogas production of 3367 L CH_4/m^3 WAS, which was 38% higher than the control. Various sludge pretreaments were carried out at temperatures of 50°C–90°C and pH of 8–11. A linear relationship was found between the rate of sludge disintegration and the applied thermochemical treatments. The best pretreatment condition was found to be 90°C and pH 11, which resulted in about 46% reduction in volatile suspended solids with methane content of 0.28 L CH_4/kg VSS (Vlyssides and Karlis, 2004).

Thermochemical method for pretreating activated sludge revealed high sCOD solubilization of 27.7%, 31.4%, and 38.3% corresponding to 25°C, 35°C, and 55°C, respectively, after 4 hours of incubation when the concentration of alkali was kept at 45 meq NaOH/L. Following anaerobic digestion based on 20-day HRT, the pretreated sludge at 25°C, 35°C, and 55°C led to a methane content of 274, 286, and 310 mL CH_4/g VS, respectively; which were found to be 66%, 73%, and 88% higher than the control (Heo et al., 2003). Shehu et al. (2012) reported the use of Box–Behnken design to optimize the sludge pretreatment by thermoalkaline method for improved biogas production. The developed quadratic model revealed the highest sludge disintegration of 61.45% and biogas yield of 36% higher than the control at an optimum temperature of 88.50°C and NaOH of 2.29 M (24.23% w/w TS). The adequacy of the model was confirmed based on its coefficient of determination (R^2) of 99.5%.

In recent days, electrochemical method has been applied in the pretreating sludge prior to anaerobic digestion and has shown a high level of flexibility, good sludge solubilization, and environmental friendliness with minimal temperature requirement (Song et al., 2010; Xu et al., 2014). Thus combined pretreatment methods are advantageous in enhancing sludge solubilization, sludge sanitation, dewaterability, and anaerobic digestion. However, insight in their wide application and economic analysis will prove their effectiveness, as some of them are still based on laboratory proof of concepts.

2.3.2.6 Biological pretreatment

This method has been advocated as a result of its environmental friendliness, where different microbial systems work synergistically in hydrolyzing complex organic matter, thereby improving anaerobic digestion (Gupta et al., 2012). In the case of organic sludge, extracellular enzyme catalyzed reactions within the system lead to sludge disintegration and solubilization. Although this process is environment-friendly and less capital-intensive compared with other methods, the process requires longer time and controlled environmental conditions for microbial growth. Hasegawa et al. (2000) reported the use of thermophilic aerobic bacteria to enhance the sludge disintegration prior to anaerobic digestion. The isolate SPT2-1 obtained from an aerobic digester growing under thermophilic condition (60°C–70°C) with a good pH tolerance of 5.0–8.5 showed increased sludge solubilization in terms of volatile suspended solids (25%–30%) by secreting hydrolytic enzymes (e.g., protease, amylase, etc.) and biogas production was found to be 1.5-fold compared with the control. Addition of Bacillus sp. to enhance anaerobic digestion showed about 95% increment in methane production compared with the control. Similarly, addition of some micronutrients (Fe^{2+}, Ni^{2+}, Co^{2+}, and Mo^{2+}) together with Bacillus sp. further improved the methane production by 167%. Following the statistical design, the actual concentrations that resulted in higher methane yield were 4.5, 0.75, 0.45, 0.09, and 12 mg/g VS for Fe^{2+}, Ni^{2+}, Co^{2+}, Mo^{2+}, and Bacillus sp., respectively (Noyola and Tinajero, 2005).

Singh et al. (2001) studied the effects of microbe activating technology, where two microbial enhancers (Aquasan and Teresan) were applied to cattle dung for anaerobic digestion. Different concentrations of Aquasan (10, 15, and 20 ppm) were applied to cattle dung prior to digestion, and biogas yield was found to be 45.1–62.1 L/kg dry matter. Following the incubation of 15 days, addition of Aquasan (15 and 20 ppm) increased the biogas yield by 15%–16%. Thus maximum biogas production was found to be 39% and 55% higher with single and double dosages of 15 ppm of Aquasan. In the case of Teresan at 10 ppm, biogas yield was 34.8% higher than the control.

Aerobic thermophilic bacteria that were found to have close similarity with *Geobacillus thermodenitrificans* were used to pretreat sludge prior to digestion. About 21% reduction in VS was obtained with 2.2-fold increment in biogas production (Miah et al., 2005). Bacterium B4 with potent hemicellulose degrading potential led to 30% improvement in the biogas potential of cattle manure, as shown by Angelidaki and Ahring (2000). Also a significant improvement in methane yield of 1100.46 mL/g VS, which was 280% higher than the control, was reported based on the biological–physicochemical pretreatment method. This was based on ultrasonication for 10 minutes, citric acid of 500 mg/L, and inoculation of Bacillus sp. at 9 wt.% prior to anaerobic digestion of oily wastewater sludge as reported by Peng et al. (2014). Generally, pretreatment using this method is found to be environmentally friendly with minimal energy consumption compared with other techniques, but offensive odor generation with limited waste reduction is one of its major drawbacks.

2.3.2.7 *Enzymatic pretreatment*

This is one of most promising biological pretreatment methods in which the rate of hydrolysis of organic waste is further enhanced prior to the digestion process. The enzyme-based pretreatment in solubilizing sludge starts immediately within the system, unlike microbial process that requires acclimatization time. However, this process is cost-intensive, which is its major drawback. Biolysis E [commercialized by Ondeo-Degremont (Suez)] made up of different enzyme systems (proteases, amylases, and lipases) shows the higher solubilization effect with 40%–80% reduction of sludge (Deleris et al., 2003). Mayhew et al. (2003) studied the enzymatic hydrolysis of waste-activated sludge, where the hydrolyzed sludge was found to be almost free from pathogens, and 10% enhancement in biogas production was recorded using a retention time of 2 days at 42°C. In addition, Parawira (2012) reported that lipase-catalyzed hydrolysis of lipid-rich sludge prior to anaerobic process enhanced both sludge disintegration and methane production. Thus using enzymatic pretreatment for biogas production holds a promising future, but this can only be a reality if studies are tailored toward producing cheap and genetically engineered enzymes.

2.3.3 Mixing or homogenizing tank

Submersible mixers are used in digesters to mix and homogenize the substrate, thus increasing the output of the plant and preventing solids from settling, which in the long run can reduce the efficiency of the plant, for example, by clogging pipes. Fig. 2.5 shows the biogas production process.

2.3.3.1 Anaerobic digester tank

An anaerobic digester is a sealed, heated tank, which provides a suitable environment for naturally occurring anaerobic bacteria to grow, multiply, and convert manure into biogas and a low-odor effluent. Typical digesters have been insulated, squat, silo-like structures or in-ground rectangular or round concrete tanks where rigid or flexible covers have been used. They are designed to hold about 20 days of manure and a small supply of biogas. Manure, added daily to the digester, remains inside for about 20 days, the retention time, before flowing to the storage facility or spreader. Because there is no volume reduction with anaerobic digestion, the same amount of material added daily to the digester is also removed daily. While manure is flowing through the digester, the bacteria convert organic matter into biogas and effluent.

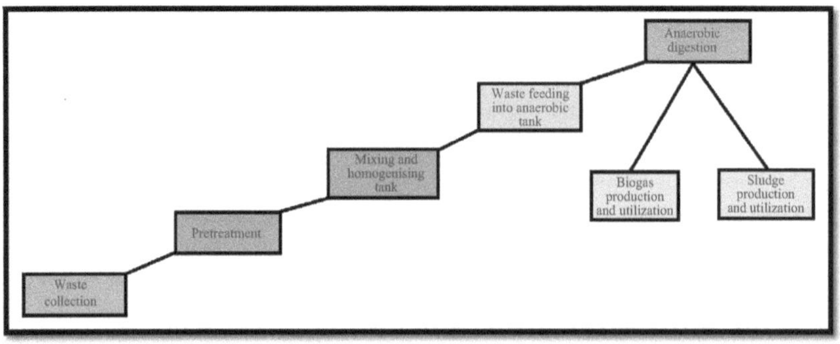

FIGURE 2.5 Biogas production process.

During the retention time, lightweight material such as bedding or animal hair can float to the top of the digester, forming a crusty scum, and heavy or insoluble material such as dirt can settle to the bottom. Settling reduces the effective volume of the digester and can cause incomplete digestion and odor problems, while crusting can keep gas from escaping the surface of the digesting manure. To control, settle, and scum formations, material in the digester can be agitated by a slurry pump, a mechanical stirrer, or strategic placement of the heating pipes. Slurry pumps are an effective way to keep material in the digester well-mixed. Mechanical mixing adds complexity to the system, but can aid thermal uniformity, reduce settling, and break up crust formation. Mechanical mixing may be necessary for certain manure handling systems such as flush systems where solid and liquid portions may separate easily into distinct layers within the digester. Strategic placement of the heating pipes will encourage thermal circulation and reduce settling problems.

The heating system is a critical part of the anaerobic digester. Heating pipes in which hot water circulates must be able to heat all material entering the digester to 95°F and to resist corrosion from manure. Adding manure to the digester as soon as possible after it is excreted from the animal will help minimize heat requirements.

- Digester size

To get an idea of the size of an anaerobic digester, consider one designed for 200 milking cows with a 20-day retention time.

Assuming each high-producing milk cow produces 2.2-ft.^{3} manure per day, the daily volume of manure from these milking cows would be:

$$200 \text{ cows} \times 2.2 \text{ ft.}^3 \text{ manure per day per cow} = 440 \text{ ft.}^3 \text{ manure per day}$$

If dilution water is needed for manure flowability or added from the milking center at a rate of 3 gallons per cow per day, the additional volume added daily would be:

$$\frac{200 \text{ cows} \times 3 \text{ gallons water per cow per day}}{7.5 \text{ gallons water/ft.}^3 \text{ water}} = 80 \text{ ft.}^3 \text{ water per day}$$

The total material added daily to the digester, therefore, would equal:

$$440 \text{ ft.}^3 \text{ manure per day} + 80 \text{ ft.}^3 \text{water per day} = 520 \text{ ft.}^3 \text{ material per day}$$

To hold 20 days worth of manure and water, the digester volume would need to be:

$$520 \text{ ft.}^3 \text{ per day} \times 20 \text{ days} = 10,400 \text{ ft.}^3$$

A digester with a rigid cover, a 3-ft. head space for gas collection, and a material volume (no bedding included) of 10,400 ft.^{3} would be approximately 15 ft. deep and 33 ft. in diameter.

2.3.4 Gas utilization equipment

Biogas is collected in the head space of the anaerobic digester (or under the flexible cover) and has about 60% of the energy density of natural gas (methane), which is about 600 British thermal unit (BTU)/$ft.^3$. With minor equipment modifications, biogas can be used in the same applications as liquid petroleum (LP) gas, propane, or natural gas. Biogas is best suited for stationary continuous operation because of its low energy density, the corrosive nature of some of the impurities, and the constant production rate. Biogas utilization equipment typically consists of either an engine-generator set with electric utility hookups, an engine operating hydraulic or air pumps, or a gas boiler. Utilization equipment should be housed in separate equipment shed apart from the digester to prevent corrosion.

Operating biogas-powered equipment continuously keeps the equipment temperature high enough to prevent condensation and sulfuric acid formation. Sulfuric acid is highly corrosive and can ruin expensive engines or boilers. Because biogas is a gas and not a liquid fuel, it is not practical for fueling vehicles. It would take 240 $ft.^3$ of biogas to produce the same energy as 1 gallon of fuel oil. Biogas cannot feasibly be compressed to a liquid fuel due to its low energy density. For electricity production, biogas is piped to an internal combustion engine. The engine drives a generator to produce electricity that can be used on the farm or sold. To maintain continuous operation, the engine throttle is adjusted to balance biogas use with production. Waste heat from the engine is used to heat the digester and for other farm heating needs. Most systems produce about 2 kWh per day per 1400 pound cow. Many utility companies in Pennsylvania pay only about 2¢ per kilowatt-hour for farm-produced electricity, much less than the consumer price for a kilowatt-hour. Therefore maximizing the replacement of purchasing energy with farm-produced energy will improve the economics of on- farm electricity generation.

2.3.5 Safety equipment

Because biogas is a potentially dangerous gas, safety devices such as gas detectors, flame traps, physical barriers, and warning signs control and minimize the hazards of biogas and manure storage. See the next section and other resources for more detailed information about required safety devices.

2.3.6 Safety hazards

Anaerobic digesters are confined spaces that pose a potential immediate threat to human life. They are designed to seal out oxygen, making death by asphyxiation possible within seconds of entry. Toxic gases such as hydrogen sulfide and ammonia accumulate inside a digester. Never enter an empty digester without extensive venting with mechanical fans, checking for toxic gases with gas detection equipment, and following safe-entry procedures. Natural ventilation is not enough to remove toxic gases from the digester or to provide sufficient breathable air. Dense hydrogen sulfide gas will sink to the bottom of the tank, lighter ammonia will linger on the top of the tank, and neither gas will escape without mechanical ventilation. Moreover, methane is explosive when mixed with air in concentrations of 5%–15%. A leak in a gas line will create a fire hazard. Anaerobic digesters are at least as dangerous, if not more so, than manure pits. For more information about safety concerns associated with anaerobic digesters, call the National Institute for Occupational Safety and Health at 1-800-35-NIOSH. See Penn State Extension Fact Sheet, Manure Storage Hazards, for an outline of safety procedures for entering manure pits.

2.3.7 Potential advantages of controlled anaerobic digestion

- There is a substantially less odor with digested manure than that is with stored liquid manure.
- Energy produced from biogas offsets the cost of the investment.
- The nutrient content of digested manure is equal to that of raw manure.
- Digested manure is more liquefied than raw manure, making it easier to pump long distances.
- Digested manure is biologically stabilized, making it easier to store for long periods without odor problems.
- Homogenous digested manure performs well in liquid application systems.
- Rodents and flies are less likely to be attracted to digested manure.
- Emissions of methane from liquid manure storage areas are reduced.

2.3.8 Potential disadvantages of anaerobic digestion

- The initial investment may be costly for a digestion system. Bankers and lenders may be wary of lending money for these systems.
- The digester requires proper care and feeding, just like an animal. Technical knowledge of the digestion process and good management are required.
- Labor is required for preventive and unscheduled maintenance. Ideally, one person will be in charge of the digester, and the digester takes precedence over that person's other farm duties.
- Daily maintenance tasks are minimal, but weekly oil changes, regular engine overhauls, and periodic digester clean out are required.
- There is no reduction in the amount of manure to be handled. If water is added to the system, the volume is increased.
- Nutrient conservation may be undesirable on a farm with excess nutrients to manage.
- Much of the nitrogen in raw manure is converted from its organic form of ammonium. Ammonium can be transformed to either ammonia or nitrate.

Ammonia can be lost from unincorporated, field-applied manure. Nitrate can be reached through the soil and may eventually reach groundwater.
- Field application and managing to reduce nitrogen losses may be more demanding for a digester event than for untreated liquid manure.
- Anaerobic digesters can be a farm safety hazard.
- *Alternatives to electric generation from anaerobic digesters*:

With minor equipment modifications, biogas can be used as a substitute for natural gas. Running a gas-fired boiler is an inexpensive and efficient method to use biogas. The obstacle will be finding uses for the heat, especially in the summer. Absorption (heat-activated) cooling systems are a promising technology for using excess heat, but currently have a high initial cost. Another option is to remove carbon dioxide and hydrogen sulfide from the biogas and sell it as natural gas. Scrubbing the gas, finding a market, providing the buyer with a dependable supply of gas, and maintaining the distribution equipment require money, time, maintenance, and management. Additionally, natural gas will be sold for a much lower price than electricity. Although other options are available for biogas utilization, electricity is the most versatile and valuable energy product from biogas.

2.3.9 Planning for future changes

If expansion of an animal production operation or a new facility is planned, but an anaerobic digestion system is not included in the layout, leaving adequate space and installing a compatible manure handling system could add to the flexibility for the future. There may be a time when investing in a digester is just the right step for a farm.

Separating solids prior to anaerobic digestion and digesting only the organic matter in the liquid portion of the manure may produce a higher quality biogas (70% methane has been observed) and typically will reduce crusting and settling problems. The solids can be field-applied, sold, or composted and used for animal bedding. Separation and marketing of solids can generate farm income. Replacing bedding with composted solids could be a money-saver if a substantial amount of bedding currently is purchased and a solids separator is owned. However, if a solids separator needs to be purchased, the savings in bedding costs may not cover the cost of solids separation.

2.4 Stages of biogas production

Biogas is produced using well-established technology in a process involving several stages:

- Biowaste is crushed into smaller pieces and slurrified to prepare it for the anaerobic digestion process. Slurrifying means adding liquid to the biowaste to make it easier to process.

- Microbes need warm conditions, so the biowaste is heated to around 37°C.
- The actual biogas production takes place through anaerobic digestion in large tanks for about 3 weeks.
- In the final stage, the gas is purified (upgraded) by removing impurities and carbon dioxide.

The heart of the biogas plant is its digestion process, which can be divided into four separate steps: hydrolysis, acidification, acetogenesis, and methanogenesis.

The important part of the digestion process outputs is the production of nutrient materials. Considering the mass balance of the whole biogas plant, approximately 5%–10% is directed out from the process as biogas flow. This leaves the remaining fraction, called a digestate, to owe more than 90% of the volume of the materials that go through the treatment process. Fig. 2.6 shows the stages of biogas production.

- *Hydrolysis*:

  In most cases, biogas is made up of large organic polymers. For the bacteria in anaerobic digesters to access the energy potential of the material, these chains must first be broken down into their smaller constituent parts. These constituent parts, or monomers, such as sugars, are readily available to other bacteria. The process of breaking these chains and

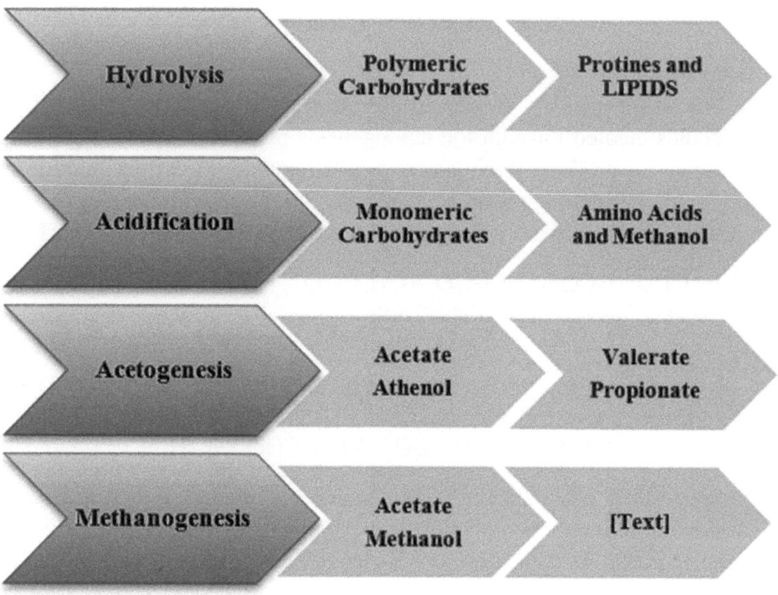

FIGURE 2.6 Stages of biogas production.

dissolving the smaller molecules into solution is called hydrolysis. Therefore hydrolysis of these high-molecular-weight polymer components is the necessary first step in anaerobic digestion. Through hydrolysis, the complex organic molecules are broken down into simple sugars, amino acids, and fatty acids. Acetate and hydrogen produced in the first stages can be used directly by methanogens. Other molecules such as VFAs with a chain length greater than that of acetate must first be catabolized into compounds that can be directly used by methanogens.

Hydrolysis is a reaction with water. Acid and base can be used to accelerate the reaction. However, this occurs in enzymes as well. In anaerobic digestion, the enzymes are exoenzymes (cellulosome, protease, etc.) from a number of bacteria, protozoa, and fungi.

$$\text{Biogas} + H_2O \rightarrow \text{monomers} + H_2$$

- *Acidogenesis*:

  The biological process of acidogenesis results in further breakdown of the remaining components of acidogenic (fermentative) bacteria. Here, VFAs are created, along with ammonia, carbon dioxide, and hydrogen sulfide, as well as other by-products. The process of acidogenesis is similar to the way milk sours. During acidogenesis, soluble monomers are converted into small organic compounds, such as short chain (volatile) acids, ketones (glycerol and acetone), and alcohols.

$$C_6H_{12}O_6 + 2H_2 \rightarrow 2CH_3CH_2COOH + 2H_2O$$

$$C_6H_{12}O_6 \rightarrow 2CH_3CH_2OH + 2CO_2$$

- *Acetogenesis*:

  The third stage of anaerobic digestion is acetogenesis. Here, simple molecules created through the acidogenesis phase are further digested by acetogens to produce largely acetic acid, as well as carbon dioxide and hydrogen. The acidogenesis intermediates are attacked by acetogenic bacteria; the products from acetogenesis include acetic acid, CO_2, and H_2.

$$CH_3CH_2COO^- + 3H_2O \rightarrow CH_3COO^- + H^+ + HCO_3^- + 3H_2$$

$$C_6H_{12}O_6 + 2H_2O \rightarrow 2CH_3COOH + 2CO_2 + 4H_2$$

$$CH_3CH_2OH + 2H_2O \rightarrow CH_3COO^- + 2H_2 + H^+$$

$$2HCO_3^- + 4H_2 + H^+ \rightarrow CH_3COO^- + 4H_2O$$

Several bacteria contribute to acetogenesis, including *Syntrophobacter wolinii*, propionate decomposer *Syntrophomonas wolfei*, butyrate decomposer *Clostridium* spp., peptococcus anaerobes, lactobacillus, and actinomyces that are acid formers.

- *Methanogenesis*:

The terminal stage of anaerobic digestion is the biological process of methanogenesis. Here, methanogens use the intermediate products of the preceding stages and convert them into methane, carbon dioxide, and water. These components make up the majority of the biogas emitted from the system. Methanogenesis is sensitive to both high and low pHs and occurs between pH 6.5 and pH 8. The remaining, indigestible material the microbes cannot use and any dead bacterial remains constitute the digestate. The last phase of anaerobic digestion is the methanogenesis phase. Several reactions take place using the intermediate products from the other phases, with the main product being methane. The following equation shows the common reactions that take place during methanogenesis:

$$2CH_3CH_2OH + CO_2 \rightarrow 2CH_3COOH + CH_4$$

$$CH_3COOH \rightarrow CH_4 + CO_2$$

$$CH_3OH \rightarrow CH_4 + H_2O$$

$$CO_2 + 4H_2 \rightarrow CH_4 + 2H_2O$$

$$CH_3COO^- + SO_4^{2-} + H^+ \rightarrow 2HCO_3 + H_2S$$

$$CH_3COO^- + NO^- + H_2O + H^+ \rightarrow 2HCO_3 + NH_4^+$$

Several bacteria contribute to methanogenesis, including Methanobacterium, methanobacillus, methanococcus, and methanosarcina. Fig. 2.7 shows stages of complex bypolymer.

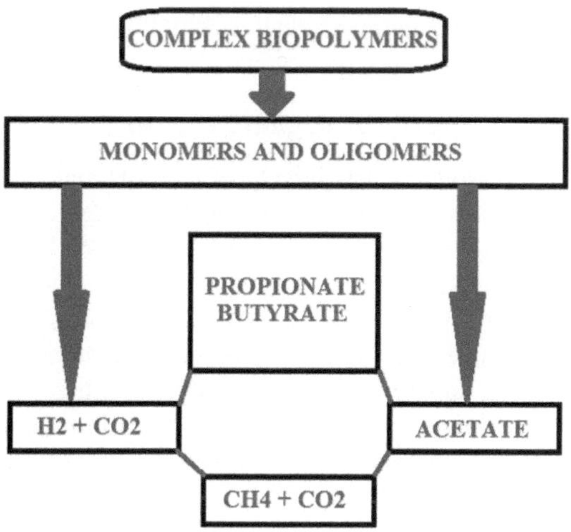

FIGURE 2.7 Stages of complex biopolymer.

2.5 Biogas production processes

Biogas is produced from organic waste (carbon), which biodegrades by means of bacteria in an anaerobic environment. This process is expedited at a process temperature of 38°C/100°F (mesophilic) or 52°C/125.6°F (thermophilic) in the plant's digester. The biogas plant receives all kinds of organic waste—typically livestock manure and organic industrial waste. The dry solid in livestock manure contains carbon, among other things, and in the process, this carbon is transformed into biogas, a compound of methane (CH_4) and carbon dioxide (CO_2).

The manure and waste are mixed in the plant's receiving tank before being heated to 38°C–52°C/100°F–125.6°F and pumped into the digester in which the biogas is produced. The biogas stays in the digester for 2–3 weeks and the fermented slurry can subsequently be used as crop fertilizer. This fertilizer has improved qualities such as less odor inconvenience when spreading the slurry and significant reduction of greenhouse gasses. Biogas is produced through the processing of various types of organic waste. It is a renewable and environmentally friendly fuel made from 100% local feedstocks that are suitable for a diversity of uses including road vehicle fuel and industrial uses. The circular-economic impact of biogas production is further enhanced by the organic nutrients recovered in the production process. Biogas can be produced from a vast variety of raw materials (feedstocks). The biggest role in the biogas production process is played by microbes feeding on the biogas. Digestion carried out by these microorganisms creates methane, which can be used as it is locally or upgraded to biogas equivalent to natural gas quality, enabling the transport of the biogas over longer distances. Material containing organic nutrients is also produced in the process, and this can be utilized for purposes such as agriculture.

2.6 Digestible property of organic matter

Organic matter in the soil consists largely of died-off material and for about 15% of living organisms. Organic matter fulfills a range of functions in the soil—it is important to a good structure and contributes to water retention, infiltration capacity, and providing nutrients. A study among farmers demonstrated that organic matter was seen as the most important indicator of soil quality and at the same time as the indicator that caused most concern. Organic matter is an umbrella term for different types of materials that largely consist of carbon. Depending on the composition, organic matter is either easy to decompose or not. Organic matter that decomposes easily provides nutrients quickly to plants and soil life and contributes to the soil structure by stimulating soil life. In a compacted soil, there may be a lack of oxygen if there is too much organic matter that decomposes easily as the microorganisms use the available oxygen to decompose the organic matter.

That risk does not exist if you have moderately stable organic matter. This produces a slowly releasing nutrition for the plants and soil life and a varied soil life. The highly stable organic matter provides better water retention and retains nutrients, such as potassium and trace elements. Highly stable organic matter also contributes to a good soil structure. The ratio between carbon and nitrogen, the C:N ratio, is an important indicator for the ease with which organic matter can be decomposed. Organic matter with a relatively low C:N ratio, such as chicken manure, slurry, and various crop remnants, digests easily. Organic matter with a relatively high C:N ratio, such as straw, will remain in the soil longer. Fresh materials with a high C:N ratio (>30) can fix nitrogen temporarily, as microorganisms use nitrogen from the soil to decompose the material. Composting reduces the C:N ratio, as carbon is converted into CO_2. The material becomes more stable due to composting, the easily digestible materials are digested first, and more stubborn materials remain. Therefore compost does contribute to building up organic matter, despite its low C:N ratio, but provides few nutrients in the short term. Decomposing organic matter is a continuous process and the rule of thumb is that each year, 2% net of the organic matter in agricultural land is decomposed. Therefore organic matter must be added to keep the organic-matter level intact. Measurements of organic matter generally measure only organic compounds or carbon, and so are only an approximation of the level of once-living or decomposed matter. Some definitions of organic matter likewise only consider "organic matter" to refer to only the carbon content, or organic compounds, and do not consider the origin or decomposition of the matter. In this sense, not all organic compounds are created by living organisms, and living organisms do not only leave behind organic material. A clam shell, for example, while biotic, does not contain much organic carbon, so may not be considered organic matter in this sense. Conversely, urea is one of many organic compounds that can be synthesized without any biological activity.

Organic matter is heterogeneous and very complex. Generally, organic matter, in terms of weight, is:

- 45%–55% carbon;
- 35%–45% oxygen;
- 3%–5% hydrogen; and
- 1%–4% nitrogen.

The molecular weights of these compounds can vary drastically, depending on if they repolymerize or not, from 200 to 20,000 amu. Up to one-third of the carbon present is in aromatic in which the carbon atoms form usually six-member rings. These rings are very stable due to resonance stabilization, so they are difficult to break down. The aromatic rings are also susceptible to electrophilic and nucleophilic attacks from other electron-donating or electron-accepting material, which explains the possible polymerization to create larger molecules of organic matter. There are also reactions that occur with organic matter and other material in the soil to create compounds never seen before. Unfortunately, it is very difficult to characterize these, because so little is known about natural organic matter in the first place. Research is currently being done to figure out more about these new compounds and how many of them are being formed. Organic matter from sewage sludge can exert significant influence on the physical, chemical, and biological properties of soils. Organic matter in general contributes greatly to soil's productive capacity. Organic matter incorporated into the soil surface can affect its structure, as denoted by porosity, aggregation, and bulk density, as well as cause an impact as expressed in terms of content and transmission of water, air, and heat, and of soil strength. Nutrients are mineralized during organic matter decomposition; C, N, and cation-exchange capacity increase following organic matter additions. Other soil chemical properties such as pH, electrical conductivity, and redox potential are changed. The soil biosystem can be altered by the addition of new energy sources for the organisms, reflected by changes in micro- and macrobiological populations, which in turn influence synthesis and decomposition of microbially produced soil humic substances, nutrient availability, interactions with soil inorganic components, and other exchanges with soil physical and biochemical properties.

- *Important factors to make the biogas digestible*:

When raw material (substrate) is digested in a container, only part of it is actually converted into methane and sludge. Some of it is indigestible to varying degrees and accumulates in the digester or passes out with the effluent and scum. The "digestibility" and other basic properties of organic matter are usually expressed in the following terms (you may also want to refer to the terminology section in the beginning of the appendix concerning fuel analysis, but be aware that the terminology is slightly different between biochemical and thermochemical applications):

Moisture: This represents the weight of water lost during drying at 105°C (220°F) until no more weight is lost.

Total solids: This is the weight of the dry material remaining after drying as above. TS weight is usually equivalent to "dry weight." (However, if you dry your material in the sun, assume that it will still contain around 30% moisture). TS is composed of digestible organic or VS, and indigestible residues or "fixed solids."

Volatile solids: This represents the weight of organic solids burned off when dry material is "ignited" (heated to around 538°C). This is a handy property of organic matter to know, since VS can be considered as the amount of solids actually converted by the bacteria.

Fixed solids: This is the weight that remains after ignition. This is biologically inert material. As an example, consider the makeup of fresh chicken manure. Starting from 100 kg of fresh chicken manure, 72–80 kg of this would be water and only 15–24 kg (75%–80% VS of the 20%–28% TS) would be available for actual digestion.

Carbon-to-nitrogen ratio: From a biological point of view, digesters can be considered as a culture of bacteria feeding upon and converting organic wastes. The elements carbon (in the form of carbohydrates) and nitrogen (as protein, nitrates, ammonia, etc.) are the chief foods of anaerobic bacteria. Carbon is utilized for energy and the nitrogen for building cell structures.

The bacteria use up carbon about 30 times faster than they use nitrogen. Anaerobic digestion proceeds best when raw material fed to the bacteria contains a certain amount of carbon and nitrogen together. The C/N represents the proportion of the two elements. A material with 15 times more carbon than nitrogen would have a C/N ratio of 15:1 (written C/N = 15/1, or simply 15).

- A C/N ratio of 30 (C/N = 30/1, 30 times as much carbon as nitrogen) will permit digestion to proceed at an optimum rate, if other conditions are favorable, of course. If there is too much carbon (high C/N ratio, e.g., 60/1) in the raw wastes, nitrogen will be used up first, with carbon left over. This will make the digester slow down.
- On the other hand, if there is too much nitrogen (low C/N ratio, e.g., 30/15, or simply 2), the carbon soon becomes exhausted and fermentation stops. The remaining nitrogen will be lost as ammonia gas (NH_3). This loss of nitrogen decreases the fertility of the effluent sludge.

There are many standard tables listing the C/N ratios of various organic materials, but they can be very misleading for at least two reasons:

1. The ratio of carbon to nitrogen measured chemically in the laboratory is often not the same as the ratio of carbon to nitrogen available to the bacteria as food.
2. The nitrogen or carbon content of even a specific kind of plant or animal waste can vary tremendously according to the age and growing conditions of the plant, and the diet, age, degree of confinement, etc. of the animal.

Nitrogen: Because nitrogen exists in so many chemical forms in nature (ammonia, NH_3; nitrates, NO_3; proteins; etc.), there are no reliable "quick" tests for measuring the total amount of nitrogen in a given material. One kind of test might measure the organic and ammonia nitrogen (the Kjeldahl test); another might measure the nitrate/nitrite nitrogen, etc. In addition, nitrogen can be measured in terms of wet weight, dry weight, or VS content of the material, all of which will give different values for the proportion of nitrogen. Finally, the nitrogen content of a specific kind of manure or plant waste can vary, depending on the growing conditions, age, diet, and so forth.

For example, one study reported a field of barley, which contained 39% protein on the 21st day of growth, 12% protein on the 49th day (bloom stage), and only 4% protein on the 86th day. You can see how much the protein nitrogen depends on the age of the plant. The nitrogen content of manure also varies a great deal. Generally, manures consist of feces, urine, and any bedding material (straw, corn stalks, hay, etc.) that may be used in the livestock shelters. Because urine is the animal's way of getting rid of excess nitrogen, the nitrogen content of manures is strongly affected by how much urine is collected with the feces.

For example, birds naturally excrete feces and urine in the same load, so that the nitrogen content of chickens, turkeys, ducks, and pigeons is the highest of the animal manures from a nitrogen content point of view. Next in nitrogen content, because of their varied diets or grazing habits, are humans, pigs, sheep, and then horses. Cattle and other ruminants (cud chewers), which rely on bacteria in their gut to digest plant foods, have a low content of manure nitrogen, because much of the available nitrogen is used to feed their intestinal bacteria. Even with the same kind of animal, there are big differences in the amount of manure nitrogen. For example, stable manure of horses may have more nitrogen than pasture manure, because feces and urine are excreted and collected in the same small place. Since there are so many variables and because anaerobic bacteria can use most forms of nitrogen, the available nitrogen content of organic materials can best be generalized and presented as total nitrogen (percentage of dry weight).

Carbon: Unlike nitrogen, carbon exists in many forms that are not directly useable by bacteria. The most common indigestible form of carbon is lignin, a complex plant compound that makes land plants rigid and decay-resistant. Lignin can enter a digester either directly with the plant material itself or indirectly as bedding or undigested plant food in manure. Thus a more accurate picture of the C part of the C/N ratio is obtained by considering the "nonlignin" carbon content of plant wastes.

- *Operating parameters to make the digester work*:

pH: Methane-producing bacteria require a neutral to slightly alkaline environment (pH 6.8–8.5) in order to produce methane. Acid-forming bacteria grow much faster than methane-forming bacteria. If acid-producing bacteria grow too fast, they may produce more acid than the methane-forming bacteria can consume. The excess acid will then build up in the system, leading to a drop in pH and the system may become unbalanced, inhibiting the activity of the methane-forming bacteria. This may stop the methane production entirely.

To prevent this type of failure, the maintenance of a large active quantity of methane-producing bacteria is crucial. Hence, retained biogas systems are inherently more stable than bacterial growth-based systems such as completely mixed and plug flow digests.

Hydraulic retention time: Most anaerobic systems are designed to retain the waste for a fixed number of days. The number of days the materials stay in the tank is called the HRT. The HRT equals the volume of the tank divided by the daily flow (HRT = V/Q). The HRT is important, since it establishes the quantity of time available for bacterial growth and subsequent conversion of the organic material to gas. A direct relationship exists between the HRT and the VS converted into gas.

Solids retention time: The SRT is the most important factor controlling the conversion of solids into gas. It is also the most important factor to maintain digester stability. Although the calculation of the SRT is often improperly stated, it is the quantity of solids maintained in the digester divided by the quantity of solids wasted each day.

$$\text{SRT} = \frac{\text{Total solids in the reactor}}{\text{Total daily solids output}} = \frac{V \cdot C_d}{Q_W \cdot C_W}$$

Here, V is the digester volume; C_d is the solids concentration in the digester; Q_w is the volume flow output each day, and C_w is the solids concentration in the output flow. In a conventional completely mixed or plug flow digester, the HRT equals the SRT. However, in a variety of retained biogas reactors, the SRT exceeds the HRT. As a result, the retained biogas digesters can be much smaller, while achieving the same solid conversion into gas. The VS conversion into gas is a function of SRT rather than HRT. At a low SRT, sufficient time is not available for the bacteria to grow and replace the bacteria lost in the effluent. If the rate of bacterial loss exceeds the rate of bacteria growth, "wash-out" occurs. The SRT at which wash-out begins to occur is the "critical SRT." Jewel established that a maximum of 65% of dairy manure's VS could be converted into gas with long SRTs. Burke established that 65%–67% of dairy farm manure COD could be converted into gas. Long retention times are required for the conversion of cellulose into gas.

Digester loading (kg/(m^3 day)): Neither the HRT nor the SRT tells the full story of the impact that the influent waste concentration has on the anaerobic digester: one waste may be diluted and the other concentrated. The concentrated waste will produce more gas per volume unit and affect the digester to a much greater extent than the diluted waste. A more appropriate measure of the waste on the digester's size and performance is the loading. The loading can be reported in kilograms of waste (influent concentration multiplied by the influent flow) per cubic meter of digester volume. A common unit is kilograms of influent waste per cubic meter of digester volume per day (kg/(m^3 day)). One kg/(m^3 day) is equal to 0.0624 lb/(ft.^{3} day).

The digester loading can be calculated if the HRT and influent waste concentration are known.

The loading in (kg/(m^3 day)) is simply:

$$L = \frac{C_I}{\text{HRT}}$$

where C_I is the influx concentration in g/m^3.

Increasing the loading will reduce the digester size but will also reduce the percentage of VS converted into gas. Underloading, the process (low substrate input rate) results in a low biogas production rate. Although this can prevent a process failure, it is uneconomical, because the capacity of the process is not fully utilized. Moreover, the process is running at a suboptimal level and the microbial populations are present in a slow and undynamic state.

Increasing the load will increase the biogas production, but risks overloading, which results in VFA (VFAs) accumulation. High concentration of VFA decreases pH and makes VFA become more toxic to the methanogens, which may again lead to a process breakdown. Sufficient nutrients are also important to microbial cell growth. Macronutrients such as carbon, hydrogen, nitrogen, and oxygen are the main components in biogas cells. Sulfur, phosphorus, potassium, calcium, magnesium, and iron are required for specific proteins. These macronutrients should be present in the cell, while the micronutrients such as nickel, cobalt, and copper are required in smaller amounts. Most nutrients can be inhibited if present in high concentrations. Sulfide and phosphate can decrease the metal ion bioavailability by precipitating. Normally, all the nutrients are present in sufficient quantities in swine and cow manure.

2.7 Undesirable gases in biogas system

There is a rapid growth in use of renewable energy globally, especially biogas. Biogas is produced in anaerobic conditions, that is, in the absence of oxygen by fermentation of organic substrates, including manure, sewerage, household waste, industrial wastewater, etc. Biogas mainly contains methane and carbon dioxide and traces of hydrogen sulfide, ammonia, hydrogen, and moisture. The biogas calorific value is around 21 MJ/m^3, which contains methane and can be used as cooking fuel in domestic kitchen or industrial canteen. If biogas is to be used as a substitute for natural gas, it needs to be cleaned and upgraded the quality equivalent of methane in the pipeline. One of the hot topics in bioenergy industries is the upgradation technology of biogas to biomethane due to rise in price of oil and natural gas. The biogas cleaning and upgradation process increases the calorific value and removes the undesirable gases, such as hydrogen sulfide, carbon dioxide, and moisture, which will be more harmful to the system to be utilized.

- *Cleaning process*:

Around 100 m^3/h of biogas with methane (CH_4), 60%–65%; carbon dioxide (CO_2), 30%–35%; and hydrogen sulfide (H_2S), 2000–2500 ppm was produced from a high rate anaerobic digester. As a value addition and to reduce greenhouse gas emissions, the biogas produced has been upgraded to biomethane (equivalent to natural gas) and is being utilized in the existing boiler and canteen as a substitute fuel to natural gas. Removal of CO_2 from biogas enhances the calorific value closer to the range of natural gas, which is nothing but CH_4, that is, natural gas. There are various options available to upgrade biogas by removing carbon dioxide, such as adsorption, absorption, cryogenic, membrane, etc. Among these upgrading technologies, a suitable site-specific technology working at low pressure, namely, "vacuum swing adsorption" (VSA) has been identified. The main advantage of the VSA process is that adsorption of carbon dioxide takes place at low pressure and reduces higher compression requirement of biogas feeding and desorption at vacuum, which completely evacuates the carbon dioxide from the adsorbents and quicker regeneration for the next cycle of operation.

Cleaning and upgradation of biogas is very important to get a natural gas equivalent fuel. Cleaning of biogas involves removal of hydrogen sulfide, water vapor, and upgradation for complete removal of carbon dioxide, which enhances the calorific value of biogas nearer to natural gas. A 100-m^3/h low-pressure VSA technology has been deployed for purification of the biogas system. Schematic arrangement of VSA-based biogas purification plant has been shown in the Fig. 2.8. The biogas produced from raw dairy effluent is stored in a biogas storage tank. A blower boosts the biogas up to 0.7-bar pressure from the storage tank and passes through the desulfurization unit, where most of H_2S present in the biogas is removed. Iron oxide is used as the packing material in H_2S scrubber. After passing through the H_2S scrubber, the gases are cooled in a chilling unit to remove the moisture present in the gas. If H_2S is not removed prior to the adsorption process, it will lead to poisoning of the composite bed of adsorbents, and with moisture, it forms a corrosive acidic solution in the downstream utilization systems. Hence, cleaning of H_2S and water vapor are very much essential during the biogas purification process. After cleaning, the biogas is passed through the twin towers packed with composite bed of molecular sieves by a low-pressure VSA process to separate CH_4 from CO_2. Adsorption of CO_2 on the molecular sieves leads to the CH_4 removal from the top of the tower and by vacuum, leading to evacuation of CO_2. The advantage of twin tower is when one tower performs adsorption, the other one is used for regeneration and the process becomes cyclic and continuous supply of biomethane of natural gas

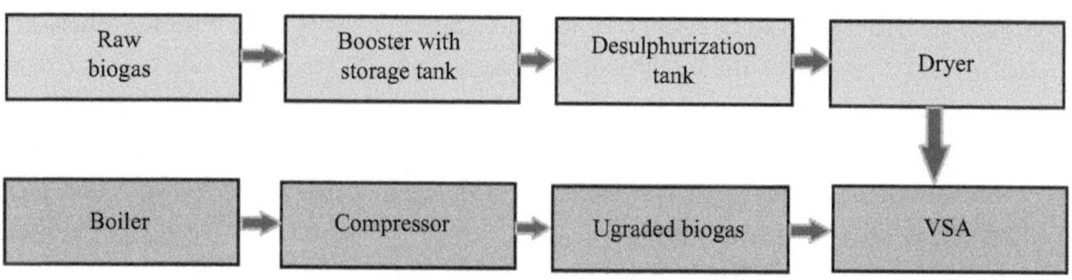

FIGURE 2.8 Stages of purification plant.

equivalent fuel is collected in a surge vessel and in biomethane storage tank. Then, with the help of a compressor, it is sent through pipelines to existing boiler or canteen for cooking purpose. Fig. 2.8 shows stages of purification plant.

2.8 Multifunctional biogas system

Multifunctional biogas systems may contribute substantially to a more efficient use of biogas resources and agricultural land, resulting in low mitigation costs of greenhouse gas emissions. Therefore in this thesis, two concepts of multifunctional biogas system's multiproduct use and cascading are investigated. In this section, the basic principle of multifunctional biogas systems, namely, the multiple use of resources, is discussed and the concepts of multiproduct use and cascading are defined. In addition, an overview of possible multifunctional biogas systems is given.

2.8.1 Multiple use of resources

Within the concept of multiple land use, land generates more than one type of product or service like the production of food, fodder, energy, and materials, the protection of the soil, wastewater treatment, recreation, or nature protection. In this thesis, however, we concentrate on multiple use of biogas resources. Hence, we exclude multiple land use from the analysis. The multiple use of biogas resources can be achieved in several ways. First, different parts of a crop may each be used for a specific purpose. Second, processing of biogas may lead to various products and by-products. Finally, biogas may be used to produce materials and energy in succession, that is, the recycling and cascading approach.

In scientific literature, the potential of using each part of a biogas resource for a specific purpose, thus increasing efficiency of biogas utilization mainly with regard to costs, has often been discussed. Producing several products and by-products from biogas is also an often-examined concept. Wright and Cushman (1997) stress the importance of increasing the use of by-products, and many authors discuss the coproduction of fuels, heat, and electricity. Another concept of multiple biogas resource use is the biorefinery. A biorefinery is an analogy to a petrochemical refinery, where crude oil is completely reverted into many different value-added products maximizing the economic benefits. In a biorefinery, the biogas resource is converted into biomaterials and/or energy carriers. Often a main product of biorefinery is a bulk chemical. Increasing the efficiency of biogas utilization by recycling of biomaterials and waste-to-energy conversion is discussed by several studies. However, in these studies, the cascading of wood products is investigated mostly in a qualitative way.

Hence, in scientific literature, the multiple use of biogas resources is seen as an important possibility to increase the efficient use of biogas for materials and energy, and reduces the costs of biogas options to mitigate green house gas (GHG) emissions. Nevertheless, only very few studies have investigated the potential of multifunctional biogas systems in a quantitative way.

- *Multiproduct use and cascading*:

The principles of multiple biogas resource use can be summarized under the term "multiproduct use" and "cascading." Multiproduct use is defined as using biogas in different applications. Cascading is the subsequent use of biogas for a number of applications. In other words, biogas is used first for a material application; next, it may be recycled for several further material applications; and finally, energy is recovered from the biomaterial waste. In this thesis, multifunctional biogas systems are systems that combine in principle biomaterial use(s) and bioenergy use(s). Fig. 2.9 gives an overview of such a multifunctional biogas system.

As many different biogas crops, biomaterial applications and bioenergy conversion steps exist; obviously, a very large number of multifunctional biogas systems are possible. The main application of the biogas resource characterizes the feasibility of a multifunctional biogas system. In the case of material production, the application determines the selection of crops, the availability of agricultural residues, the supply of by-products, potential recycling steps, and waste-to-energy conversions. The biomaterial application in a multifunctional biogas system should at least offer a good possibility for

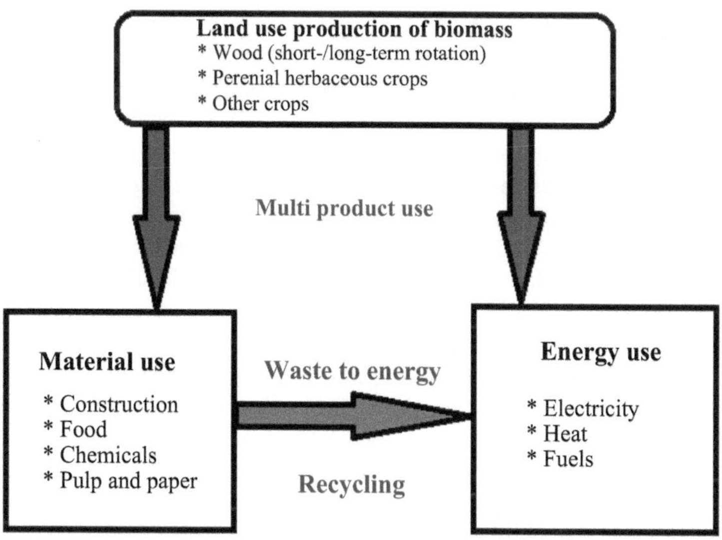

FIGURE 2.9 Multifunctional biogas system.

multiproduct use or cascading. In the case of multiproduct use, this means that residues or by-products from material production may be used and, in the case of cascading, that the material may be recycled.

Main sectors of biomaterial uses are pulp and paper, chemicals, and construction. The pulp and paper industry currently uses a large part of woody biogas. The additional potential for pulp and paper production to substitute the use of other materials, plastics, and, thus, to reduce GHG emissions is probably low. Contrarily, the chemical industry is an interesting sector, offering new opportunities for biogas utilization. Especially, bulk chemicals from biogas have a large potential to substitute fossil fuel feedstock in the chemical industry. Finally, biomaterials for construction have already a large market share. In this case, particularly, a large cascading potential exists, based on a variety of recycling options.

Obviously, biogas resources should have a suitable quality of the application selected. Furthermore, biogas crops in a multifunctional biogas system should either have a high total biogas yield for bulk material and energy applications or have a high yield of a specific plant component required for a high-value added material application, for example, fibers, sugar, or starch. Finally, bioenergy conversion technologies in a multifunctional biogas system should be able to convert efficiently the available agricultural residues, processing residues, and biomaterial wastes. Summarizing, multiple biogas resource use may be a promising concept to decrease the costs and land use of GHG emission mitigation by biogas material and energy uses. In this thesis, we will focus on multifunctional biogas systems with multiproduct use and cascading. Material application that seems promising for these systems is bulk chemical and construction materials.

Multifunctional biogas-refinery system is largely used in the production of biogas energy. This includes the use of agricultural residues for energy production and the use of waste to energy for recovery. The level of system performance in carrying out these tasks is quite exceptional, and it does not take a lot of time and energy to produce the desired energy. This means that a fast energy production involves low costs of production in terms of the materials used per kilogram, which saves on production costs and land use. The market prices of land may be high, but since the system can operate faster, less materials and products and the products that are used are maximized so that they can produce the desired energy.

All biorefinery multifunctional biogas systems can offer net savings of nonrenewable energy consumption of 70–220 GJ/hectare every year and net savings on greenhouse gas emissions in 3–17 mg every year per hectare. These are great net benefits that the system offers, and that is why it is very important to ensure proper viability of biogas systems, for increased production of energy and more opportunities for savings. Multifunctional biogas should also be introduced in large scale so that good own price elasticity is achieved, because high demand of energy production materials affects the overall costs of biogas refinery system. With the large-scale introduction of biogas material, then, there will be production of an energy, which can be used for various applications within and outside the house.

- *Savings of nonrenewable energy consumption*:

Investigating energy balances of different biogas systems is a common way to compare especially bioenergy systems. Furthermore, energy balances can be an important indicator of GHG emission balances as most GHG emissions in biogas systems are related to fossil energy uses and savings. In order to set up an energy balance, generally, energy inputs and outputs of a biogas system have to be accounted for. These inputs are directing energy uses, for example, diesel use for tractor in biogas production, or indirect energy uses, for example, energy uses for the production of a biogas combustion plant. In scientific literature, the energy inputs are compared often with the energy content of the bioenergy carrier resulting in net energy outputs or net energy ratios. Taken into account that a material or energy carrier produced from biogas replaces a reference product with the same function, energy balances can also be used to investigate primary or nonrenewable energy savings.

2.9 Exercise

1. What is the meaning of prefeasibility analysis of biogas power plant?
2. Write different conditions that are required to be satisfied during the prefeasibility assessment of biogas power plant.
3. Which type of critical issues is raised during the prefeasibility assessment of biogas power plant?
4. What is the purpose of prefeasibility assessment of biogas power plant?
5. What are the different steps of conceptual design of prefeasibility assessment of biogas power plant?
6. What is the meaning of decomposition of biogas?
7. Write short note on the anaerobic digestion.
8. Explain different steps of anaerobic digestion with the help of block diagram.
9. Explain anaerobic digestion system with the help of block diagram.
10. What is the significance of the biogas production system in biogas power plant?
11. Write short note on the following:
    a. Thermal pretreatment
    b. Mechanical pretreatment
    c. Chemical pretreatment
    d. Combined pretreatment
12. Explain potential advantages of controlled anaerobic digestion system.
13. Explain potential disadvantages of controlled anaerobic digestion system.

14. Explain different stages of biogas production.
15. Write short note on digestible property of organic matter.
16. Explain different important factor to make the biogas digestible.
17. Explain different operating parameters of anaerobic digester.
18. Explain multifunctional biogas system with its block diagram.

References

Ak, M.S., Muz, M., Komesli, O.T., Gokcay, C.F., 2013. Enhancement of bio-gas production and xenobiotics degradation during anaerobic sludge digestion by ozone treated feed sludge. Chem. Eng. J. 230, 499–505.

Angelidaki, I., Ahring, B.K., 2000. Methods for increasing the biogas potential from the recalcitrant organic matter contained in manure. Water Sci. Technol. 41, 189–194.

Appels, L., Van Assche, A., Willems, K., Degreve, J., Van Impe, J., Dewil, R., 2011. Peracetic acid oxidation as an alternative pre-treatment for the anaerobic digestion of waste activated sludge. Bioresour. Technol. 102, 4124–4130.

Appels, L., Houtmeyers, S., Degreve, J., Van Impe, J., Dewil, R., 2013. Influence of microwave pre-treatment on sludge solubilization and pilot scale semi-continuous anaerobic digestion. Bioresour. Technol. 128, 598–603.

Apul, O.G., Sanin, F.D., 2010. Ultrasonic pretreatment and subsequent anaerobic digestion under different operational conditions. Bioresour. Technol. 101, 8984–8992.

Azbar, N., Keskin, T., Yuruyen, A., 2008. Enhancement of biogas production from olive mill effluent (OME) by co-digestion. Biogas Bioenergy 32, 1195–1201.

Bougrier, C., Delgenes, J.P., Carrere, H., 2008. Effects of thermal treatments on five different waste activated sludge samples solubilisation, physical properties and anaerobic digestion. Chem. Eng. J. 139, 236–244.

Bougrier, C., Albasi, C., Delgenes, J.P., Carrere, H., 2006a. Effect of ultrasonic, thermal and ozone pre-treatments on waste activated sludge solubilisation and anaerobic biodegradability. Chem. Eng. Process. Process Intensif. 45, 711–718.

Bougrier, C., Delgenes, J.P., Carrere, H., 2006b. Combination of thermal treatments and anaerobic digestion to reduce sewage sludge quantity and improve biogas yield. Process. Saf. Environ. Protect. 84, 280–284.

Carballa, M., Manterola, G., Larrea, L., Ternes, T., Omil, F., Lema, J.M., 2007. Influence of ozone pre-treatment on sludge anaerobic digestion: removal of pharmaceutical and personal care products. Chemosphere 67, 1444–1452.

Carrere, H., Dumas, C., Battimelli, A., Batstone, D.J., Delgenes, J.P., Steyer, J.P., et al., 2010. Pretreatment methods to improve sludge anaerobic degradability: a review. J. Hazard. Mater. 183, 1–15.

Castrillon, L., Fernandez-Nava, Y., Ormaechea, P., Maranon, E., 2011. Optimization of biogas production from cattle manure by pre-treatment with ultrasound and co-digestion with crude glycerin. Bioresour. Technol. 102, 7845–7849.

Comino, E., Riggio, V.A., Rosso, M., 2012. Biogas production by anaerobic co-digestion of cattle slurry and cheese whey. Bioresour. Technol. 114, 46–53.

Connaughton, S., Collins, G., O'Flaherty, V., 2006. Psychrophilic and mesophilic anaerobic digestion of brewery effluent: a comparative study. Water Res. 40, 2503–2510.

Deleris, S., Larose, A., Geaugey, V., Lebrun, T., 2003. Innovative strategies for the reduction of sludge production in activated sludge plant: BIOLYSIS® O and BIOLYSIS® E. In:

Proceedings of the International Water Association (IWA) Specialist Conference, June 23–25, 2003, Norwegian University of Science and Technology (NTNU), Trondheim, Norway, pp. 55–61.

Devlin, D.C., Esteves, S.R.R., Dinsdale, R.M., Guwy, A.J., 2011. The effect of acid pretreatment on the anaerobic digestion and dewatering of waste activated sludge. Bioresour. Technol. 102, 4076–4082.

Dewil, R., Appels, L., Baeyens, J., Degreve, J., 2007. Peroxidation enhances the biogas production in the anaerobic digestion of biosolids. J. Hazard. Mater. 146, 577–581.

Divya, D., Gopinath, L.R., Christy, P.M., 2015. A review on current aspects and diverse prospects for enhancing biogas production in sustainable means. Renew. Sustain. Energy Rev. 42, 690–699.

Elbeshbishy, E., Aldin, S., Hafez, H., Nakhla, G., Ray, M., 2011. Impact of ultrasonication of hog manure on anaerobic digestability. Ultrason. Sonochem. 18, 164–171.

Eskicioglu, C., Kennedy, K.J., Droste, R.L., 2009. Enhanced disinfection and methane production from sewage sludge by microwave irradiation. Desalination 248, 279–285.

Eskicioglu, C., Terzian, N., Kennedy, K.J., Droste, R.L., Hamoda, M., 2007. Athermal microwave effects for enhancing digestibility of waste activated sludge. Water Res. 41, 2457–2466.

Eskicioglu, C., Droste, R.L., Kennedy, K.J., 2006. Performance of continuous flow anaerobic sludge digesters after microwave pretreatment. In: Proceedings of the 79th Water Environment Federation Annual Technical Exhibition and Conference, October 22–25, 2006, Dallas, TX, USA., pp. 526–540.

Esposito, G., Frunzo, L., Giordano, A., Liotta, F., Panico, A., Pirozzi, F., 2012. Anaerobic codigestion of organic wastes. Rev. Environ. Sci. Bio/Technol 11, 325–341.

Ferrer, I., Ponsa, S., Vazquez, F., Font, X., 2008. Increasing biogas production by thermal (70 °C) sludge pre-treatment prior to thermophilic anaerobic digestion. Biochem. Eng. J. 42, 186–192.

Fuerstenau, D.W., Abouzeid, A.Z.M., 2002. The energy efficiency of ball milling in comminution. Int. J. Miner. Process. 67, 161–185.

Gonzalez-Fernandez, C., Leon-Cofreces, C., Garcia-Encina, P.A., 2008. Different pretreatments for increasing the anaerobic biodegradability in swine manure. Bioresour. Technol. 99, 8710–8714.

Gupta, P., Singh, R.S., Sachan, A., Vidyarthi, A.S., Gupta, A., 2012. A re-appraisal on intensification of biogas production. Renew. Sustain. Energy Rev. 16, 4908–4916.

Hasegawa, S., Shiota, N., Katsura, K., Akashi, A., 2000. Solubilization of organic sludge by thermophilic aerobic bacteria as a pretreatment for anaerobic digestion. Water Sci. Technol. 41, 163–169.

Haug, R.T., Stuckey, D.C., Gossett, J.M., McCarty, P.L., 1978. Effect of thermal pretreatment on digestibility and dewaterability of organic sludges. J. Water Pollut. Control. Fed. 50, 73–85.

Heo, N.H., Park, S.C., Lee, J.S., Kang, H., 2003. Solubilization of waste activated sludge by alkaline pretreatment and biochemical methane potential (BMP) tests for anaerobic codigestion of municipal organic waste. Water Sci. Technol. 48, 211–219.

Izumi, K., Okishio, Y.K., Nagao, N., Niwa, C., Yamamoto, S., Toda, T., 2010. Effects of particle size on anaerobic digestion of food waste. Int. Biodeterior. Biodegrad. 64, 601–608.

Jones, D.A., Lelyveld, T.P., Mavrofidis, S.D., Kingman, S.W., Miles, N.J., 2002. Microwave heating applications in environmental engineering—a review. Resour. Conserv. Recycl. 34, 75–90.

Kafle, G.K., Kim, S.H., Sung, K.I., 2013. Ensiling of fish industry waste for biogas production: a lab scale evaluation of biochemical methane potential (BMP) and kinetics. Bioresour. Technol. 127, 326–336.

Kalloum, S., Bouabdessalem, H., Touzi, A., Iddou, A., Ouali, M.S., 2011. Biogas production from the sludge of the municipal wastewater treatment plant of Adrar city (Southwest of Algeria). Biogas Bioenergy 35, 2554–2560.

Kameswari, K.S.B., Kalyanaraman, C., Thanasekaran, K., 2011. Effect of ozonation and ultrasonication pretreatment processes on co-digestion of tannery solid wastes. Clean. Technol. Environ. Policy 13, 517–525.

Kavacik, B., Topaloglu, B., 2010. Biogas production from co-digestion of a mixture of cheese whey and dairy manure. Biogas Bioenergy 34, 1321–1329.

Kim, J., Park, C., Kim, T.H., Lee, M., Kim, S., Kim, S.W., et al., 2003. Effects of various pretreatments for enhanced anaerobic digestion with waste activated sludge. J. Biosci. Bioeng. 95, 271–275.

Lehne, G., Muller, A., Schwedes, J., 2001. Mechanical disintegration of sewage sludge. Water Sci. Technol. 43, 19–26.

Li, Y., Noike, T., 1989. The effect of thermal pretreatment and retention time on the degradation of waste activated in anaerobic digestion. Jpn. J. Water Pollut. Res. 12, 112–121 (in Japanese).

Lin, J.G., Chang, C.N., Chang, S.C., 1997. Enhancement of anaerobic digestion of waste activated sludge by alkaline solubilization. Bioresour. Technol. 62, 85–90.

Lin, Y., Wang, D., Wu, S., Wang, C., 2009. Alkali pretreatment enhances biogas production in the anaerobic digestion of pulp and paper sludge. J. Hazard. Mater. 170, 366–373.

Luostarinen, S., Luste, S., Sillanpaa, M., 2009. Increased biogas production at wastewater treatment plants through co-digestion of sewage sludge with grease trap sludge from a meat processing plant. Bioresour. Technol. 100, 79–85.

Mata-Alvarez, J., Dosta, J., Romero-Guiza, M.S., Fonoll, X., Peces, M., Astals, S., 2014. A critical review on anaerobic co-digestion achievements between 2010 and 2013. Renew. Sustain. Energy Rev. 36, 412–427.

Mata-Alvarez, J., Mace, S., Llabres, P., 2000. Anaerobic digestion of organic solid wastes. An overview of research achievements and perspectives. Bioresour. Technol. 74, 3–16.

Mayhew, M., Le, M., Brade, C., Harrison, D., 2003. The united utilities enzymic hydrolysis process-validation of phased digestion at full-scale to enhance pathogen removal. In: Proceedings of the WEF/AWWA/CWEA Joint Residuals and Biosolids Management Conference and Exhibition, February 19–22, 2003, Baltimore, MD, USA, pp. 1000–1013.

Miah, M.S., Tada, C., Yang, Y., Sawayama, S., 2005. Aerobic thermophilic bacteria enhance biogas production. J. Mater. Cycles Waste Manage. 7, 48–54.

Montusiewicz, A., Lebiocka, M., Rozej, A., Zacharska, E., Pawlowski, L., 2010. Freezing/thawing effects on anaerobic digestion of mixed sewage sludge. Bioresour. Technol. 101, 3466–3473.

Mottet, A., Steyer, J.P., Deleris, S., Vedrenne, F., Chauzy, J., Carrere, H., 2009. Kinetics of thermophilic batch anaerobic digestion of thermal hydrolysed waste activated sludge. Biochem. Eng. J. 46, 169–175.

Mudhoo, A., Sharma, S.K., 2011. Microwave irradiation technology in waste sludge and wastewater treatment research. Crit. Rev. Environ. Sci. Technol. 41, 999–1066.

Muller, C.D., Abu-Orf, M., Blumenschein, C.D., Novak, J.T., 2009. A comparative study of ultrasonic pretreatment and an internal recycle for the enhancement of mesophilic anaerobic digestion. Water Environ. Res. 81, 2398–2410.

Nah, I.W., Kang, Y.W., Hwang, K.Y., Song, W.K., 2000. Mechanical pretreatment of waste activated sludge for anaerobic digestion process. Water Res. 34, 2362–2368.

Noyola, A., Tinajero, A., 2005. Effect of biological additives and micronutrients on the anaerobic digestion of physicochemical sludge. Water Sci. Technol. 52, 275–281.

Ogejo, J.A., Li, L., 2010. Enhancing biomethane production from flush dairy manure with turkey processing wastewater. Appl. Energy 87, 3171–3177.

Ormeci, B., Vesilind, P.A., 2001. Effect of dissolved organic material and cations on freeze-thaw conditioning of activated and alum sludges. Water Res. 35, 4299–4306.

Oz, N.A., Yarimtepe, C.C., 2014. Ultrasound assisted biogas production from landfill leachate. Waste Manage. 34, 1165–1170.

Parawira, W., 2012. Enzyme research and applications in biotechnological intensification of biogas production. Crit. Rev. Biotechnol. 32, 172–186.

Park, W.J., Ahn, J.H., 2011. Effects of microwave pretreatment on mesophilic anaerobic digestion for mixture of primary and secondary sludges compared with thermal pretreatment. Environ. Eng. Res. 16, 103–109.

Patil, P.N., Gogate, P.R., Csoka, L., Dregelyi-Kiss, A., Horvath, M., 2016. Intensification of biogas production using pretreatment based on hydrodynamic cavitation. Ultrason. Sonochem. 30, 79–86.

Peng, L., Bao, M., Wang, Q., Wang, F., Su, H., 2014. The anaerobic digestion of biologically and physicochemically pretreated oily wastewater. Bioresour. Technol. 151, 236–243.

Perez-Elvira, S.I., Diez, P.N., Fdz-Polanco, F., 2006. Sludge minimisation technologies. Rev. Environ. Sci. Bio/Technol. 5, 375–398.

Pilli, S., Bhunia, P., Yan, S., LeBlanc, R.J., Tyagi, R.D., Surampalli, R.Y., 2011. Ultrasonic pretreatment of sludge: a review. Ultrason. Sonochem. 18, 1–18.

Qiao, W., Yan, X., Ye, J., Sun, Y., Wang, W., Zhang, Z., 2011. Evaluation of biogas production from different Biogas wastes with/without hydrothermal pretreatment. Renew. Energy 36, 3313–3318.

Rai, C.L., Rao, P.G., 2009. Influence of sludge disintegration by high pressure homogenizer on microbial growth in sewage sludge: an approach for excess sludge reduction. Clean. Technol. Environ. Policy 11, 437–446.

Saidu, M., Yuzir, A., Salim, M.R., Azman, S., Abdullah, N., 2013. Influence of palm oil mill effluent as inoculum on anaerobic digestion of cattle manure for biogas production. Bioresour. Technol. 141, 174–176.

Schroder, P., Herzig, R., Bojinov, B., Ruttens, A., Nehnevajova, E., et al., 2008. Bioenergy to save the world: producing novel energy plants for growth on abandoned land. Environ. Sci. Pollut. Res. Int. 15, 196–204.

Shehu, M.S., Manan, Z.A., Alwi, S.R.W., 2012. Optimization of thermo-alkaline disintegration of sewage sludge for enhanced biogas yield. Bioresour. Technol. 114, 69–74.

Singh, S., Kumar, S., Jain, M.C., Kumar, D., 2001. Increased biogas production using microbial stimulants. Bioresour. Technol. 78, 313–316.

Solyom, K., Mato, R.B., Perez-Elvira, S.I., Cocero, M.J., 2011. The influence of the energy absorbed from microwave pretreatment on biogas production from secondary wastewater sludge. Bioresour. Technol. 102, 10849–10854.

Song, L.J., Zhu, N.W., Yuan, H.P., Hong, Y., Ding, J., 2010. Enhancement of waste activated sludge aerobic digestion by electrochemical pre-treatment. Water Res. 44, 4371–4378.

Tyagi, V.K., Lo, S.L., 2011. Application of physico-chemical pretreatment methods to enhance the sludge disintegration and subsequent anaerobic digestion: an up to date review. Rev. Environ. Sci. Bio/Technol 10, 215–242.

Vlyssides, A.G., Karlis, P.K., 2004. Thermal-alkaline solubilization of waste activated sludge as a pre-treatment stage for anaerobic digestion. Bioresour. Technol. 91, 201–206.

Weemaes, M., Grootaerd, H., Simoens, F., Verstraete, W., 2000. Anaerobic digestion of ozonized biosolids. Water Res. 34, 2330–2336.

Wilson, C.A., Novak, J.T., 2009. Hydrolysis of macromolecular components of primary and secondary wastewater sludge by thermal hydrolytic pretreatment. Water Res. 43, 4489–4498.

Xu, J., Yuan, H., Lin, J., Yuan, W., 2014. Evaluation of thermal, thermal-alkaline, alkaline and electrochemical pretreatments on sludge to enhance anaerobic biogas production. J. Taiwan Inst. Chem. Eng. 45, 2531–2536.

Chapter 3

Biogas digester plant

Chapter outline

Objectives

- To provide knowledge about general description, component, and classification of biogas power plant.
- To provide knowledge about functioning and installation of biogas power plant.
- To provide knowledge about operation and maintenance of biogas power plant.
- To provide knowledge about MATLAB simulation of biogas power plant.
- To provide knowledge about design of biomass power plant by HOMER software.

3.1 General description and types

Biogas plant (BGP) is an airtight container that facilitates fermentation of material under anaerobic condition. The other names given to this device are "biogas digester," "biogas reactor," "methane generator," and "methane reactor." The recycling and treatment of organic wastes (biodegradable material) through

Design and Optimization of Biogas Energy Systems. DOI: https://doi.org/10.1016/B978-0-12-822718-3.00003-4

anaerobic digestion (fermentation) technology provide biogas not only as a clean and convenient fuel but also as an excellent and enriched biomanure. Thus the BGP also acts as a miniature biofertilizer factory; hence, some people prefer to refer to it as "Biogas Fertilizer Plant" or "Biomanure Plant." The fresh organic material (generally in a homogenous slurry form) is fed into the digester of the plant from one end, known as inlet pipe or inlet tank. The decomposition (fermentation) takes place inside the digester due to bacterial (microbial) action, which produces biogas and organic fertilizer (manure) rich in humus and other nutrients. There is a provision for storing biogas on the upper portion of the BGP. There are some BGP designs that have floating gas-holder and others have a fixed gas storage chamber (GSC). The other end of the digester outlet pipe or the outlet tank (OT) provides for the automatic discharge of the liquid digested manure. A biogas digester is the apparatus used to control anaerobic decomposition. In general, it consists of a sealed tank or a pit that holds the organic material and some means to collect the gases that are produced.

Many different shapes and styles of BGPs have been experimented with:

i. horizontal;
ii. vertical;
iii. cylindrical;
iv. cubic; and
v. dome shaped.

There are two basic parts to the design: a tank that holds the slurry (a mixture of manure and water) and a gas cap or drum on the tank to capture the gas released from the slurry. To get these parts to do their jobs, of course, requires provision for mixing the slurry, piping off the gas, drying the effluent, etc. In addition to the production of fuel and fertilizer, a digester becomes the receptacle for animal, human, and organic wastes. This removes from the environment possible breeding grounds for rodents, insects, and toxic bacteria, thereby producing a healthier environment in which to live.

The feed material is mixed with water in the effluent collecting tank. The fermentation slurry flows through the inlet into the digester. The bacteria from the fermentation slurry are intended to produce biogas in the digester. For this purpose, they need time. The digester must be designed in a way that only fully digested slurry can leave it. The bacteria are distributed in the slurry by stirring (with a stick or stirring facilities). The fully digested slurry leaves the digester through the outlet into the slurry storage pits. The biogas is collected and stored until the time of consumption in the gasholder. The gas pipe carries the biogas to the place where it is consumed by gas appliances. Condensation collecting in the gas pipe is removed by a water trap. Depending on the available building material and type of plant under construction, different variants of the individual components are possible. The following components of a BGP can also play an important role and are described separately

3.2 Component of biogas plant

The main component of BGP describes the following:

1. *Foundation*: The BGP has been built within the surface of the earth underground and it is on the foundation that the whole BGP is based. The foundation forms the base of the digester where the most important processes of BGP occur. The foundation base of the digester is made up of cement concrete (CC) and brick ballast. The construction should be built in such a manner that it should be able to provide a stable foundation for the digester walls and be able to sustain the full load of slurry filled in it. The foundation should be waterproof so that there are no percolation and leakage of water. Fig. 3.1 shows the structure of different parts of biogas power plant.
2. *Digester*: It is either an underground cylindrical-shaped or an ellipsoidal shaped structure where the digestion (fermentation) of substrate takes place. The digester is also known as "Fermentation Tank or Chamber." In a simple rural household (RHh) BGP working under ambient temperature, the digester (fermentation chamber) is designed to hold slurry equivalent to 55, 40, or 30 days of daily feeding. This is known as hydraulic retention time (HRT) of BGP. The designed HRT of 55, 40, and 30 days is determined by the different temperature zones in the country—the states and regions falling under the different temperature zones are already defined for different countries. The digester can be constructed of brick masonry, CC or reinforced cement concrete, or stone masonry or prefabricated cement concrete blocks or ferrocement (ferroconcrete) or steel- or rubber- or bamboo-reinforced cement mortar (BRCM). In the case of smaller capacity floating gasholder plants of 2 and 3 m^3, no partition wall is provided inside the digester, whereas the BGPs of 4-m^3

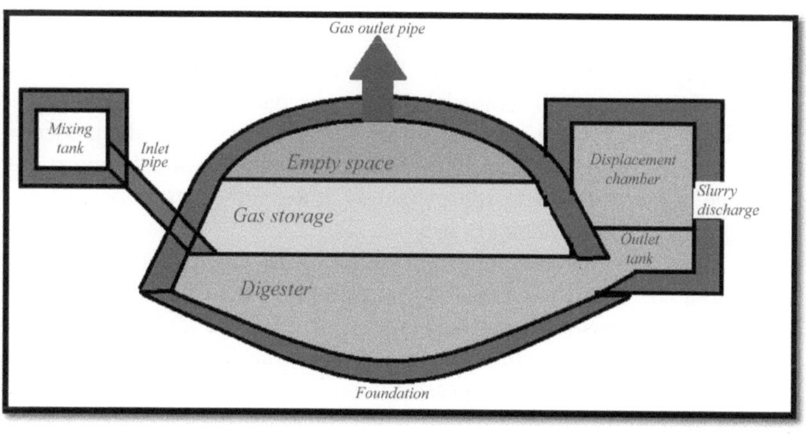

FIGURE 3.1 Parts of a biodigester.

capacity and above have been provided partition wall in the middle. This is provided for preventing short-circuiting of slurry and promoting better efficiency. This means the partition wall also divides the entire volume of the digester (fermentation chamber) into two halves. As against this, no partition wall is provided inside the digester of a fixed-dome design. The reason for this is that the diameter of the digesters in all the fixed-dome models is comparatively much bigger than the floating-drum BGPs, which takes care of the short-circuiting problems to a satisfactory level, without adding to additional cost of providing a partition wall.

3. *Gasholder or gas storage chamber (dome)*: In the case of floating gasholder BGPs, the gasholder is a drum-like structure, fabricated of either of mild steel sheets or ferrocement (ferroconcrete) or high-density plastic (HDP) or fiber glass-reinforced plastic (FRP). It fits like a cap on the mouth of digester where it is submerged in the slurry and rests on the ledge, constructed inside the digester for this purpose. The drum collects gas, which is produced from the slurry inside the digester, as it gets decomposed and rises upward, being lighter than air. To ensure that there is enough pressure on the stored gas so that it flows on its own to the point of utilization through pipeline when the gate valve is open, the gas is stored inside the gasholder at a constant pressure of 8–10 cm of water column. This pressure is achieved by making the weight of biogas holder as 80–100 kg/cm^2. In its up and down movements, the drum is guided by a central guide pipe. The gas formed is otherwise sealed from all sides except at the bottom. The scum of the semidried mat formed on the surface of the slurry is broken (disturbed) by rotating the biogas holder, which has a scum-breaking arrangement inside it. The gas storage capacity of a family-sized floating biogas holder BGP is kept as 50% of the rate capacity (daily gas production in 24 hours). This storage capacity comes to approximately 12 hours of biogas produced every day.

   In the case of fixed-dome designs, the biogas holder is commonly known as GSC. The GSC is the integral and fixed part of the main unit of the plant (MUP) in the case of fixed-dome BGPs. Therefore the GSC of the fixed-dome BGP is made of the same building material as that of the MUP. The gas storage capacity of a family-sized fixed-dome BGP is kept as 33% of the rate capacity (daily gas production in 24 hours). This storage capacity comes to approximately 8 hours of biogas produced during the night when it is not in use.

4. *Inlet chamber*: The cow dung is supplied to the digester of the BGP via inlet chamber, which is made at the ground level so that the cow dung can be poured easily. It has bell mouth sort of shape and is made up of bricks, cement, and sand. The outlet wall of the inlet chamber is made inclined so that the cow dung easily flows to the digester.
   a. In the case of floating biogas holder pipe, the inlet is made of a CC pipe. The inlet pipe reaches the bottom of the digester well on one

side of the partition wall. The top end of this pipe is connected to the mixing tank (MT).

b. In the case of the first approved fixed-dome models (Janata model), the inlet is like a chamber or tank—it is a bell mouth-shaped brick masonry construction and its outer wall is sloppy. The top end of the outer wall of the inlet chamber has an opening connecting the MT, whereas the bottom portion joins the inlet gate. The top (mouth) of the inlet chamber is kept covered with a heavy slab. The inlet of the other fixed-dome models (Deenbandhu and Grameen Bandhu) has asbestos CC pipes of appropriate diameters.

5. *Outlet chamber*: Through the outlet chamber (OC), the digested slurry from which the biogas has been generated is removed from the BGP. A few steps are usually provided in the OC so that some person can go into the pit and clean it. The opening of the OC is also at the ground level. The slurry from the OC flows to the pit made especially for this purpose.
   a. In the case of floating gasholder pipe, the outlet is made of a CC pipe standing at an angle, which reaches the bottom of the digester on the opposite side of the partition wall. In smaller plants (2- and 3-m^3 capacity BGPs), which have no partition walls, the outlet is made of a small (approximately 2-ft. length) CC pipe inserted on the topmost portion of the digester, submerged in the slurry.
   b. In the two fixed-dome plants, the outlet is made in the form of rectangular tank. However, in the case of Grameen Bandhu model, the upper portion of the outlet [known as outlet displacement chamber (ODC)] is made hemispherical in shape, designed to save in the material and labor cost. In all the three fixed-dome models (Janata, Deenbandhu, and Grameen Bandhu models), the bottom end of the OT is connected to the outlet gate. There is a small opening provided on the outer wall of the OC for the automatic discharge of the digested slurry outside the BGP, equal to approximately 80%–90% of the daily feed. The top mouth of the OC is kept covered with heavy slab.
6. *Mixing tank*: It is the first part of the biogas generator, where the water and cow dung are mixed together in the ratio of 1:1 to form the slurry, which is fed into the inlet chamber. This is a cylindrical tank used for making homogenous slurry by mixing the manure from domestic farm animals with appropriate quantity of water. Thoroughly mixing of slurry before releasing it inside the digester, through the inlet, helps in increasing the efficiency of digestion. Normally, a feeder fan is fixed inside the MT for facilitating easy and faster mixing of manure with water for making homogenous slurry.
7. *Gas outlet pipe and valve*: The gas outlet pipe (GOP) is located at the top of the dome where the biogas produced in the digester is collected. The flow of the gas through the dome via gas pipe can be controlled by

valve. The gas taken from the pipe can be transferred to the point of use. The GOP is made of GI pipe and fixed on the top of the drum at the center in the case of a floating biogas holder BGP and on the crown of the fixed-dome BGP. From this pipe, the connection to gas pipeline is made for conveying the gas to the point of utilization. A gate valve is fixed on the GOP to close and check the flow of biogas from plant to the pipeline.

3.3 Classification of biogas plant

The anaerobic digester design has continued to evolve over the years, but systems are generally variations around the theme of the floating-dome and the fixed-dome design. Often construction materials vary, or loading positions differ. Concerning the construction, two main types of simple BGPs can be distinguished:

a. fixed-dome plants; and
b. floating-drum plants.

But also other types of plants play a role, especially in past developments. In developing countries, the selection of appropriate design is determined largely by the prevailing design in the region. Typical design criteria are space, existing structures, cost minimization, and substrate availability. The designs of BGPs in industrialized countries reflect a different set of conditions. Table 3.1 shows some of the most common BGPs that are recognized by the different agencies in all over the world.

The most important types of BGPs are described below:

a. fixed-dome plants;
b. floating-drum plants;
c. balloon plants;

TABLE 3.1 Different types of biogas plant recognized by Ministry of Nonconventional Energy Sources.

1	Floating-drum plant with a cylinder digester (Khadi and Village Industries Commission model).
2	Fixed-dome plant with a brick-reinforced, molded dome (Janata model).
3	Floating-drum plant with a hemisphere digester (Pragati model).
4	Fixed-dome plant with a hemisphere digester (Deenbandhu model).
5	Floating-drum plant made of angular steel and plastic foil (Ganesh model).
6	Floating-drum plant made of prefabricated-reinforced concrete compound units.
7	Floating-drum plant made of fiber glass-reinforced polyester.

d. horizontal plants;
e. anaerobic baffled reactor (ABR); and
f. toilet linked.

Typical designs in industrialized countries and appropriate design selection criteria have also been considered. BGPs can be classified on the basis of various design, construction, and operational factors as discussed below:

A. *According to gas storage*: The design of biogas digester may vary accordingly to suit the requirements of the owner. This can be divided into three groups, namely:
   i BGP with fixed roof;
   ii BGP with floating gasholder;
   iii BGP with separate gasholder; and
   iv flexible bag BGPs.
B. *According to geometrical shapes*: Biogas digester can be constructed in various geometrical shapes: vertical cylinder, spherical, rectangular, square, pipe-shaped, oval, spindle-shaped, elliptical, arch, oblate, etc. Fig. 3.2 shows the classification according to the layout of geometrical shapes.
C. *According to orientations of inlet and outlet*: The arrangement of the different components of biogas system can be varied according to what is suitable to the condition of the area. The different orientations of inlet and outlet are shown in Fig. 3.3 for design flexibility.

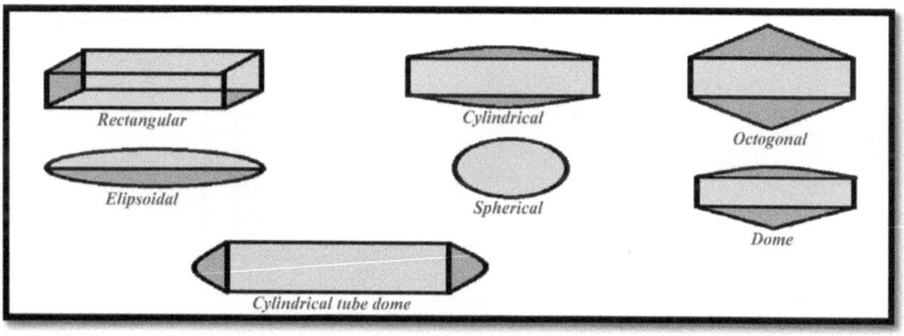

FIGURE 3.2 Classifications according to the layout geometrical shapes.

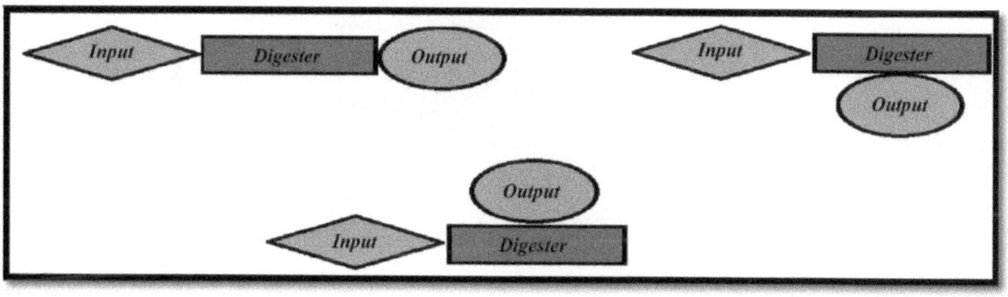

FIGURE 3.3 Design according to the orientation of inlet and outlet.

D. *According to buried position*: Biogas digesters can be erected either of the following ways, which is shown in Figs. 3.4–3.6. Fermentation chamber is connected with the gas storage tank, and in that case, biogas is entered through gas pipe. In semiburied position, gas storage and slurry tank are interconnected with the inlet, outlet, and gas pipe.

3.3.1 Fixed-dome biogas plant with fixed and integrated gas storage chamber

The plant is constructed underground, protecting it from physical damage and saving space. While the underground digester is protected from low

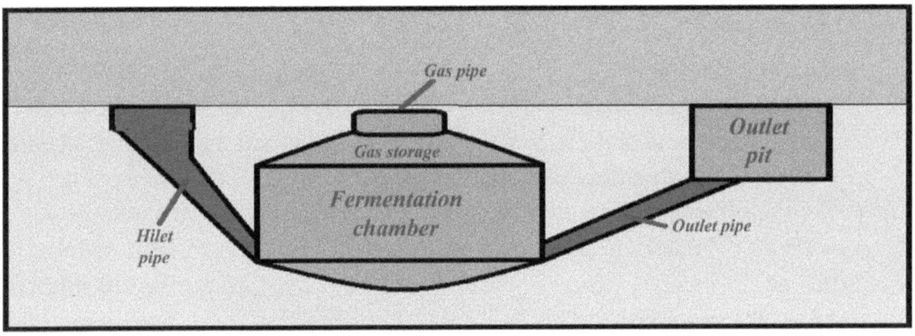

FIGURE 3.4 Layout of fermentation.

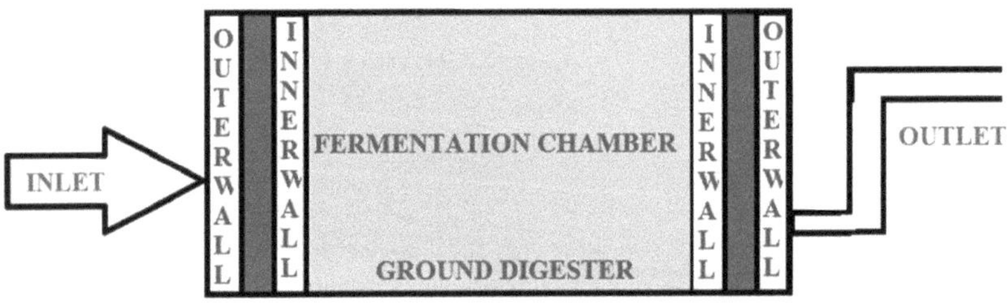

FIGURE 3.5 Ground digester.

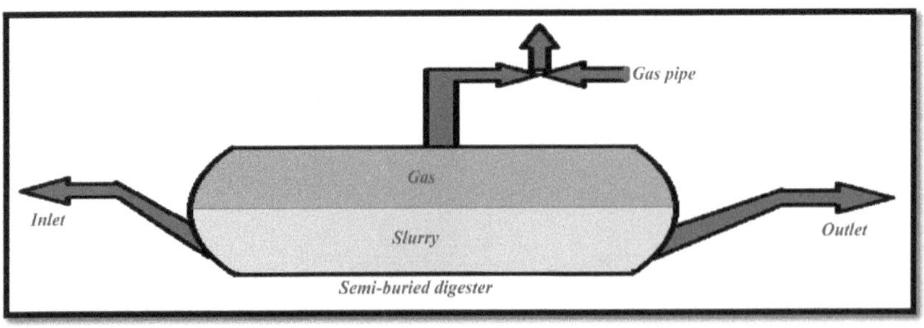

FIGURE 3.6 Design according to the buried position.

temperatures at night and during cold seasons, sunshine and warm seasons take longer to heat up the digester. The costs of a fixed-dome BGP are relatively low. It is simple as no moving parts exist. There are also no rusting steel parts, and hence, a long life of the plant (20 years or more) can be expected. No day/night fluctuations of temperature in the digester positively influence the bacteriological processes. They are as follows:

- Consist of a digester with a fixed, nonmovable gasholder, which sits on top of the digester.
- When gas production starts, the slurry is displaced into the compensation tank; gas pressure increases with the volume of gas stored.
- The height difference between the slurry level in the digester and the slurry level in the compensation tank is the most important point of design.
- All concrete constructions are, hence, durable and lifelong investment.
- Simple structure and least cost.
- No moving parts and metal components and, thus, easy to maintain.
- Capable of generating higher gas pressure (on the average 10 times higher than floating gasholder type) and does not use floating tank.

The construction of fixed-dome plants is labor-intensive, thus creating local employment. Fixed-dome plants are not easy to build. They should only be built where construction can be supervised by experienced biogas technicians. Otherwise, plants may not be gastight (porosity and cracks). The basic elements of a fixed-dome plant (here the Nicarao design) are shown in Fig. 3.7.

The fixed-dome digester is the most common type of design. The four major components of the digester, which are gas storage, fermentation

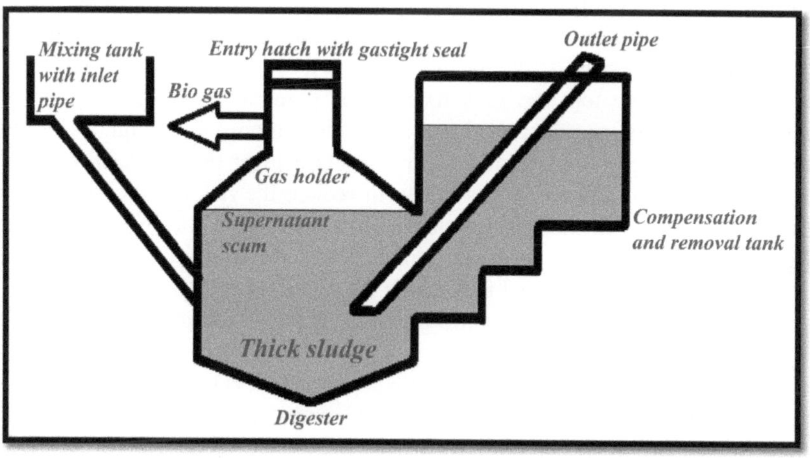

FIGURE 3.7 Fixed-dome plant a/c Nicarao design.

chambers, hydraulic tank, and inlet tanks, are integrated into one structure. Their distinct advantages over the other designs are as follows:

1. all concrete constructions, hence, are durable and lifelong investment, of simple structure, and of least cost.
2. No moving parts and metal components and, thus, easy to maintain.
3. Capable of generating higher gas pressure (on the average 10 times higher than floating gasholder type) and does not use floating tank.
4. Completely constructed underground and, thus save land space. Input materials flow easily into the digester by gravity, hence simplifying operation.

Function: A fixed-dome plant comprises of a closed, dome-shaped digester with an immovable, rigid gasholder and a displacement pit, also named "compensation tank." The gas is stored in the upper part of the digester. When gas production commences, the slurry is displaced into the compensating tank. Gas pressure increases with the volume of gas stored, that is, with the height difference between the two slurry levels. If there is little gas in the gasholder, the gas pressure is low.

Digester: The digesters of fixed-dome plants are usually masonry structures, and structures of cement and ferrocement exist. The main parameters for the choice of material are as follows:

1. technical suitability (stability, gastightness, and liquidtightness);
2. cost-effectiveness;
3. availability in the region and transport costs; and
4. availability of local skills for working with the particular building material.

Fixed-dome plants produce just as much gas as floating-drum plants, if they are gastight. However, utilization of the gas is less effective as the gas pressure fluctuates substantially. Burners and other simple appliances cannot be set in an optimal way. If the gas is required at constant pressure (e.g., for engines), a gas pressure regulator or a floating gasholder is necessary.

Gasholder: The top part of a fixed-dome plant (the gas space) must be gastight. Concrete, masonry, and cement rendering are not gastight. The gas space must, therefore, be painted with a gastight layer (e.g., "waterproofer," latex, or synthetic paints). A possibility to reduce the risk of cracking of the gasholder consists in the construction of a weak ring in the masonry of the digester. This "ring" is a flexible joint between the lower (waterproof) and the upper (gas-proof) part of the hemispherical structure. It prevents cracks that develop due to the hydrostatic pressure in the lower parts to move into the upper parts of the gasholder.

The plants based on fixed-dome concept were developed in India in the middle of 1970, after a team of officers visited China. The Chinese fixed-dome plants use seasonal crop wastes as the major feedstock for feeding; therefore, their design is based on the principle of "Semibatch-fed digester."

However, the Indian fixed-dome BGP designs differ from that of Chinese designs, as the animal manure is the major substrate (feedstock) used in India. Therefore all the Indian fixed-dome designs are based on the principle of "semicontinuous-fed digester." While the Chinese designs have no fixed storage capacity for biogas due to use of variety of crop wastes as feedstock, the Indian household BGP designs have fixed storage capacity, which is 33% of the rated gas production per day. The Indian fixed-dome plant designs use the principle of displacement of slurry inside the digester for storage of biogas in the fixed GSC. Due to this, Indian fixed-dome designs have Displacement Chamber(s), either on both the inlet and outlet sides (like Janata model) or only one the outlet side (like Deenbandhu or Grameen Bandhu model). Fig. 3.8 shows the fixed-dome-type BGP.

Therefore in Indian fixed-dome design, it is essential to keep the combined volume of inlet displacement chamber (IDC) and ODC equal to the volume of the fixed GSC; otherwise, the desired quantity of biogas will not be stored in the plant. The pressure developed inside the Chinese fixed-dome BGP ranges from a minimum of 0 to a maximum of 150 cm of water column. And the maximum pressure is normally controlled by connecting a simple manometer on the pipeline near the point of gas utilization. On the other hand, the Indian fixed-dome BGPs are designed for pressure inside the plant, varying from a minimum of 0 to a maximum of 90 cm of water column. The discharge opening located on the outer wall surface of the ODC automatically controls the maximum pressure in the Indian design.

Climate and size of fixed-dome BGPs: Fixed-dome plants must be covered with earth up to the top of the gas-filled space to counteract the internal pressure (up to 0.15 bars). The earth cover insulation and the option for internal heating make them suitable for colder climates. Due to economic parameters, the recommended minimum size of a fixed-dome plant is 5 m^3. Digester volumes up to 200 m^3 are known and possible.

Advantages of fixed-dome BGPs: The advantages are low initial costs and long useful life span; no moving or rusting parts involved; basic design is compact, saves space, and is well insulated; construction creates local employment. Table 3.2 shows the advantages and disadvantages of fixed-dome biogas power plant.

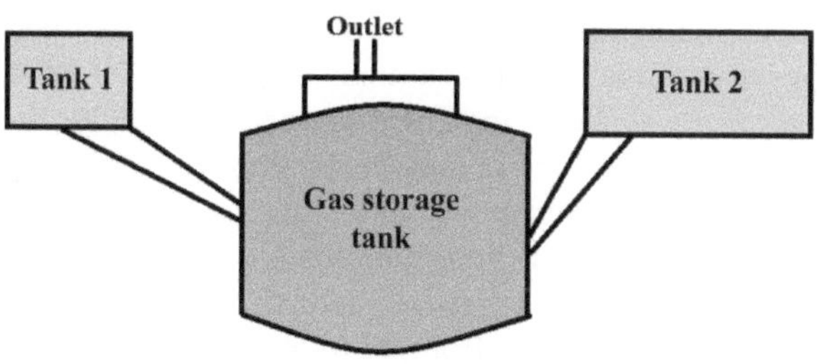

FIGURE 3.8 Fixed-dome-type biogas plant.

TABLE 3.2 Advantages and disadvantages of fixed-dome biogas plants.

Advantages	Limitations
Relatively low construction costs	Frequent problems with the gastightness of the brickwork gasholder (a small crack in the upper brickwork can cause heavy losses of biogas)
Absence of moving parts and rusting steel parts	Gas pressure fluctuates substantially depending on the volume of the stored gas
Long life span if well constructed	Even though the underground construction buffer temperature extremes, digester temperatures are generally low
Underground construction saves space and protects the digester from temperature changes	
Construction provides opportunities for skilled local employment	

Disadvantages of fixed-dome BGPs: Masonry gasholders require special sealants and high technical skills for gastight construction; gas leaks occur quite frequently; fluctuating gas pressure complicates gas utilization; amount of gas produced is not immediately visible, plant operation is not readily understandable; fixed-dome plants need exact planning of levels; excavation can be difficult and expensive in bedrock. Fixed-dome plants can be recommended only where construction can be supervised by experienced biogas technicians.

3.3.1.1 Types of fixed-dome plants

Chinese fixed-dome plant: Chinese fixed-dome plant is the archetype of all fixed-dome plants. Several million have been constructed in China. The digester consists of a cylinder with round bottom and top. Round bottom and top is the specialty of this type. Fixed-dome Chinese model BGP (also called drumless digester) was built in China as early as 1936. It consists of an underground brick masonry compartment (fermentation chamber) with a dome on the top for gas storage. In this design, the fermentation chamber and gasholder are combined as one unit. This design eliminates the use of costlier mild steel gasholder, which is susceptible to corrosion. The life of fixed-dome-type plant is longer (from 20 to 50 years) than that of the Khadi and Village Industries Commission (KVIC) plant. The concrete dome is the

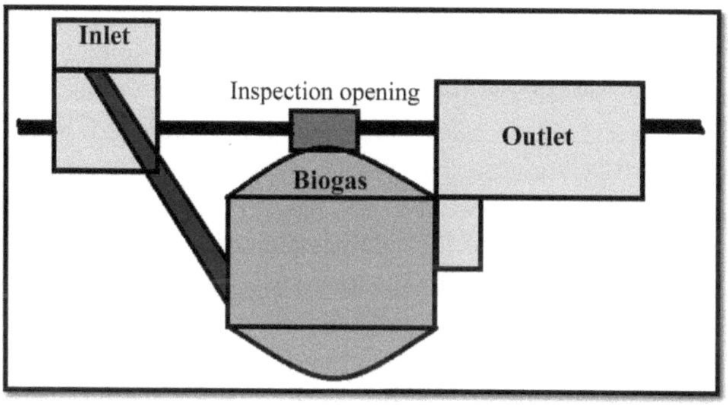

FIGURE 3.9 Schematic diagram of Chinese fixed-dome biogas plant.

main characteristic of GGC design; several million have been constructed in China. The digester consists of a cylinder with round bottom and top. Fig. 3.9 shows the schematic diagram of Chinese fixed-dome power plant.

Janata model: The design of this plant is of Chinese origin, but it has been introduced under the name "Janata biogas plant" (JBP) by Gobar Gas Research Station, PRAD, Ajitmal PRI, Lucknow, Uttar Pradesh, in 1978, in view of its reduced cost. This is a fixed roof (dome) BGP where no steel is used; there is no moving part in it and maintenance cost is low. The plant can be constructed by village mason taking some preexplained precautions and using all the indigenously available building materials. Good quality of bricks and cement should be used to avoid the afterward structural problems like cracking of the dome and leakage of gas. This model have a higher capacity than the KVIC model; hence, it can be used as a community BGP. This design has longer life than KVIC models. Substrates other than cattle dung such as municipal waste and plant residues can also be used in Janata-type plants. The plant consists of an underground well sort of digester made of bricks and cement having a dome-shaped roof that remains below the ground level, as shown in the figure. At almost middle of the digester, there are two rectangular openings facing each other and coming up to a little above the ground level, and they act as an inlet and outlet of the plant. Dome-shaped roof is fitted with a pipe at its top, which is the gas outlet of the plant. The principle of gas production is the same as that of KVIC model. The biogas is collected in the restricted space of the fixed dome; hence, the pressure of gas is much higher, which is around 90 cm of water column. Janata model was the first fixed-dome design in India, as a response to the Chinese fixed-dome plant. With a brick-reinforced, molded dome, it is not constructed anymore. The mode of construction leads to cracks in the gas-holder, and very few of these plants had been gastight.

This is also a semicontinuous flow hydraulic digester BGP. The main features of JBP are that the digester and the GSC (gasholder) are the integral part of a composite unit made of bricks and cement mortar. The JBP has a cylindrical digester with dome-shaped roof. A large "Inlet Chute" and an "OT" is attached, respectively, with the Inlet and Outlet Gate of the Digester. The top end of the Inlet Chute and OT is attached, respectively, to the lower end of "IDC" and to the "ODC" of equal volume (capacities). The upper end of the IDC is connected to the "Slurry Mixing Tank" with a channel. Whereas the outlet wall of the ODC has a small opening (known as discharge opening) for directing the digested slurry automatically to the slurry storage pits (or the compost pits) each day when JBP is under regular operation as per the prescribed guidelines. Construction of JBP requires shuttering, formwork, and mud mold for making a gas leak-proof dome-shaped roof; therefore, skilled and properly trained master masons are essential for construction of this model. The JBP costs between 20% and 30% less than the KVIC model BGP. Fig. 3.10 shows the schematic diagram and structure design of Janata model biogas power plant.

Deenbandhu model: The successor of the Janata plant in India, with improved design, was more crack-proof and consumed less building material than the Janata plant with a hemisphere digester. Deenbandhu model was developed in 1984, by Action for Food Production, a voluntary organization based in New Delhi, India. Schematic diagram of a Deenbandhu biogas plant (DBP) shown in Figs. 3.11 and 3.12 entire biogas programmed of India as it reduced the cost of the plant half of that of KVIC model and brought biogas technology within the reach of even the poorer sections of the population. The cost reduction has been achieved by minimizing the surface area through joining the segments of two spheres of different diameters at their bases. The DBP has a hemispherical fixed-dome type of gasholder, unlike the floating dome of the KVIC design and the dome is made from prefabricated ferrocement or reinforced concrete and attached to the digester, which has a curved

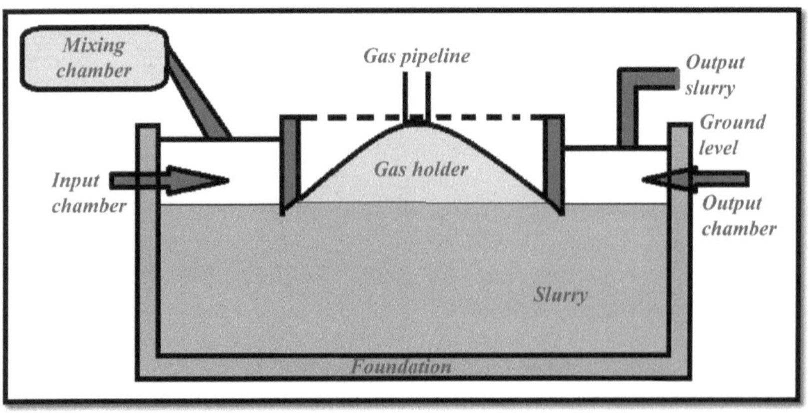

FIGURE 3.10 Janata model biogas plant.

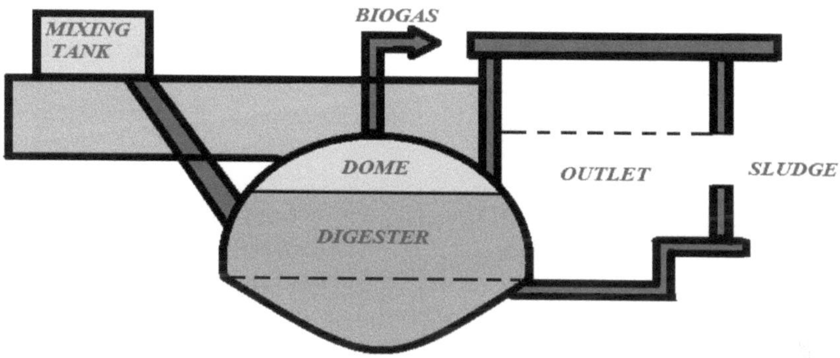

FIGURE 3.11 Schematic diagram Deenbandhu model biogas plant.

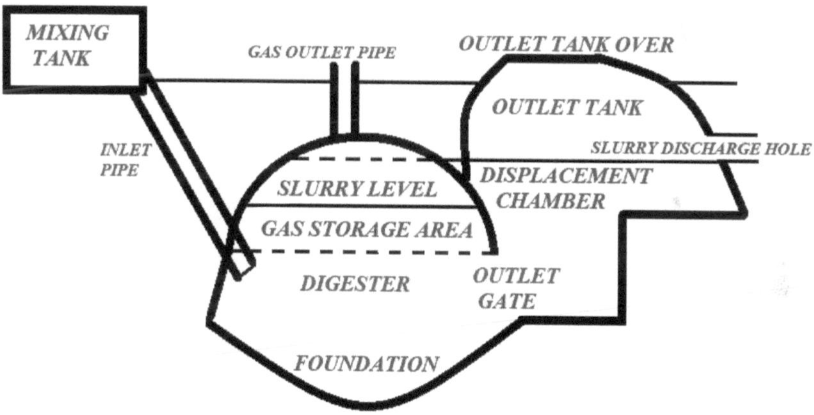

FIGURE 3.12 Schematic diagram of Deenbandhu model biogas plant with slight variations.

bottom. The slurry is fed from an MT through an inlet pipe connected to the digester. After fermentation, the biogas collects in the space under the dome. It is taken out for use through a pipe connected to the top of the dome, while the sludge, which is a by-product, comes out through an opening in the side of the digester. About 90% of the BGPs in India are of the Deenbandhu type.

Deenbandhu, the successor of the Janata plant in India, with improved design, was more crack-proof and consumed less building material than the Janata plant with a hemisphere digester.

1. It is either built of CC or brick-walled. The gas is stored in the upper dome.
2. With increase in gas, pressure in the dome rises up. It pushes the digested material out of the tank. The costs of a fixed-dome BGP are relatively low.

3. It is simple as no moving parts exist. There are also no rusting steel parts, and hence, a long life of the plant (20 years or more) can be expected.
4. Fixed-dome plants are not easy to build. They should only be built where construction can be supervised by experienced biogas technicians.
5. Due to the porosity and cracks, the gas generated may leak.

The objective of action R&D for the development of this model was to reduce further the cost of RHh plants to make biogas technology within the reach of wider sections of rural society, as the cost of building materials had started going up. The team was successful in designing and fabricating a new low-cost fixed-dome model BGP, which was not only around 20% cheaper than the Janata biogas power plant but also more sturdier and simpler to construct, after the master masons were given systematic, brick-by-brick training on the construction techniques of this technology. The reduction in cost was brought about without affecting the strength and efficiency of this new model. This plant was christened as the Deenbandhu as the design was a step closer to making this technology within the reach of poorer sections of the community. The Deenbandhu model is also a semicontinuous flow hydraulic digester plant.

After intensive trial and testing under controlled conditions as well as, field evaluating it under farmers field conditions, the design and drawings of 1- 2-, 3-, 4-, and 6-m^3 DBPs were standardized for promotion as RHh digesters. A manual on Deenbandhu model was prepared and the design of DBP model was submitted to the Department of Nonconventional Energy Sources (DNES) in 1986 [which later became the Ministry of Non-Conventional Energy Sources (MNES) and have recently been renamed the Ministry of New and Renewable Energy], Government of India. The DNES approved this model for transfer, promotion, and extension under the National program for biomass development (NPBD) from the financial year 1987–88 onward.

The DBP is built manly with locally available building materials such as brick, sand, and local skills, in the form of rural master masons. The cement is the only building material that comes from the factory at a distant place but is easily available throughout the country. As the construction of Deenbandhu requires no formwork and shuttering materials, labor requirement is also reduced. The requirement of cement and bricks is also less as compared with JBP. There is considerable saving in the construction time as compared with JBP due to the simpler and less time-consuming construction techniques. The DBP is the most popular plant in India at present, among all the approved designs of MNES, constituting over 75% of the share of annual target of over 100,000 plants constructed each year under NPBD.

Camartec model: It has a simplified structure of a hemispherical dome shell based on a rigid foundation ring only and a calculated joint of fraction, the so-called weak/strong ring. Camartec (Fig. 3.13) was developed in the late 1980s in Tanzania.

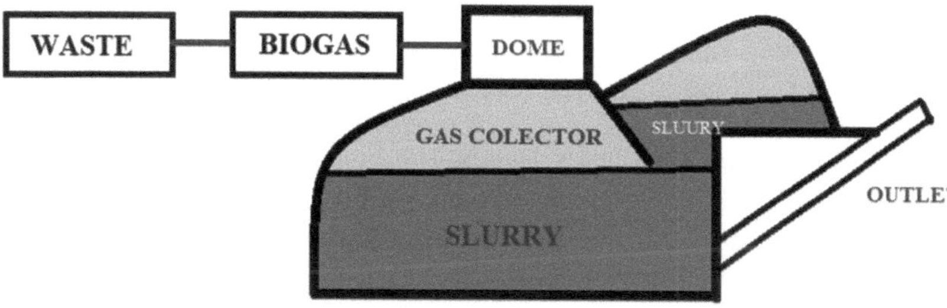

FIGURE 3.13 CAMARTEC fixed-dome plant.

FIGURE 3.14 Floating gasholder-type biogas plant.

3.3.2 Floating-drum plants

This is one of the common designs in India and comes under the category of semicontinuous-fed plant. It has a cylindrical-shaped floating biogas holder on the top of the well-shaped digester. As the biogas is produced in the digester, it rises vertically and gets accumulated and stored in the biogas holder at a constant pressure of 8–10 cm of water column. The biogas holder is designed to store 50% of the daily gas production. Therefore if the gas is not used regularly, then the extra gas will bubble out from the sides of the biogas holder. The floating gasholder digester makes use of a floating tank for gas storage. Fig. 3.14 shows the floating gasholder-type biogas power plant. This can be further subdivided into the following:

1. *Top floating gasholder digester*: The floating tank for gas storage is directly installed on the top of the digester. This is usually employed for small-sized digester.
2. *Separate floating gasholder digester*: The application of this style is for medium-to-large-sized digester. There are two tanks involved: one is the fermentation tank and the other is the storage tanks.

The drum: In the past, floating-drum plants were mainly built in India. A floating-drum plant consists of a cylindrical- or dome-shaped digester and a moving, floating gasholder, or drums. The gasholder floats either directly in the fermenting slurry or in a separate water jacket. The drum in which the biogas collects has an internal and/or external guide frame that provides stability and keeps the drum upright. If biogas is produced, the drum moves up, and if gas is consumed, the gasholder sinks back.

Size: Floating-drum plants are used chiefly for digesting animal and human faces on a continuous feed mode of operation, that is, with daily input. They are used most frequently by small-to-middle-sized farms (digester size: 5–15 m^3) or in institutions and larger agroindustrial estates (digester size: 20–100 m^3).

3.3.2.1 Water-jacket floating-drum plants

Water-jacket plants (Fig. 3.15) are universally applicable and easy to maintain. The drum cannot get stuck in a scum layer, even if the substrate has high solid content. Water-jacket plants are characterized by a long useful life and a more aesthetic appearance (no dirty gasholder). Due to their superior sealing of the substrate, they are recommended for use in the fermentation of night soil. The extra cost of the masonry water jacket is relatively modest.

Material of digester and drum: The digester is usually made of brick, concrete, or quarry-stone masonry with plaster. The gas drum normally consists of 2.5-mm steel sheets on the sides and 2-mm sheets on the top. It has welded braces, which break up surface scum when the drum rotates and the drum must be protected against corrosion. Suitable coating products are oil paints, synthetic paints, and bitumen paints. There must be at least two preliminary coats and one topcoat. In coastal regions, repainting is necessary at least once a year and at least every other year in dry uplands. Gas production will be higher if the drum is painted black or red rather than blue or white, because the digester temperature is increased by solar radiation. Gas drums made of 2-cm wire-mesh-reinforced concrete or fiber-cement must receive a gastight internal coating. The gas drum should have a slightly sloping

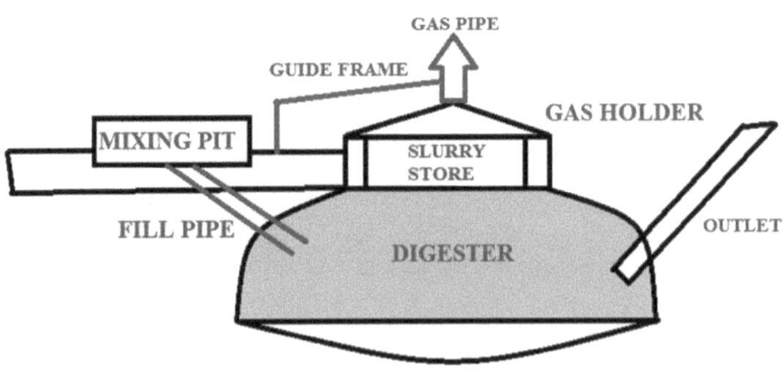

FIGURE 3.15 Water-jacket plant with external guide frame. 1: Mixing pit; 11: fill pipe; 2: digester; 3: gasholder; 31: guide frame; 4: slurry store; 5: gas pipe.

roof; otherwise, rainwater will be trapped on it, leading to rust damage. An excessively steep-pitched roof is unnecessarily expensive and the gas in the tip cannot be used, because when the drum is resting on the bottom, the gas is no longer under pressure. Floating drums made of glass-fiber-reinforced plastic and high-density polyethylene (HDPE) have been used successfully, but the construction costs are higher than using steel. Floating drums made of wire-mesh-reinforced concrete are liable to hairline cracking and are intrinsically porous. They require a gastight, elastic internal coating. Polyvinyl chloride (PVC) drums are unsuitable, because they are not resistant to ultraviolet (UV).

Guide frame: The side wall of the gas drum should be just as high as the wall above the support ledge. The floating drum must not touch the outer walls. It must not tilt; otherwise, the coating will be damaged, or it will get stuck. For this reason, a floating drum always requires a guide. This guide frame must be designed in a way that allows the gas drum to be removed for repair. The drum can only be removed if air can flow into it, either by opening the gas outlet or by emptying the water jacket. The floating gas drum can be replaced by a balloon above the digester. This reduces construction costs, but in practice, problems always arise with the attachment of the balloon to the digester and with the high susceptibility to physical damage.

Advantages of floating-drum BGPs: Floating-drum plants are easy to understand and operate. They provide gas at a constant pressure, and the stored gas volume is immediately recognizable by the position of the drum. Gastightness is no problem, provided the gasholder is derusted and painted regularly.

Disadvantages of floating-drum BGPs: The disadvantages are high construction cost of floating drum, many steel parts liable to corrosion, resulting in short life (up to 15 years; in tropical coastal regions about 5 years for the drum), and regular maintenance costs due to painting. In spite of these disadvantages, floating-drum plants are always to be recommended in the cases of doubt. Water-jacket plants are universally applicable and especially easy to maintain. The drum would not stick, even if the substrate has high solid content. Floating drums made of glass-fiber-reinforced plastic and HDPE have been used successfully, but the construction cost is higher than that with steel. Floating drums made of wire-mesh-reinforced concrete are liable to hairline cracking and are intrinsically porous. They require a gastight, elastic internal coating. PVC drums are generally unsuitable, because they are not resistant to UV. The floating gas drum can be replaced by a balloon above the digester. This reduces construction costs (channel-type digester with folia), but in practice, problems always arise with the attachment of the balloon at the edge. These types of plants are still being tested under practical conditions. Table 3.3 shows the advantages and disadvantages of floating-drum biogas power plant.

TABLE 3.3 Advantages and disadvantages of floating-drum biogas plants.

Advantages	Limitations
Simple, easily understood operation.	High material costs of the steel drum.
The volume of stored gas is directly visible.	Susceptibility of steel parts to corrosion (because of this, floating-drum plants have a shorter life span than fixed-dome plants).
The gas pressure is constant (determined by the weight of the gasholder).	Regular maintenance costs for the painting of the drum.
Construction is relatively easy.	If fibrous substrates are used, the gasholder shows a tendency to get "stuck" in the resultant floating scum.
Construction mistakes do not lead to major problems in operation and gas yield.	

3.3.2.2 *Types of floating-drum plants*

Floating-drum plant is made of prefabricated-reinforced concrete compound units and fiber-glass-reinforced polyester. There are different types of floating-drum plants:

1. KVIC model with a cylindrical digester, the oldest and most widespread floating-drum BGP from India; and
2. Pragati model with a hemisphere digester.

Khadi and Village Industries Commission model: With a cylindrical digester, this is the oldest and most widespread floating-drum BGP from India. In 1961–62, the KVIC decided to undertake Gram Laxmi-III model, designed by the Late Dr. Jasbhai Patel, the then Director (Biogas), KVIC, in 1956–57, for popularization. Since then, this plant has been associated with KVIC and is now popularly known as KVIC model. It is a semicontinuous flow hydraulic digester BGP. The KVIC design consists of a deep well-shaped underground digesters connected with inlet and outlet pipes at the bottom, just opposite to each other but are separated by a partition wall dividing the digester into two equal parts. The height of the partition walls is three-fourth of the total height of the digester. A mild steel gas storage drum (gasholder) is inverted in the digester over the slurry, which goes up and down around a guide with the accumulation and withdrawal of gas. In KVIC model, the cost of steel drum gasholder itself constitutes around 40% of the total cost of the plant. The gasholders of KVIC model can now be made of FRP, ferrocement, and HDP, in place of steel gasholder, but till date have not been widely accepted in rural areas. Fig. 3.16 shows the schematic diagram of KVIC biogas power plant.

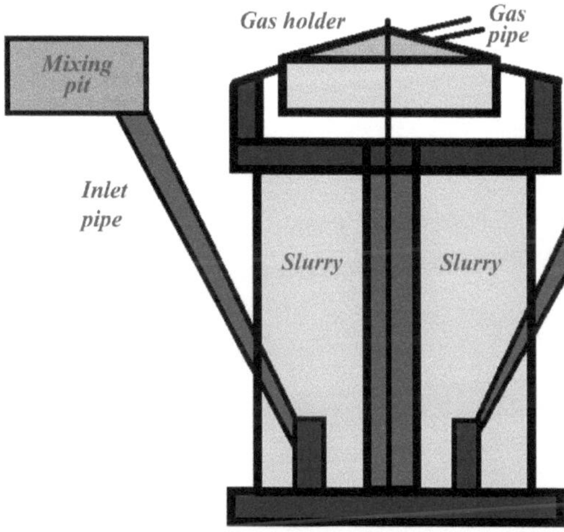

FIGURE 3.16 Schematic diagram of Khadi and Village Industries Commission model biogas plant.

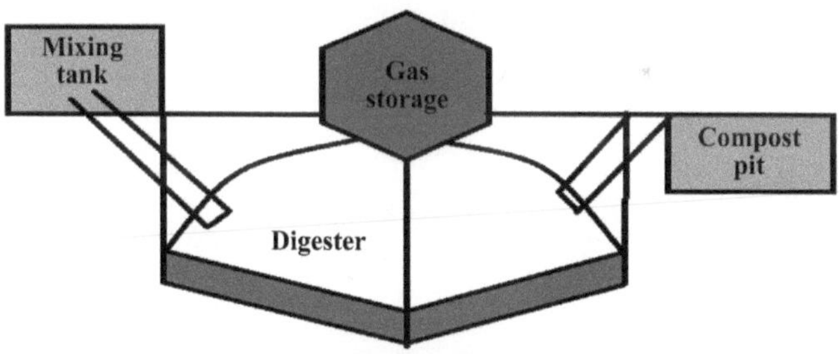

FIGURE 3.17 Schematic diagram of Pragati model biogas plant.

Pragati model: Pragati model (Fig. 3.17) is the combination of Deenbandhu and KVIC designs. The lower part of the digester is semispherical with conical bottom. Floating drum acts as gas storage. The Pragati model of BGP combines the merits of both the floating gasholder-type and fixed-dome-type BGPs. The foundation of this type of BGP is conical and the digester is shell-shaped. The inlet and outlet assemblies are almost the same as in the case of the floating gasholder-type plant. There is a slight modification in the central guide frame and steel gasholder, so that the least load falls on the shell of the digester.

3.3.3 Flexible bag biogas plant (balloon plants)

A balloon plant (Fig. 3.18) consists of a plastic or rubber digester bag, in the upper part of which the gas is stored. The inlet and outlet are attached directly to the skin of the balloon. When the gas space is full, the plant

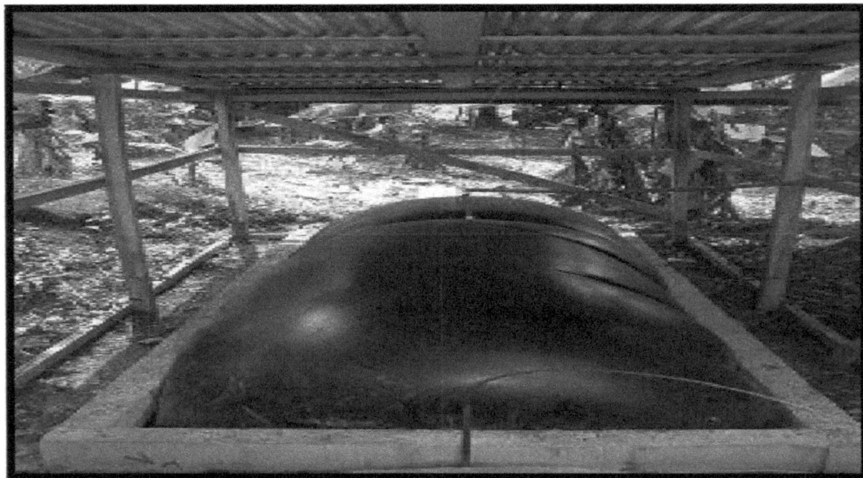

FIGURE 3.18 Balloon-digester biogas plant.

works like a fixed-dome plant, and the balloon is not inflated; it is not very elastic. The fermentation slurry is agitated slightly by the movement of the balloon skin. This is favorable to the digestion process. Even difficult feed materials, such as water hyacinths, can be used in a balloon plant. The balloon material must be UV-resistant. Materials that have been used successfully include red mud plastic (RMP), Trevira, and butyl.

The bag digester is a type of digester with a separate bag for gas storage. The entire MUP including the digester is fabricated out of rubber, high-strength plastic, neoprene, or RMP. The Inlet and Outlet is made of heavy duty PVC tubing. A small pipe of the same PVC tubing is fixed on the top of the plant as GOP. The Flexible Bag BGP is portable and can be easily erected. Being flexible, it needs to be provided support from outside, up to the slurry level, to maintain the shape as per its design configuration, which is done by placing the bag inside a pit dug at the proposed site. The depth of the pit should as per the height of the digester (fermentation chamber) so that the mark of the initial slurry level is in line with the ground level. The outlet pipe is fixed in such a way that its outlet opening is also in line with the ground level. Some weight has to be added on the top of the bag to build the desired pressure to convey the generated gas to the point of utilization. The advantage of this plant is that the fabrication can be centralized for mass production, at the district or even at the block level. Individuals or agencies having land and some basic infrastructure facilities can take up fabrication of this BGP with small investment, after some training. However, the cost of good quality plastic and rubber is high, which increases the comparative cost of fabricating it. Moreover, the useful working life of this plant is much less, compared with other Indian simple household BGPs, therefore in spite of having good potential; the Flexible Bag BGP has not been taken up seriously for promotion by the field agencies. Fig. 3.19 shows bag-balloon gasholder-type biogas power plant.

Advantages: The various advantages are standardized prefabrication at low cost, low construction sophistication, ease of transportation, and shallow installation suitable for use in areas with a high groundwater table; high digester temperature in warm climates; uncomplicated cleaning, emptying

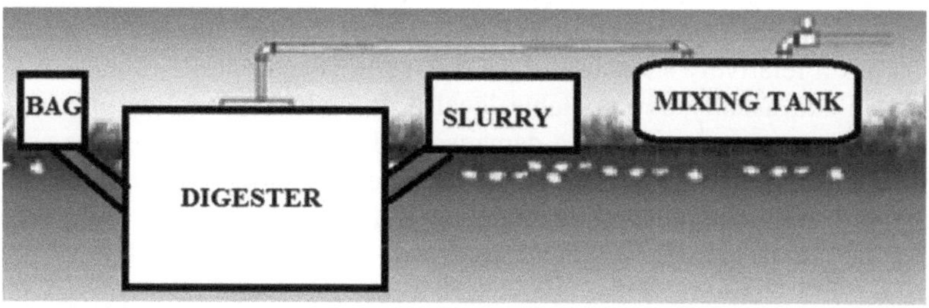

FIGURE 3.19 Bag-balloon gasholder-type biogas plant.

TABLE 3.4 Advantages and disadvantages of balloon-digester biogas plants.

Advantages	Limitations
Low cost	Relatively short life (about 5 years)
Low construction sophistication	Little creation of local employment
High digester temperatures	Limited self-help potential
Uncomplicated cleaning, emptying, and maintenance	Little knowledge for repairing by local craftsmen
Ease of transportation	Susceptibility to damage
Low-cost system suitable for rural and semiurban communities	Only applicable where land is available and cheap
High community participation in construction and O&M possible	Permanent overload leads to breakdown of biological cleansing processes
Simple operation and maintenance	Misuse of system leads to public health hazard
Resistant against shock load and variable inflow volume, if lagoon size is big enough	Sullage is in the open and, thus, poses a potential health threat
	Low treatment efficiency, effluent is still infectious
	Not suitable where there is a high groundwater table due to infiltration of sullage
	Public health hazard if system is overloaded

and maintenance; and difficult substrates like water hyacinths can be used. Table 3.4 shows the advantages and disadvantages of balloon-digester biogas power plant.

Disadvantages: The disadvantages are low gas pressure may require gas pumps; scum cannot be removed during operation; and the plastic balloon has a relatively short useful life span and is susceptible to mechanical damage and usually not available locally. In addition, local craftsmen are rarely in a position to repair a damaged balloon. There is only little scope for the creation of local employment and, therefore, limited self-help potential. Balloon BGPs are recommended if local repair is or can be made possible and the cost advantages are substantial. It can be used as sanitary latrine with BGP and toilet-linked BGPs for conversion of night soil into biogas.

3.3.4 Anaerobic baffled reactor

In this type of biodigester, mechanical and anaerobic cleansing processes are applied in sequence. This reactor consists of different chambers in which mode of flow is upstream. Wastewater is intensively mixed up with the sludge and supplied to digester chamber for digestion. It has integrated sedimentation chamber for pretreatment. The prime motive for installation of this reactor is wastewater treatment, and therefore its sizing and design consideration are based on the availability and amount of wastewater and, therefore, biogas production consideration is secondary (Table 3.5).

3.3.5 Toilet-linked biogas plants

Night-soil-based or toilet-linked BGPs are widely used in Asia for the codigestion of human excreta along with animal manure (e.g., cattle or buffalo dung, etc.) The hygienically safe on-site treatment of toilet water and recovery of valuable energy in the form of biogas are to be used as a substitute to liquefied petroleum gas (LPG) in cooking and lighting. The prime motive for installation of this reactor is for wastewater treatment, and therefore its sizing and design consideration are based on the availability and amount of wastewater and, therefore, biogas production consideration is secondary. Fig. 3.20 shows the schematic diagram of toilet-linked biogas power plant. Table 3.6 shows the advantages and limitations of toilet-linked digester BGPs.

LPG, Liquefied petroleum gas.

TABLE 3.5 Advantages and disadvantages of baffled reactor biogas plants.

Advantages	Limitations
Suitable for small and large settlements	Experts required for design and supervision
Little space required due to underground construction	Master mason required for high-quality plastering work
Very low operation and maintenance costs	Infectious organisms are not sufficiently removed
Low investment costs	Well-organized combined biomass (CBO) or service provider needed for operation and maintenance
Standardized designs and standard operation procedures available	Manual desludging of the tank is highly hazardous and an inhumane task
Simple operation and maintenance	Mechanical desludging (vacuum trucks) requires sophisticated instruments
High treatment efficiency	

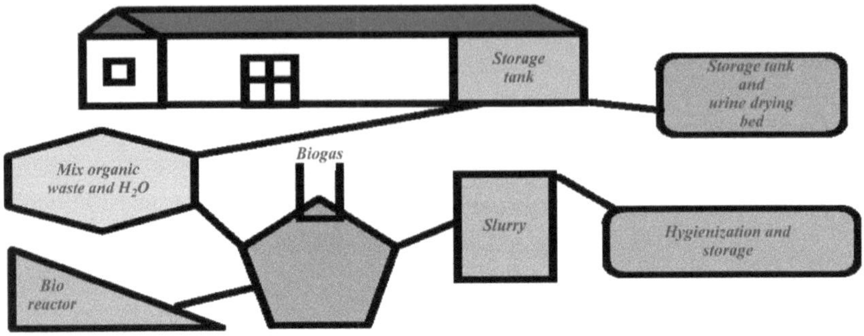

FIGURE 3.20 Schematic diagram of toilet-linked biogas plant.

TABLE 3.6 Advantages and limitations of toilet-linked digester biogas plants.

Advantages	Limitations
No handling of raw (unprocessed) toilet water	Limited biogas production if only toilet water is treated
Increased biogas production if additional feed material (e.g., animal manure, etc.) is available for codigestion	
Biogas may be used as a substitute to LPG in cooking	
Application of digested slurry as soil amendment to agricultural plots possible	

3.4 Functioning of biogas plant

The BGP detailed in this manual consists of five main structures or functions of components:

1. inlet tank;
2. digester vessel;
3. dome;
4. OC; and
5. compost pits.

The required quantity of feedstock and water is mixed in the inlet tank, and the slurry is discharged to the digester vessel for digestion. The fresh organic material (generally in a homogenous slurry form) is fed into the digester of the plant from one end, known as Inlet. Fixed quantity of fresh material fed each day (normally in one lot at a predetermine time) goes down at the bottom of the digester and forms the "bottommost active layer," being heavier than the previous day and older material. The decomposition (fermentation) takes place inside the digester due to bacterial (microbial) action, which produces biogas and digested or semidigested organic material.

As the organic material ferments, biogas is formed through methanogenesis in the digester, which rises to the top and gets accumulated in the gasholder (in the case of floating gasholder BGPs) or GSC also called the dome (in the case of fixed-dome BGPs). A GOP is provided on the topmost portion of the gasholder (GSC) of the BGP. Alternatively, the biogas produced can be taken to another place through pipe connected on the top of the GOP and stored separately. The gas is supplied from the dome to the point of application through a pipeline. The slurry then flows through the overflow opening in the OT to the compost pit. The slurry (semidigested and digested) occupies the major portion of the digester and the sludge (almost fully digested) occupies the bottommost portion of the digester. The digested slurry (also known as effluent) is automatically discharged from the other opening, known as outlet, to the OT through the manhole and is an excellent biofertilizer, rich in humus. The anaerobic fermentation increases the ammonia content by 120% and quick acting phosphorous by 150%. Similarly, the percentage of potash and several micronutrients useful to the healthy growth of the crops also increases. The nitrogen is transformed into ammonia that is easier for plant to absorb. This digested slurry can either be taken directly to the farmer's field along with irrigation water or stored in slurry pits (attached to the BGP) for drying or directed to the compost pit for making compost along with other waste biomass. The slurry and the sludge contain a higher percentage of nitrogen and phosphorous than the same quantity of raw organic material fed inside the digester of the BGP.

When a BGP is underfed, the gas production will be low; in this case, the pressure of the gas might not be sufficient to displace fully the slurry in the OC. It is important to design the plant keeping hydrostatic pressure higher at the inlet tank than the OT. The hydrostatic pressure from slurry in the inlet and OTs will pressurize the biogas accumulated in the dome. If too much material is fed into the digester and the volume of gas is consumed, the slurry may enter the gas pipe and to the appliances. Starting an anaerobic digester is relatively easy if a few points are kept in mind.

- Need to have the appropriate bacteria present. In cattle and pig wastes, there are usually the right bacteria but poultry waste and vegetable waste may need inoculation, so you may have to start with cattle or pig manure and gradually introduce the other waste form.
- Start-up will be slow and the actual process is fairly slow and proceeds in a series of steps, just like AD itself. First, the oxygen in headspace air has to be used up to get anaerobic conditions, so the first gas collected will be carbon dioxide. Acetogenesis then generates more CO_2 and acetic acid, so if this stage gets too advanced, methanogenesis can be inhibited.

Once there is some acetic acid, methanogenesis can start, but of course, some initial bacteria must be present.

- Any sudden changes can upset the process, and if you need to change the amount of waste added per day, the concentration, or the type of waste, it is better to make the change gradual, so the bacterial population has time to adjust. While the change is underway monitor gas quality, a drop in quality (more CO_2 and/or less CH_4) is often the first indication of possible digester failure. If the gas quality does deteriorate, it is best to stop feeding and give the digester a chance to recover.
- The preferred way to start a digester is to part fill with water (so you can check operation and make any alterations necessary without having to deal with effluent) and then put in half the working volume of dairy or piggery effluent (or active digester sludge if available). Wait until the digester is producing flammable gas, which may take a few weeks if temperatures are cool, before starting feeding at half the design rate. Once operation has settled down, increase to full rate, still using the initial waste and wait for operation to settle down again. If you want to use different wastes, use a 50/50 mix for at least one retention time (RT) before changing over completely, again monitor the digester performance, and wait until you are satisfied that the operation has adjusted before making any further change.
- *Initial filling*: The initial filling of a new BGP should, if possible, consist of either digested slurry from another plant or cattle dung. The age and quantity of the inoculants (starter sludge) have a decisive effect on the course of fermentation. It is advisable to start collecting cattle dung during the construction phase in order to have enough by the time the plant is finished. When the plant is being filled for the first time, the substrate can be diluted with more water than usual to allow a complete filling of the digester.

Type of substrate: Depending on the type of substrate in use, the plant may need from several days to several weeks to achieve a stable digesting process. Cattle dung can usually be expected to yield good gas production within 1 or 2 days. The breaking-in period is characterized by:

1. low-quality biogas containing more than 60% CO_2;
2. very odorous biogas;
3. sinking pH; and
4. erratic gas production.

Potential feedstock for biodigesters: The raw materials that can be considered as suitable substances for the production of biogas through the bioconversion process in a digester are as follows.

3.4.1 What type of waste produces biogas?

a. urban waste (garbage):
   i urban refuse (human excreta);

- ii rural and agricultural waste;
- iii cow dung; and
- iv animal waste from butchery.

b. agricultural waste includes:

- i straw of rice, wheat, other cereals, or crops;
- ii bagasse of sugar cane;
- iii groundnut shell, rise husk, wheat husk, etc.;
- iv unused food grain of all types.

c. fruit waste and fruit tree waste includes:

- i Waste of all types of fruits, for example, coconut tree waste, coconut husk, coconut shells, unused and spoilt grapes, cashew, banana, mango, etc.

d. rural animal waste includes:

- i cow dung, horse dung, and sheep dung;
- ii slaughterhouse waste;
- iii droppings and poultry waste;
- iv piggery dung and piggery waste;
- v urban waste and municipal garbage.

e. aquatic waste and aquatic crops:

- i waste from fishery;
- ii harvested algae;
- iii water hyacinth, etc.

1. *Crop Residue*: This includes sugarcane trash, weed, corn and crop stubble, straw, spoiled fodder, etc. Studies have shown that the content of water-soluble substances such as sugars, amino acids, proteins, and mineral constituents decreases with the age of the plant and it is low enough to limit the rate of the digestion process. Thus the decomposition of crop residues takes a longer time than manures due to their fibrous content and larger particle sizes.
2. *Manure*: Cattle-shed waste (dung, urine, and litter), sheep and goat droppings, slaughterhouse waste (blood and meat), fishery waste, leather, etc. from the agricultural community represent a significant source of feedstock for the biogas production. Manure is a source of carbon and nitrogen required for the successful operation of the fermentation process. The quantity and composition of animal waste is dependent on the type of animal. Poultry, for example, produces more volatile solids (VS), nitrogen, and phosphorus, per unit weight. In addition, the composition of manure depends to a large extent on the feed ration of the animal. Animals feeding on grass alone show much lower nitrogen content in their manure and urine.
3. *Human waste*: Human faces and urine are raw materials that can be used for biogas production. No work using this type of feedstock is currently visible in Guyana.
4. *By-products and waste from agriculture-based industries and aquatic growth*: Oil cakes, bagasse, rice bran, seeds, water from fruit and

vegetable processing, filter press mud from sugar factories, marine algae, seaweed, etc. can also be sued as feedstock form biodigesters. Table 3.7 shows the amount of human and animal wastes discharged per day. Table 3.8 shows the compositional analysis of biogas produced from each substrate (cm^3)

3.4.2 What type of waste does not produce biogas?

i. Fiber-rich waste such as wood, leaves, etc. are difficult to digest;
ii. heavy metals; and
iii. inorganic materials in high concentration (nitrate, sodium, sulfate, sodium, potassium, calcium, magnesium, etc.).

Stabilization of the process: The digesting process will stabilize more quickly if the slurry is agitated frequently and intensively. Only if the process shows extreme resistance to stabilization, should lime or more cattle dung be added in order to balance the pH value. No additional biomass should be put into the BGP during the remainder of the starting phase. Once the process has stabilized, the large volume of unfermented biomass will result in a high rate of gas production. Regular loading can commence after gas production has dropped off to the expected level.

Gas collection and storage: The digester itself provides the simplest gas storage, but supply pressure will fall as gas is used. You may need some weights to place on the digester to develop more pressure, but be careful of puncturing the digester body. If the use is a fair way from the digester, better burner performance may be obtained by having a separate gas storage near the use point. This can be either a flexible, gastight bag or floating-drum-type storage. With the second storage, it is a good idea to allow low-pressure collection of the gas (to minimize the possibility of gas leaks) and to use weight to increase pressure during burner operation, but you need the "back flow preventer" to make best use of this. The biogas in a floating-drum-type digester is collected in an inverted drum. The walls of the drum extend down into the slurry to provide a seal. The drum is free to move to accommodate more or less gas as needed. The weight of the drum provides the pressure on the gas system to create flow. The biogas flows through a small hole in the roof of the drum. A nonreturn valve here is a valuable investment to prevent air being drawn into the digester, which would destroy the activity of the bacteria and provide a potentially explosive mixture inside the drum. Larger plants may need counterweights of some sort to ensure that the pressure in the system is correct. The drum must obviously be slightly smaller than the tank, but the difference should be as small as possible to prevent loss of gas and tipping of the drum. Table 3.9 shows biogas compression and storage pressure and material.

The storage should be sized to hold one day's use of gas if possible, which should be as closely matched as possible to digester gas production.

TABLE 3.7 Amount of human and animal wastes discharged per day.

Kinds	Body weight (kg)	Daily amount of excrement (kg)	Daily amount of urine (kg)	Annual amount of excrement discharged (kg)	Annual amount of excrement collection (kg)	Daily yield of biogas per capita (m^3)
Pig	50	6	15	2190	1752	0.18–0.25
Ox	500	34	34	12,410	9928	0.36–0.96
Horse	500	10	15	3650	2920	
Sheep	15	1.5	2	548	438.4	
Chicken	1.5	0.10	0	36.80	29.44	0.0076–0.0112
Human	50	0.50	1	182.50	146.00	0.028

TABLE 3.8 Compositional analysis of biogas produced from each substrate (cm^3).

Substrate	Total biogas (cm^3)	CH_4 (cm^3)	CH_4 (%)	CO_2 (cm^3)	CO_2 (%)	H_2S (cm^3)	H_2S (%)
Cow dung	8540	5640	65.9	2829	32.9	74.6	1
Millet Husk	6522	3789	58.00	2656	41.00	52.12	1
Rice husk	1375	899	65.00	461	33.00	22.39	2
Saw dust	965	669	69.00	289	29.89	14.23	1.5
Paper waste	469	342	72.60	116	24.28	14.23	3.12

TABLE 3.9 Biogas compression and storage pressure and material.

Pressure	Storage device	Material
Low	Water sealed gasholder	Steel
Low	Gas bag	Rubber, plastic, and vinyl
Medium	Propane and butane tank	Steel
High	Commercial gas cylinder	Alloy

If production does exceed use, then any excess gas needs to be flared, as the methane in biogas is about 23 times worse as a greenhouse gas than carbon dioxide produced during combustion. As methane is very hard to compress, I see its best use as for stationary fuel, rather than mobile fuel. It takes a lot of energy to compress the gas (this energy is usually just wasted) plus you have the hazard of high pressure. Variable volume storage (flexible bag or floating drum are the two main variants) is much easier and cheaper to arrange than high-pressure cylinders, regulators, and compressors.

Gas quality: The required quantity of feedstock and water is mixed in the inlet tank and the slurry is discharged to the digester vessel for digestion. The gas produced through methanogenesis bacteria in the digester is collected in the dome. The digested slurry flows to the OT through the manhole. The slurry then flows through the overflow opening in the OT to the compost pit. The gas is supplied from the dome to the point of application through a turret and pipeline. When a BGP is underfed, the gas production will be low; in this case, the pressure of the gas might not be sufficient to displace fully the slurry in the OC. It is important to design the plant keeping hydrostatic pressure higher at the inlet tank than the OT. The hydrostatic pressure from slurry in the inlet and OTs will pressurize the biogas accumulated in the dome. If too much material is fed into the digester and the volume of gas is consumed, the slurry may enter the gas pipe and to the appliances. As soon as the biogas becomes reliably combustible, it can be used for the intended purposes. Less-than-optimum performance of the appliances due to inferior gas quality should be regarded as acceptable at first. However, the first two gasholder fillings should be vented unused for reasons of safety, since residual oxygen poses an explosion hazard.

Substrate input: For a simple, small-scale biogas system, only a minimum amount of time and effort must be spent on procuring the feedstock and preparing it for fermentation. The technical equipment is relatively inexpensive. Theoretically, any organic material can be digested. Substrate preprocessing and conveying depend on the type of material to be used. One of the most important problems in substrate management to be considered is the problem of scum.

Effluent sludge: The sludge resulting from the digestion process represents a very valuable material for fertilization. The following aspects of sludge treatment and use are considered here:

1. sludge storage;
2. composition of sludge;
3. fertilizing effect of effluent sludge; and
4. sludge application and slurry-use equipment.

3.4.2.1 Stratification (layering) of digester due to anaerobic fermentation

In the process of digestion of feedstock in a BGP, many by-products are formed. They are biogas, scum, supernatant, digested slurry, digested sludge, and inorganic solids. If the content of biogas digester is not stirred or disturbed for a few hours, then these by-products get formed into different layers inside the digester. The heaviest by-product, which is inorganic solids, will be at the bottommost portion, followed by digested sludge, and so on, as shown in the three diagrams of Fig. 3.21 for three different types of digester.

a. *Biogas*: Biogas is a combustible gas produced from the anaerobic digestion of organic matter comprising 55%–70% methane, 30%–45% carbon dioxide, 1%–2% of hydrogen sulfide, and traces of other gases.
b. *Scum*: Mixture of coarse fibrous and lighter material that separates from the manure slurry and floats on the topmost layer of the slurry is called scum. The accumulation and removal of scum is sometimes a serious problem. In moderate amount, scum do any harm and can be easily broken by gentle stirring, but in large quantity can lead to slowing down of biogas production and even shutting down of the BGPs.

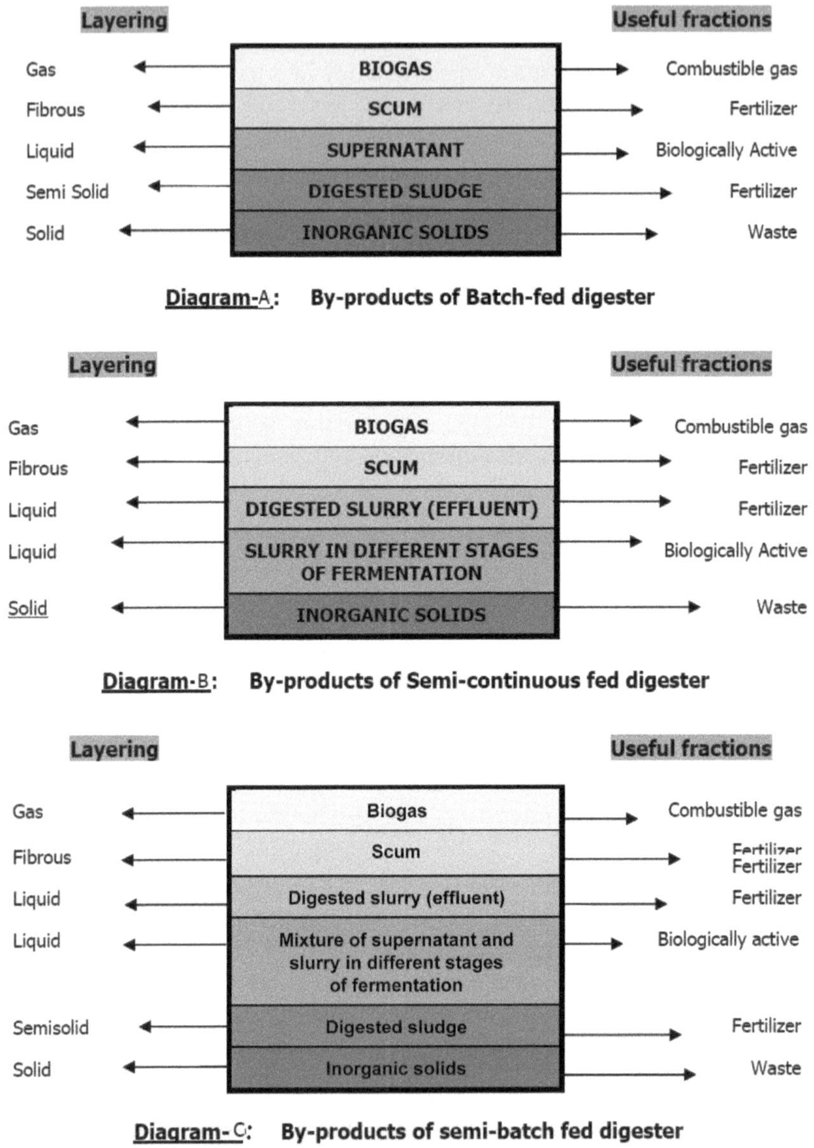

FIGURE 3.21 Schematic diagram of stratification for three types of biodigestion processes.

c. *Supernatant*: The spent liquid of the slurry (mixture of manure and water) layering just above the sludge, in the case of batch-fed and semibatch-fed digester, is known as supernatant. Since supernatant has dissolved solids, the fertilizer value of this liquid (supernatant) is as great as that of effluent (digested slurry). Supernatant is a biologically active by-product and, therefore, must be sun dried before using it in agricultural fields.

d. *Digested slurry (effluent)*: The effluent of digested slurry is in liquid form and has its solid content [total solid (TS)] reduced to approximately 10%–20% by volume of the original (influent) manure (fresh) slurry, after going through the anaerobic digestion cycle. Out of the three types of digestion processes mentioned above, the digested slurry in effluent-form comes out only in semicontinuous BGP. Digested slurry effluent, either in liquid form or after sun drying in slurry pits, makes excellent biofertilizer for agriculture, horticulture, and aquaculture.

Sludge: In the batch-fed or semibatch-fed digester where the plant wastes and other solid organic materials are added, the digested material contains less of effluent and more of sludge. The sludge precipitates at the bottom of the digester and is formed mostly of the solid substances of plant wastes. The sludge is usually composted with chemical fertilizers as it may contain a higher percentage of parasites and pathogens and hookworm eggs, etc., especially if the semibatch digesters are either connected to the pigsty or latrines. Depending on the raw materials used and the conditions of the digestion, the sludge contains many elements essential to the plant life, for example, nitrogen, phosphorous, potassium, plus a small quantity of salts (trace elements), indispensable to the plant growth—the trace elements such as boron, calcium, copper, iron, magnesium, sulfur, and zinc. The fresh digested sludge, especially if the night soil is used, has high ammonia content and in this state may act like a chemical fertilizer by forcing a large dose of nitrogen than required by the plant and, thus, increasing the accumulation of toxic nitrogen compounds. For this reason, it is probably best to let the sludge age for about 2 weeks in open place. The fresher the sludge, the more it needs to be diluted with water before application to the crops; otherwise, very high concentration of nitrogen may kill the plants.

e. *Inorganic solids*: In village situation, the floor of the animal shelters is full of dirt, which gets mixed with the manure. Added to this, the collected manure is kept on the unlined surface that has plenty of mud and dirt. Due to all these, the feedstock for the BGP always has some inorganic solids, which goes inside the digester along with the organic materials. The bacteria cannot digest the inorganic solids, and therefore settles down as a part of the bottommost layers inside the digester. The inorganic solids contain mud, ash, sand, gravel, and other inorganic materials. The presence of too many inorganic solids in the digester can adversely affect the efficiency of the BGP. Therefore to improve the efficiency and enhance the life of a semicontinuous BGP, it is advisable to empty even it in a period of 5–10 years for thoroughly cleaning and washing it from inside and then reloading it with fresh slurry.

Monitoring digester operation: One of the best indicators of proper digester operation is that the gas volume produced per day is consistent with the waste input and that the gas will support combustion (which indicates at least 50% methane). As the digester is a well-buffered system, a simple pH measurement will really only tell you that the digester is in trouble, usually well after the gas quality/volume has dropped.

A syringe body fitted with some flexible tube and some dilute sodium hydroxide solution can be used to estimate carbon dioxide percentage, as NaOH absorbs CO_2 but not methane. Draw up 20–30 mL of biogas and put the end of the tube into the NaOH solution, and then push out excess gas to get a 10-mL gas sample (you have to allow for the gas in the tube, which may be 4–5 mL). Now draw up approximately 20 mL of solution and keep the end of the tube submerged while you shake the syringe for 30 seconds. Point the syringe downward and push out excess liquid, so the syringe plunger reaches 10 mL. Now read the volume of liquid, which should be 3–4 mL indicating about 30%–40% of gas absorbed, so we assume the balance is methane. If you get over 50% methane (a reading of less than 5 mL of liquid) and the flame will still not burn properly, you must have nitrogen or some other gas present. To measure pressure in a digester, simply put a clear plastic tube into a container of water (a glass container is easiest) and connect the tube to the digester gas line. The digester pressure will push down the water surface in the tube and the difference between inside and outside levels is known as "water gauge." If the pressure is too high, gas will bubble out the bottom of the tube. You can also make a "manometer" by bending a plastic tube into a U and putting water in—when one end is connected to the gas line and the other is open to atmosphere (you may need to restrict this to a small hole so the levels settle quickly), you can measure the difference in water heights.

Digester performance: When looking at an anaerobic digester as a biogas production unit the performance depends on the type of waste used, the amount of VS put in per day, that is, loading rate, the operating temperature (T), and the RT. Since VS reduction is directly related to both biogas production and chemical/biological oxygen demand (the pollution load), the same variables apply to the digester as a pollution treatment unit anyway. The various parameters affecting the digester performance are discussed in Table 3.10:

C/N, Carbon–nitrogen; *HRT*, hydraulic retention time.

Efficiency of biogas production: Other parameters being the same, the overall efficiency of biogas production for a given plant design depends on three things. They are:

- the optimum yield from a given substrate for a given HRT or solid detention time;
- the gas production per unit time (usually 24 hours in a day); and

TABLE 3.10 Physiochemical conditions for optimum production of biogas.

S. no.	Character	Range	Comments
1	pH	6.8–7.2	
2	C/N ratio	1:25–30	
3	Temperature	32–35	Can be compensated with HRT
4	Solid concentration	8–10	
5	Toxic substances	0–25 ppm	

- the daily gas production per unit of digester volume (m^3/m^3/day or L/m^3/day). These three are very important criteria for ascertaining overall efficiency of a simple BGP, which also results in cutting down the cost, especially in a simple RHh, that is, family-sized BGP.

  Gas yield:

i. *Total gas yield*: It is the maximum potential of biogas, which can be given by a particular organic, feedstock (substrate) under anaerobic conditions. This yield is found out by fermenting (digesting) 1 kg of substrate under laboratory conditions and optimum mesophyllic range of temperature until the time the maximum feasible gas releasing capacity of that particular substrate is more or less exhausted.
ii. *Optimal gas yield*: It is computed from the gas yield curve for a given RT (residence time), which can be taken as a practical figure for design of a digester. For example, a simple RHh family-sized Bio-gas (BG) model operating as a semicontinuous hydraulic digester plant, using cattle manure with 10% TS, will give an average biogas yield of 0.04 m^3 (or 40 L) per kg of fresh manure (dung) for an HRT of either 30, 40, or 55 days for the three different temperature (ambient) zones in India, respectively.

Gas production rate: The gas production rate is the quantity of biogas produced per unit time—it is normally expressed in terms of cubic meter per day (m^3/day) in 24 hours or liters/day or ft^3/day.

Gas production per unit digester volume: It is the biogas production in m^3 per cubic meter effective digester (i.e., fermentation chamber) volume per day (24 hours). It is commonly referred as m^3/cum/day or liter/cum/day or ft^3/cu ft/day. The comparative efficiency of two or more BGPs is ascertained in terms of their ability to produce maximum possible gas from the least possible volume from their digesters (fermentation chambers) for the same substrate and under similar conditions, etc. In the case of simple RHh semicontinuous hydraulic digester, the cost of the entire biogas unit also needs to be taken into account to ascertain the biogas production efficiency vis-à-vis the capital investment to build (install) the unit. Combination of these two types of information can also be used as a thumb rule by a layperson to choose an appropriate biogas model based on the least investment per unit digester volume out of the various options of plant designs, available to him/her for making selection.

Loading rate: Loading rate is the amount of raw material fed to the digester per day per unit volume of digester capacity. In the case of cow-dung plant, the thumb rule is to put 6 kg of fresh dung in $1/m^3$ size of biodigester. For example, if the size of BGP is $10\ m^3$, about 60 kg of dung is required to be loaded per day for optimum gas production. In fact, the correct rate of loading is essential for efficient gas production. If the plant is overfed, acidity will accumulate, and methane production will be inhibited; if the loading rate is lower, the gas production will not be sufficient. Fig. 3.22 shows the schematic diagram of stratification for three types of biodigestion processes (Table 3.11.

Seeding or bacterial population: Acetogenic (acid forming bacteria) and methanogenic bacteria are naturally present in cow dung. However, their number is quite small. Acid forming bacteria proliferate faster and increase their number, while methanogenic bacteria develop very slowly. Therefore for the initial reaction, small amount of sludge of another digester is generally used as seeding or inoculums. This sludge contains a high concentration of acetogenic and methanogenic bacteria, which could enhance the process of anaerobic digestion of organic materials. Some study has shown that the seeding materials can be mixed with the input slurry up to the ratio of 30%–50%. If inoculums are increased further, less volume of gas is obtained due to reduced inputs fed to the digester.

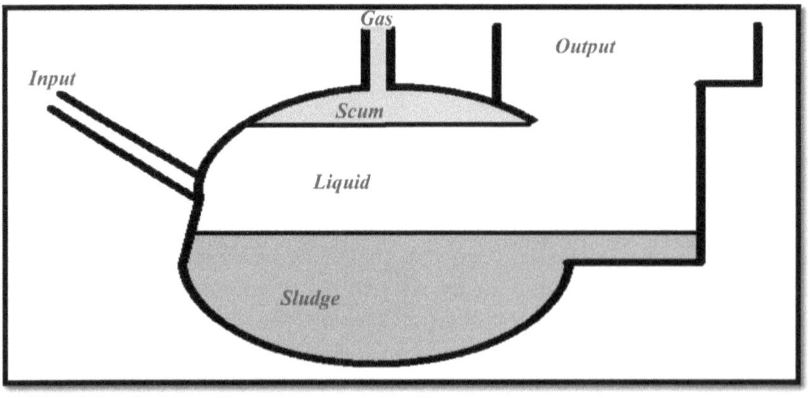

FIGURE 3.22 Changes in dry matter concentration inside the digester.

TABLE 3.11 Loading rate for various biogas plant sizes.

Plant size (m^3)	Daily loading rate (kg)	
	Hills	Tarai
4	24	30
6	36	45
8	48	60
10	60	75
15	90	110
20	120	150

Dilution and consistency of inputs: Before feeding the digester, the excreta, especially fresh cattle dung, has to be mixed with water at the ratio of 1:1 on unit volume basis (i.e., the same volume of water for a given volume of dung). However, if the dung is in dry form (that has to be crushed before putting into the digester), the quantity of water has to be increased accordingly to arrive at the desired consistency of the inputs (e.g., ratio could vary from 1:1.25 to even 1:2). The dilution should be made to maintain the TS from 5% to 10%. Sometime, the particles impede the flow of gas formed at the lower part of digester. In both cases, gas production will be less than optimum. A survey made by BSP reveals that the farmers often overdilute the slurry. For thorough mixing of the cow dung and water (slurry), GGC has devised a slurry mixture machine that can be fitted in the inlet of a digester. For proper solubilization of organic materials, the ratio between solid and water should be 1:1 when the domestic wastes are used. Fig. 3.22 shows changes in dry matter concentration inside the digester.

Acidity: Anaerobic digestion occurs best within a pH range of 6.8–8.0. More acidic or basic mixtures will ferment at a lower speed. The introduction of raw material will often lower the pH (make the mixture more acidic). Digestion will stop or slow dramatically until the bacteria have absorbed the acids. A high pH will encourage the formation of ammonia, which makes sludge more alkaline.

PH value is crucial for a good result:

1. optimal production when pH 7.0–7.2;
2. inhibition (due to acids) if $pH < 6.2$;
3. inhibition (due to ammonia) if $pH > 7.6$; and
4. deviation from the optimum range results in: lower gas yield and inferior gas quality.

Activity of enzyme is very sensitive according to pH level. The acidic condition lowers down methane formation. The optimum biogas production is achieved when the pH value of input mixture in the digester is between 6 and 7. The pH in a biogas digester is also a function of the relation time. In the initial period of fermentation, as large amounts of organic acids are produced by acid forming bacteria, the pH inside the digester can decrease to below 5. This inhibits or even stops the digestion or fermentation process. Methanogenic bacteria are very sensitive to pH and do not thrive below a value of 6.0. Later, as the digestion process continues, concentration of NH_4 increases due to digestion of nitrogen, which can increase the pH value to above 8. When the methane production level is stabilized, the pH range remains buffered between 7.2 and 8.2. However, some of the feeding materials especially industrial waste have the tendency of decreasing the pH of the digestion slurry. In such cases, pH can be adjusted by the addition of lime ($CaCO_3$), but this amount needs to be calculated. Overliming harms the bacteria. When high nitrogen content materials are used for feeding, during the process of methane formation, nitrogen is liberated to the slurry and forms ammonium hydroxide. This increases the pH value of the slurry. If this condition appears, addition of straw would help amelioration of pH.

Carbon–nitrogen ratio: The bacteria responsible for the anaerobic process require both elements but they consume carbon roughly 30 times faster than nitrogen. Assuming all other conditions are favorable for biogas production, a carbon–nitrogen (C/N) ratio of about 30:1 is ideal for the raw material fed into a BGP. A higher ratio will leave carbon still available after the nitrogen has been consumed, starving some of the bacteria of this element. These will in turn die, returning nitrogen to the mixture, but slowing the process. Too much nitrogen will cause this to be left over at the end of digestion (which stops when the carbon has been consumed) and reduce the quality of the fertilizer produced by the BGP. The correct ratio of carbon to nitrogen will prevent loss of either fertilizer quality or methane content.

Necessary elements such as carbon, hydrogen, nitrogen, phosphorus, and many other microelements must be present in adequate quantities for the normal growth of the microorganisms. It has been recognized that all living organisms need nitrogen for the synthesis of protein. In the absence of sufficient nitrogen, the bacteria would not be able to utilize all the carbon present and the process would be less efficient. In general, a ratio of around 20–30:1 is considered best for anaerobic digestion. C/N ratio should never be more than 35, with an optimum of 30. If the C/N ratio is very high, nitrogen will be consumed rapidly and the rate of reaction will decrease. On the other hand, if the C/N ratio is very low, nitrogen will be liberated and accumulated in the form of ammonia, which is toxic under certain conditions. Animal waste, particularly cattle dung, has an average C/N ratio of about 24. The plant materials such as straw and sawdust contain a higher carbon.

TABLE 3.12 Carbon–nitrogen ratio of some organic matters.

Raw material	Carbon–nitrogen ratio
Duck dung	8
Human excreta	8
Chicken dung	10
Goat dung	12
Pig dung	18
Sheep dung	19
Cow/buffalo dung	24
Water hyacinth	25
Elephant dung	43
Straw (maize)	60
Straw (rice)	70
Straw (wheat)	90
Saw dust	>200

The human excreta have a C/N ratio as low as 8. Table 3.12 shows the C/N ratio of some organic matters.

Toxicity: Mineral ions, heavy metals, and the detergents are some of the toxic materials that inhibit the normal growth of pathogens in the digester. Small quantity of mineral ions (e.g., sodium, potassium, calcium, magnesium, ammonium, and sulfur) also stimulates the growth of bacteria, while very heavy concentration of these ions will have toxic effect. For example, presence of NH_4 from 50 to 200 mg/L stimulates the growth of microbes, whereas its concentration above 1500 mg/L produces toxicity.

Addition of succulent plant or algae: For the effective and high production of biogas from cow dung and animal dung, many succulent plants or algae are added. Green algae, water hyacinth, and lemna grass are added in the practice. The amount of biogas produced from the algae was twice (344 mL/g dry algae) of that obtained from cow dung (179/g dry cow dung) alone. In addition, the duration of gas evolution increased with increasing the proportion of slurry. The caloric value of the gas was 4800 K Cal/m^3 and the percentage of methane was 55.4%.

Temperature: Anaerobic breakdown of waste occurs at temperatures lying between 0°C and 69°C, but the action of the digesting bacteria will decrease sharply below 16°C. Production of gas is most rapid between 29°C and 41°C or between 49°C and 60°C. This is due to the fact that two different types of bacteria multiply best in these two different ranges, but the high temperature bacteria are much more sensitive to ambient influences. A temperature between 32°C and 35°C has proven most efficient for stable and continuous production of methane. Biogas produced outside this range will have a higher percentage of carbon dioxide and other gases than within this range.

Substrate temperature in the digester: Anaerobic fermentation can work in an ambient temperature between 30°C and 70°C and, if colder, the reactor has to be insulated and/or heated. Methane production is very sensitive to changes in temperature shown in Table 3.13.

Temperature control: In hot regions, it is relatively easy to simply shade the digester to keep it in the ideal range of temperature, but cold climates present more of a challenge. The first action is, naturally, to insulate the digester with straw or wood shavings. A layer about 50–100-cm thick, coated with a waterproof covering, is a good start. If this still proves to be insufficient in winter, then heating coils may have to be added to the biogas digester. It is relatively simple to keep the digester at the ideal temperature if hot water, regulated with a thermostat, is circulated through the system. Usually, it is sufficient to circulate the heating for a couple of hours in the morning and again in the evening. Naturally, the biogas produced by the digester can be used for this purpose. The small quantity of gas "wasted" on heating the digester will be more than that compensated for by the greatly increased gas production.

Stirring: Some method of stirring the slurry in a digester is always advantageous. If not stirred, the slurry will tend to settle out and form a hard scum on the surface, which will prevent release of the biogas. This problem is much greater with vegetable waste than that with manure, which will tend to remain in suspension and have better contact with the bacteria as a result. Continuous feeding causes fewer problems in this direction, since the new charge will break up the surface and provide a rudimentary stirring action.

Stirring improves the efficiency of digestion by:

1. removing metabolites (gas removal);
2. bringing fresh material in contact with bacteria;
3. reducing scum formation and sedimentation;
4. preventing temperature gradients in the digester; and
5. avoiding the formation of blind spots (short cuts).

TABLE 3.13 Common temperature ranges for bacteria.

Psychrophillic bacteria	Below 20°C
Mesophillic bacteria	20°C–40°C
Thermophillic bacteria	Above 40°C

Simple biogas units normally do not have mechanical stirring devise. Covered lagoon digesters are the most popular ones; however, their operation depends on the climatic conditions and has very low performance. High failure rate has been encountered: up to 70% in complete-mix and plug flow digesters. Reasons for failure of anaerobic digesters are not understood. Improper or insufficient mixing can be one of the reasons. Despite anaerobic digestion being a slow reaction, mixing plays an important role.

6. Enhances microorganisms and substrate contact and distribution.
7. Ensures uniform pH and temperature.
8. Prevents deposition of denser solids at the bottom and flotation of lighter solids at the top.
9. Helps to release biogas bubbles.

The information available on mixing is contradictory, and thus, the extent of mixing in digesters is not understood properly. More reliable information of impact of mixing on digester performance can be obtained by systematic and carefully planned experiments. Fig. 3.23 shows types of stirring.

Mixing is important to maintain a uniform environment, and thus the effect of mixing on digester performance needs to be evaluated.

Performance of anaerobic digesters: effect of mixing in lab-scale digesters: Systematic lab-scale performance studies were carried out to study the effect of the following variables (Table 3.14) on the performance of 6-in. diameter (3.78 L) digesters (with the same power input per unit volume, 8 W/m^3):

1. geometry of digester;
2. intensity of mixing (1–3 lpm gas flow rate); and
3. percentage TS in the feed (5% and 10%).

And the conclusions are as follows.

1. Mode/intensity of mixing or the geometry does not affect the performance of laboratory-scale digester.

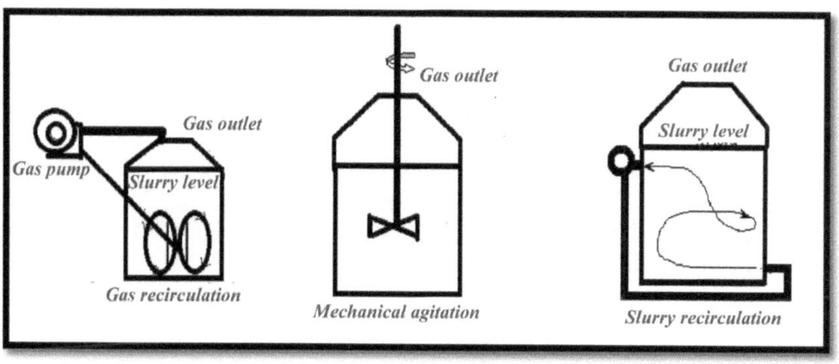

FIGURE 3.23 Types of stirring.

TABLE 3.14 Effect of mixing (stirring) on biogas production.

Type of mixing	Biogas production rate (L/L/day)	Methane yield (L/gmVS loaded)	%TS reduction	%VS reduction
Unmixed	0.92	0.19	41	35
Gas mixed	1.07	0.21	49	39
Impeller mixed	1.14	0.23	47	41
Slurry recirculation	1.20	0.24	45	35

2. Studies at large-scale digester needs to be performed to evaluate the true effect of mixing on its performance.
3. Performance of laboratory-scale digesters does not depend on mixing. Is it true at all scales of operation?

Retention time: RT (also detention time) is the average duration of time a sample remains in the digester. In a cow-dung plant, the RT is calculated by dividing the total volume of the digester by the volume of slurry added daily. Usually, for a cow-dung plant, an RT of 40–60 days is required depending on the temperature. Thus the fermenting pit should have a volume of from 40 to 60 times the slurry added daily. However, for a night-soil digester, a longer RT (70–90 days) is needed in order to kill the pathogens present in human faces.

Production of biogas in cold season: During the winter season and at the higher altitude, production of biogas is drastically reduced due to decreased temperature. In order to cope with this problem, various attempts are made to increase the biogas production in cold season through physical, chemical, and biological methods.

Heat loss from the digesters:

A. The total heat transfer coefficient from dome to slurry and vice versa during the day and off-sunshine hours through gas: Few assumptions are:
   I. For day, Ts = 25°C and TD = 60°C, and the height of exposed surface of dome from slurry (characteristic length) = 5 m.
   II. For off-sunshine hours, Ts = 40°C and TD = 25°C.
   III. During the day time, evaporation heat loss is assumed to be zero because of the large heat capacity of the slurry.

   Therefore the radioactive heat transfer coefficient from dome to slurry will be calculated as:

$$h_{\text{rds}} = 0.8 \times 5.67 \times 10^{-8} \left[\frac{(60+273)^4 - (25+273)^4}{(60-35)} \right] = 5.74 \ \text{W/m}^2 - {}^\circ\text{C}.$$

In the present case, the hot plate is facing downward; hence, $Nu = 0.27(Gr^*Pr)^{0.25}$, where Gr and Pr are calculated at an average temperature of dome and slurry. The values of different parameters at an average temperature are $\rho = 1.12\ kg/m^3$

$$\beta = \frac{1}{\Delta T} = 3.169 \times 10^{-3}/K, \mu = 1.98 \times 10^{-5}\ Ns/m^2$$

$$K = 0.027\ W/m^{\circ}C \text{ and } Pr = 0.7$$

$$Gr = \frac{(9.8)(3.169 \times 10^{-3})(5^3)(1.12^2)(35)}{(1.98 \times 10^{-5})} = 4.34 \times 10^{11}.$$

Now, calculating the convective heat transfer coefficient (h_{cds}) from dome to slurry, we get:

$$h_{cds} = \frac{Nu \cdot K}{L} = \frac{0.27(Gr \cdot Pr)^{0.25} \cdot K}{L}$$

$$h_{cds} = \frac{0.27(4.34 \times 10^{11} \times 0.7)^{0.25} \cdot 0.027}{5} = 1.08\ W/m^2 - {}^{\circ}C.$$

Therefore the total heat transfer coefficient from dome to slurry during day time is:

$$= 5.74 + 1.08 = 6.82\ W/m^{2\circ}C.$$

B. *During the night time*: The process will be revered, and heat transfer will take place from the slurry to the dome and can be calculated as follows.

The radioactive heat transfer coefficient from the slurry to the dome will be:

$$h_{rsd} = 0.8 \times 5.67 \times 10^{-8} \left[\frac{(40+273)^4 - (25+273)^4}{(40-35)}\right] = 5.20\ W/m^{2\circ}C.$$

The convective heat transfer can be determined by using vapor pressure at the slurry and the dome temperature, as 7335 and 3166 N/m^2, and we get:

$$h_{csd} = 0.884 \times \left[(40-25)\frac{(7335-3166)(40+273)}{268 \times 10^3 - 7335}\right]^{1/3} = 2.39\ W/m^{2\circ}C.$$

The evaporative heat transfer is determined as:

$$h_{esd} = 0.016 \times h_{csd} \left[\frac{P_s - P_d}{T_s - T_d}\right]$$

$$h_{esd} = 0.016 \times 2.39 \times \left[\frac{(7335 - 3166)}{40 - 25}\right] = 10.62\ W/m^{2\circ}C$$

Hence, the total transfer from the slurry to the dome during the night is:

$$\begin{aligned} &h_{rsd} + h_{csd} + h_{esd} \\ &= 5.20 + 2.39 + 10.62. \\ &= 18.21 \ \ \mathrm{W/m^2\ ^\circ C} \end{aligned}$$

Physical method (heating and insulating digesters): To reach optimum operating temperatures (30°C–37°C or 85–100 °F), some measures must be taken to insulate the digester, especially in high altitudes or cold climates. Straw or shredded tree bark can be used around the outside of the digester to provide insulation. Other forms of heating can also be used such as solar water heaters or the burning of some of the methane produced by the digester to heat water that is circulated through copper coils on the inside of the digester. Solar or gas heating will add to the cost of the digester, but in cold climates, it may be necessary. Few techniques for heating are discussed below.

Solar hut:

1. Solar hut is a simple way of preserving solar energy inside the simply made green house, especially for the cold climate.
2. A black plastic hut is built over the dome of the biogas digester in order to absorb and conserve the solar heat on the dome.
3. In view of increasing the temperature inside the digester.

Integration of solar energy:

1. Integration of solar energy in the biodigester for the production of biogas during winter time has been practiced in some parts of the world.
2. The solar water heating system is applied into the biogas digester through a long coil of pipe inserted inside the digester.
3. However, during cold season, when water start freezing, this system can create a problem due to water freezing inside the pipe.
4. Therefore for such areas, a low boiling liquid has to be used instead of water.

Through the integration of solar system in the digester, the temperature inside the digester can be raised to 25°C–30°C, which is optimum or above to enhance biogas production.

However, cost of the system can become a barrier as a result of addition of solar system into the biodigester. Fig. 3.24 shows integrated solar collector water heater with biodigester.

Black paint:

1. Some people have practiced the painting of the dome of biogas digester with black paint for absorption of solar heat.
2. This method also helps increase the temperature of the digester to some extent.
3. Instead of paint, some people also used charcoal to cover the dome.

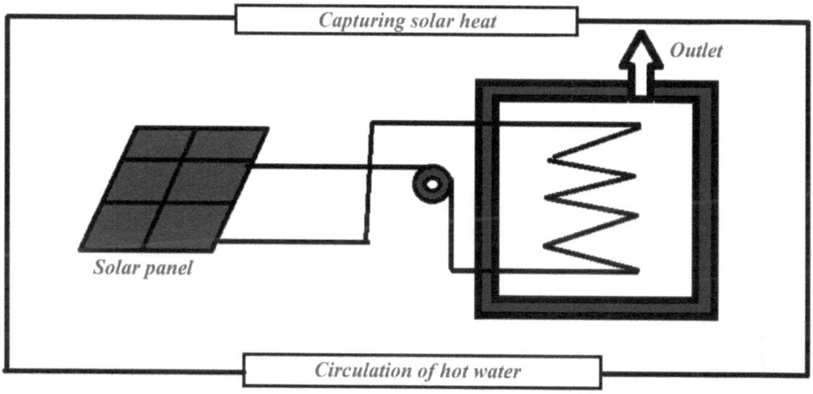

FIGURE 3.24 Integrated solar collector water heater with biodigester.

4. Charcoal also helps retain the heat for some time. This method had been reported to be in practice by some institute in Solan, India.

Composting pile:

1. A composting pile is also built on the top of the digester dome in order to conserve heat loss from the dome surface. During the composting, heat is generated due to metabolic process.
2. This heat also helps conserve the temperature inside the dome. Mostly, farm residues are piled on the top of the digester. It is then covered with the layer of straw to conserve the heat inside the digester. The pile is then covered with black plastic sheet, which not only conserve the heat generated through composting metabolism but also help absorb solar heat during sunny days.

3.5 Installation of a biogas plant

Before a BGP can be built at a particular site, many factors need to be studied. These include the quantity of feedstock available on daily basis, the type of waste, the availability of water, the availability of space, the type of soil, and the level of the groundwater table on the site. The installation of a BGP, especially the fixed-dome model, requires well trained and experienced masons and the use of good quality materials. If the masons follow the design measurements strictly, the plant will be long-lasting. The success or failure of a BGP mainly depends on how and by whom it has been installed, what type and quality of materials have been used, and how well the plant has been maintained. Fig. 3.25 shows the layout diagram of biodigester construction.

A mason has to carry out the following activities in sequential order while installing BGP in farmer's premises:

1. selection of correct size of biodigester;

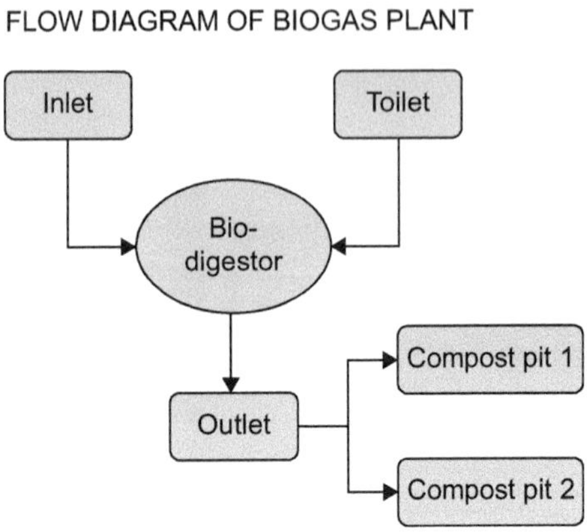

FIGURE 3.25 Layout diagram of biodigester construction.

2. selection of construction site;
3. collection of construction materials that meet the quality standards;
4. layout of plant;
5. digging of the pit (excavation);
6. fixing the diameter and laying of collar (base layer for brick/stone work) for digester and manhole walls;
7. construction of digester walls and manhole;
8. installation of inlet pipes;
9. backfilling the empty spaces outside the digester wall;
10. construction of the top of manhole (usually called as beam);
11. construction of gasholder (preparation of mold, concreting, and fixing of dome gas pipe);
12. constructing inlet chamber;
13. constructing OC and outlet covers;
14. plastering of the inside of dome;
15. construction of turret;
16. installation of pipeline, fittings, and appliances;
17. testing for leakages;
18. filling the plant with feeding;
19. construction of slurry pit(s);
20. filling the top of dome and sides of OT with earth;
21. cleaning the site; and
22. orienting the users on simple operation and maintenance activities.

3.5.1 Site selection (location of BGPs)

There are several points to keep in mind before actual construction of the digester begins. The most important consideration is the location of the digester. The following procedures should be followed while site selection for a BGP is made:

- Be sure there is enough space to construct the digester. The area available should be adequate to accommodate all the units of the plant. A plant that produces 3 m^3 of methane requires an area of approximately 2 m × 3 m. If a larger plant is required, figure space needs accordingly.
- Plan for a site that is open and exposed to the sun. Care should be taken that the site receives full sunlight without any obstruction from other surrounding structures or vegetation. The digester should receive little or no shade during the day. The digester operates best and gives better gas production at high temperatures (35°C or 85–100 °F).
- Do not select low lying areas for the plant as water logging will create problems.
- Check the water table in the existing wells close to the plant location before site selection. If the water table is above the bottom level of the digester, an alternative appropriate site should be selected. If the water table is reached when digging, it will be necessary to cement the inside of the digester pit. This increases the initial expense of building the digester, but prevents contamination of the drinking supply.
- The plant should be located at least 20 m away from the water sources, such as wells, springs, tube wells, etc., to avoid possible contamination of water sources. Arrange to have water readily available for mixing with the manure.
- The site selected should be away from trees or tree stumps to mitigate the root hazard in the pre- or postconstruction phase.
- To make plant operation easy and to avoid wastage of raw materials especially the waste substrate, the plant must be as close as possible to the waste source (cattle shed, poultry waste collection chamber, kitchen waste, and night soil pipe). Try to locate the digester near the stable, so excessive time is not spent to transport the manure. Remember, the fresher the manure, the more methane is produced as the final product and the fewer problems with biogas generation will occur. To simplify collection of manure, animals should be confined. Although the gas plant itself takes up a very small area, the slurry should be stored either as is or dried. The slurry pits should be large and expandable.
- The nearest water source should not be at a distance of more than 20-minute walk. Otherwise, more time in fetching water from the source to the plant will bring unnecessary burden to the owner during the operation of the plant.
- Locate the gas plant as close as possible to the point of gas consumption. This tends to reduce costs and pressure losses in piping the gas. Methane can be stored fairly close to the house as there are few flies or mosquitoes or odor associated with gas production. Furthermore, longer pipe increases the risk of gas leakage due to more joints in it.

- The main valve has to be opened and closed before and after use. Therefore the plant should be as close as possible to the point of use so that the above problems are eliminated.
- The edge of the foundation of the plant should be at least 2 m away from the house or any other building to avoid risk of damages.

Layout of plant and digester:

1. Calculate the tentative length and breadth of the plant required as per the available drawing.
2. Level the ground before a plant layout could be started.
3. Mark with white powder on the center line for inlet, digester, outlet, and compost pits on the ground so that all are accommodated in the same plane.
4. Fix two wooden pegs 2 m away from the end points of the plant as reference points during the construction.
5. Fix a small wooden peg in the ground, which will act as the center for the digester.
6. One end of a cord is attached to this peg with the length equal to the internal radius of the digester including plaster thickness, wall thickness, and the footing offset.
7. The other end of the cord is held tight without disturbing the position of the wooden peg and moved along a circular path. This circular mark made on the ground is covered with white powder (lime).

3.5.2 Selection of construction materials

If the construction materials to be used in the plant construction, such as cement, sand, aggregate, etc., are not of good quality, the quality of plant will be poor even if design and workmanship involved are excellent. Domestic fixed-dome BGP should be constructed by stone round wall and outlet, dome with plain concrete and slabs with reinforced concrete, and inlet with either stones or bricks. In order to select these materials of best quality, their brief description regarding the specifications has been given hereunder.

a. *Cement*:
   1. The cement brand selected must have Indian Standard.
   2. The cement shall be 43 grade ordinary portland cement.
   3. The cement bags shall be properly sealed and bear the Nepal Standard on the cover.
   4. It shall be stored in a dry place, in regular piles not exceeding 10 bags high and in such a manner that it will be efficiently protected from moisture and contamination.
   5. Set cement should immediately be removed from the work and replaced.

b. *Sand*:
1. Fine aggregates or sand shall be clean and free from coagulated lumps.
2. The quantity of impurities especially the mud in the sand can be determined by a simple test using a bottle. This is called the "bottle test."
3. Coarse and granular sand can be used for concreting work, but fine sand will be better for plastering work.

c. *Gravel*:
1. Aggregate shall be stone crushed. They shall be hard, strong, and clean.
2. It should be free from other materials such as plastics, papers, brickbats, plants, etc.
3. If it is dirty, it should be washed with clean water.

d. *Water*:

Water is needed for construction, that is, for preparing the mortar for masonry work, concreting work, and plastering. It is also used to soak bricks/stones and for washing sand and aggregates before using them. Irrigation canal water or groundwater is best, as long as it does not contain mud. Piped water can be used, but it often contains chlorine, used to kill bacteria. This water should be left in a container, so the chlorine can come out of solution. It can also be aerated. pH value of water should not exceed 7.

e. *Bricks*:

Well kiln burnt mud bricks are the best type to use for a fixed-dome BGP. In many places, hollow bricks are more easily available, as they are light weight. These tend to weaker and more easily broken. They have smaller surface areas onto which cement can bind. Solid bricks are better. A typical brick size is 220 mm × 100 mm × 70 mm. Locally available bricks should be in regular shape. However, the sizes available are varying from hilly to plain areas.

To check the quality of bricks, two bricks are randomly selected from the delivered quantity. If the bricks are well burnt, they produce a distinct sound when hit to each other.

Auxiliary parts:

a. *Mixing device*:

This device is used to prepare good quality water–dung solution in the inlet tank when cattle dung is used as feeding material. Usually, for household biogas digesters, vertical mixing devices are installed. The device should be of good quality, as per the design, and the mixing blades have to be well galvanized. The blade should be properly aligned for the effective mixing.

b. *The central guide mechanism*:

As the volume of gas that is generated increases, it pushes the gasholder upward, which later retracts back into the digester pit as the gas is used up. This up and down movement of the gasholder requires a central guide mechanism to prevent the gas drum from jamming onto the sides of the digester. The central guide mechanism consists of a mild steel rod of 30 mm coated with one layer of primer and two layers of oil paint onto which a layer of grease is applied to lubricate the system and also to protect the rod against corrosion.

c. *Slurry influent and effluent pipes*:

Many types of pipe can be used for this purpose, made from plastic, concrete, or stone ware, which are usually used for drainage. A diameter of 15 cm should be used. If plastic drain pipes are used, such as PVC, a thick wall thickness should be used, between 4 and 6 mm. Standard pipes of PVC should be used for slurry flowing into and out of the digester. PVC is recommended, since it does not corrode in the alkaline conditions prevailing in the digester and is, therefore, more durable than galvanized iron pipes.

d. *Main gas pipe*:

Gas stored in the gasholder is conveyed to the pipeline through this pipe, which is placed in the topmost portion of the dome. The joint of reduction elbow with this pipe should be perfect and gastight; otherwise, gas leakage from this joint cannot be stopped easily. Therefore it is recommended that the reduction elbow has to be fitted in a workshop to ensure perfect airtightness of the joint. Steel pipe is required, usually ½-in. diameter galvanized iron, with the top end threaded. Two steel wire pieces, each 300-mm long, are welded either side of the pipe, about 30 mm from the bottom. These hold the GOP in the top of the dome. Class "B" galvanized iron can be used, and a local welding shop can be contracted to do the welding. This pipe should be made up of light quality iron, and Municipal solid (MS) rod has to be welded at one end to embed it with the concrete during installation. The length of this pipe should be at least 60 cm. Steel pipe lengths are required to link the HDPE pipe to the main gas valve at one end and the rubber hose at the other. These are usually ½" diameter and about 150-mm long.

e. *Gas transfer pipe*:

A pipe is needed to connect the digester and the biogas burner. The pipe is placed above ground; so it needs to be strong and long-lasting. HDPE is most suitable. A pipe with an outside diameter of about 20 mm and a wall thickness of 3 mm is suitable.

f. *Main gas valve to control the gas flow*:

A gate valve is required at the outlet of the gas drum to regulate the flow of the gas depending on the consumer requirements. It controls the flow of biogas in the pipeline from the gasholder. It is opened when gas is to be used and closed after each use. If substandard quality of main gas valve is used, there is always risk of gas leakage. This valve should be of high quality and approved by the concerned quality control authorities. A ½" diameter valve is screwed onto the thread of the GOP. A ball valve is better than other types of valve, but it can be made of a range of materials. PVC valves with nylon sealing rings are good, although they are fragile and easily broken. Metal valves are more liable to leak. In addition, if a metal valve is knocked, it can disturb the GOP in the top of the dome, which is difficult to fix. A broken plastic valve can be replaced easily.

g. *Rubber hosepipe*:

It is used to convey gas from the gas tap to the stove. This pipe should be made up of high-quality neoprene rubber and should not develop cracks when folded. It should have 15-mm outer and 9-mm inner diameters. The minimum wall thickness of the pipe should be 2.5 mm. A length of rubber hose is used to connect a nipple in the end of the HDPE pipe to the biogas burner. The size of the hose depends on the size of the spigot on the biogas burner.

h. *Pipe nipple*:

A brass nipple, with a spigot at one end and a pipe thread at the other, is required to link the galvanized pipe length to the rubber hose. The spigot should be the same size as that of the biogas burner.

i. *Water trap*:

When the gas collected flows along the GOP, some water condenses along the pipe, necessitating the use of a water trap. It drains the water condensed inside the pipeline when biogas comes in contact with the cool pipe. This is an important component of BGP, and therefore its quality should be controlled carefully. It should be easy to operate, and threads in it should be perfect. It should be ensured that the hole in the screw nut is bored properly and is located at the right place. The thickness of the nylon washer has to be 4 mm, and either a 4-cm long handle pin or a properly knurled opener should be used. This appliance should be approved by the concerned authorities. Standard water traps are available and should be installed along the gas pipe and just before the consumer point. Water traps encourage condensation, subsequent retention, and eventual discharge of water in the biogas.

j. *Mild steel bars*:

MS bars are used to construct the covers of OT and water drain chamber. It should meet the engineering standard generally adopted. For plants of 4, 6, and 8 cum, MS rods of 8-mm diameter are recommended, and for plant of 10 cum capacity, 10-mm diameter is recommended. MS bar should be free from heavy rust.

k. *Acrylic emulsion paint*:

It is used to make the gasholder (dome) of biodigester airtight. Paint of this type should meet quality standard, and they must be approved from concerned quality control authority.

l. *Biogas burner*:

Biogas needs specialized burners. LPG burners are not suitable. A supplier of biogas burners is needed, usually from either India or China. It is possible to import some parts that are specifically made for biogas and use them to adapt LPG burners.

m. *Gas stove*:

Gas stoves can be found with single and double burners. In general, a single burner gas stove used for household purpose consumes 350–400 L of gas per hour. The efficiency of gas stove is very important for the successful functioning of the biodigester. The stove should be of good quality and strong enough to firmly rest in ground. The primary air intake should be easily adjustable, and the holes should be properly placed. The jet and pipe leading to the burner should be straight and aligned properly. The holes in the burner cap should be evenly spread across it.

n. *Gas lamp*:

Gas lamp is another important appliance used in biodigesters. Often users complain about the malfunctioning of these lamps. These lamps should be of high quality with the efficiency of more than 60%. Usually, a biogas lamp consumes 150–175 L of biogas per hour. Lamps to be used in biodigesters have to be approved by the concerned quality control authority.

o. *Gas tap*:

Gas tap is used for regulating flow of gas to the gas stove. Care should be taken to install gas tap of high quality. It has been often complained by the users that this taps are becoming problematic with gas leakage through them. It is important that the "o" ring is placed properly and is greased thoroughly and regularly. The gas tap should not be too tight or loose to operate. The taps to be used in biodigesters should be approved by concerned quality control authority.

p. *Gas pressure gauge*:

U-shaped pressure gauge (manometer) made up of a transparent plastic or glass tube and filled with colored water or a clock-type digital or analog pressure meter has to be installed in the conveyance system to monitor the pressure of gas. Whatever may be the type, this device should be best among those available in the local market and should meet set quality standards, if any.

3.5.3 Construction work

Reinforced cement concrete work:

1. Select the proper size of the rebar as required.
2. Clean, straighten, cut, and bend the rebar as per the requirement.

3. Mix cement, sand, and aggregate in the ratio of cement 1: sand 1.5: aggregate 3 (1:1.5:3) on a clean dry surface.
4. Add water gradually in it to make a uniform concrete mix.
5. The quantity of water for mixing will be equal to 50% of the cement quantity used for the mix.

Plain cement concrete work:

1. Mix cement, sand, and aggregate in the ratio of cement 1: sand 2: aggregate 4 (1:2:4) on a clean dry surface.
2. Add water gradually in it to make a uniform concrete mix.
3. The quantity of water for mixing will be equal to 50% of the cement quantity used for the mix.

Brick work:

1. Immerse bricks in a clean water tank for at least 6 hours for proper soaking.
2. Mix cement and sand in the ratio of cement 1: sand 4 (1:4) on a clean dry surface.
3. Add water gradually to it to make a uniform cement mortar.
4. The quantity of water for mixing will be equal to 50% of the cement quantity used for the mix.
5. 10-mm cement mortar shall be laid on each layer of brick wall.
6. The brick joint gap for cement mortar should be in the range of 10–15 mm.
7. The brick joints of the first layer and the second layer should never fall in a vertical line.
8. The verticality of the wall should be checked with a plumb bob for each layer of the brick wall during the laying of bricks.
9. The wall height shall be above the ground level so that no rain water enters the pits. Fig. 3.26 shows the construction of correct brick works.

Cement plasters work:

1. Mix cement and sand in the ratio of cement 1: sand 4 (1:4) on a clean dry surface.
2. Add water gradually in it to prepare a uniform cement mortar.

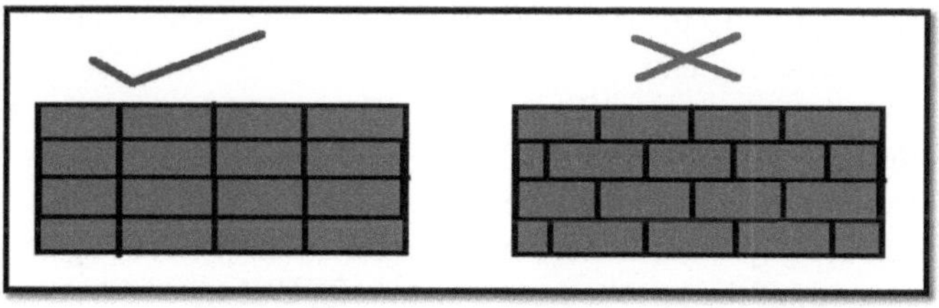

FIGURE 3.26 Diagram showing the correct brick works.

3. The quantity of water for mixing will be equal to 50% of the cement quantity used for the mix.
4. Sprinkle water on all the walls.
5. 12.5-mm-thick plaster is applied on all the walls.
6. Sprinkle water 4–5 times a day for 7 days on these plastered walls for curing.

3.6 Operation and maintenance of biogas power plant

The maintenance and daily care needs to be done easily by looking after by the rural housewife or even the teen age children by devoting only 15–30 minutes each day. Some of the simple guidelines and tips for general care and maintenance given below, if followed regularly, will increase working efficiency and the operational life of the BGPs several folds.

a. The gate valve should be opened only when the gas has to be actually used.
b. Before opening the valves, one must ensure that all the preparations for cooking have been made, to avoid unnecessary wasteful consumption of biogas.
c. The air injector should not be closed very tightly on the side of the burner. The inflow of the air should be adjusted properly in the injector.
d. The manhole provided on the top of the ODC of plant should never be left uncovered.

Specific care and maintenance:

In addition to the above, the daily, weekly, monthly, yearly, and five yearly cares and maintenance should be done as per the schedule given below.

a. *Daily maintenance*:
   1. Charge your biodigester with the necessary recommended quantity of feed material daily. Use proper slurry mixture. Use clean feed material, free from soil, straw green biomass, and other floating material.
   2. Clean MT thoroughly before and after use.
   3. Check the inlet and outlet 10-in. PVC pipes to ensure that the level of water in the bag is adequate.
   4. Check the pressure release valve to ensure that the bottle is filled with water up to the small water hole. Bubbling water is an indication of a functioning unit.
   5. Check the inlet and outlet 10-in. PVC pipes to be sure that no air is entering.
   6. Check for damage to the digester bag.
   7. Clean off any mud, stones, or foreign material on the bag and around the mouth of the inlet and outlet 10-in. PVC pipes.

b. *Weekly care and maintenance*:

  1. Do gentle stirring of the slurry inside the digester for about 5–10 minutes every week. Use a long bamboo pole with a piece of cloth or jute mat tied properly at its one end so that it can conveniently go inside the 10-cm or 100-mm (4-in.) diameter inlet pipe and act as a piston10. This will also ensure proper and effective stirring of the slurry inside the digester.

  2. Open the tap of the manually operated moisture trap to drain off moisture condensed in the pipeline.

  3. The nozzle of the biogas lamps should be properly cleaned.

c. *Monthly care and maintenance*:

  1. Remove digested slurry from the slurry collection tank to the compost pit.

  2. If compost pits are provided next to the ODC, check the level of slurry in it. If filled, divert the slurry to the next compost pit.

  3. Check gate valve, GOP, and gas pipe fittings for leakage. The "pot scrub" inside the PVC "T" in the release valve should be replaced at least every 6 months or when necessary.

  4. Check the biogas pipeline and the moisture trap (water removal system) for any possible leakage.

d. *Annual care and maintenance*:

  1. Check for gas and water leakages from pipeline and appliances.

  2. Repair the worn out accessories including pipes (if polyethylene pipes of cheaper quality are used, there are chances of developing cracks in them).

  3. Replace damaged or nonworking accessories. Open the gate valve and remove all the gas from the plant. After this, check the level of slurry in the OC. If the slurry level is above the second step counted from the bottom in the OC (above the initial slurry level), remove it up to the second step.

e. *Five yearly care and maintenance*:

  1. Empty the plant completely and clean the sludge and inorganic materials from the bottom of the plant.

  2. Give a thorough check to the gas distribution system for possible leakage.

  3. Repaint the ceiling of the dome, free space area, and GSC with black enamel paint.

  4. Recharge (reload) the plant with fresh slurry.

General safety precautions and important things to remember:

Safety concerns related to biogas generation include health hazards and risks of fire or explosion. Biogas is flammable and can be explosive when mixed with air.

1. Keep the digester away from naked flames and electrical equipment that may spark.
2. Buildings should be well ventilated.
3. Explosion-proof motors, wiring, and lights should be used.
4. Perform periodic system checks for gas leaks.
5. Follow Bunsen burner safety rules (hair tied back and no loose clothes) when testing the gas with a flame.
6. Utilize gas detection and alarm devices in enclosures.
7. Take care with sharp scissors when cutting the tubing.
8. Do not divert the effluent from the unit directly into lakes or streams.

Do's and don'ts and general precautions:

Some of the important DOs, DON'Ts, and General Precautionary Measures essential for the operation of biogas Grameen Bandhu plant (GBP) are given as follows.

a. *Do's*:
   1. Select the size of the GBP depending on the quantity of dung available with the beneficiaries.
   2. Install the GBP at a place near the kitchen as well as the cattle shed as far as possible.
   3. Ensure that the plant is installed in an open space, and gets plenty of sunlight for the whole day, all round the year.
   4. Ensure that the outer side of the plant is firmly compacted with soil.
   5. Feed the BGP with right proportion of fresh slurry mixture prepared from animal manure and water, for example, when the bovine (cattle and/or buffalo) manure is used as feedstock, then add one part of cattle dung to one part of water by weight for making a homogenous slurry mixture.
   6. Ensure that the fresh slurry (mixture of dung and water) is free from soil, straw, etc.
   7. For efficient gas utilization, use good quality and approved burners and biogas lamps.
   8. Only use appropriate appliances.
   9. Open the gas regulator/cock only at the time of its actual use.
   10. Adjust flame by turning air regulator till a blue flame is obtained—this will give maximum heat.
   11. Light the match before opening the gas cock.
   12. Cover the manhole of ODC with BRCM manhole cover (MhC), to avoid accidental falling of cattle and children inside the plant.
   13. Check that the grating (made of bamboo sticks) is properly placed and fixed on the horizontal opening at the level of second step (from bottom end) of OC.
   14. Purge air from all delivery lines, allowing gas to flow for some time prior to first use.

15. Adjust the flame by regulator, provided on the biogas burner/stove, till it is blue in color.

b. *Don'ts*:
   1. Do not install a bigger size of GBP if sufficient cattle dung or any other feedstock to be used for gas production is/are not available on regular basis.
   2. Do not install the GBP at a long distance from the point of gas utilization to save the cost on pipeline.
   3. Do not install a plant under or very close to a tree, especially a big tree.
   4. Do not allow soil or sand particles to enter into the digester.
   5. Do not allow the scum to form in the digester; otherwise, the production of gas might be affected, and biogas generation may even stop completely.
   6. Do not burn the gas directly from the GOP even for the testing purpose as it can be dangerous.
   7. Do not use burner in the open; otherwise, there will be enormous loss of heat.
   8. Do not leave the gas regulator (valve) open when the gas is not in use.
   9. Do not use the biogas if the flame is yellow.
   10. Do not let any water accumulate in the gas pipeline; otherwise, the required pressure of gas will not be maintained, and the flame will sputter.
   11. Do not make digested slurry pit more than 1.0 m (3.0 ft) deep.
   12. Do not inhale the biogas as it may be hazardous.
   13. Do not add any foreign material in the plant to enhance the gas production.
   14. Do not hurry to get biogas after initial loading of slurry, as it may take 10–25 days for gas production in freshly loaded plants.
   15. Do not allow building a maximum pressure of above 80 cm or 800 mm (800 kg/m^2) of water column to avoid any damage to the GBP.
   16. Do not allow any person to enter the GBP when it has slurry inside to avoid accidental fall due to slipping, which may cause even death.

Common problems and suggested solutions:

The biodigester does not seem to be producing any gas.

1. Gas production may drop or cease for many reasons including the entrance of air into the bag, changes in temperature, water pH, and contamination in the wastes used to charge the digester.
2. Check to be sure that no air is entering the biodigester from the inlet or outlet bucket.

3. Check the digester for any bag damage from foreign objects or animals that may allow gas to enter. If necessary, increase the water level inside the bag.
4. Some producers have noted a drop in biodigester gas production during long periods of rain.

Soil around the biodigester is washing onto and compressing the bag.

1. When soil or mud falls on the biodigester, they can deflate the bag and cause sedimentation to occur inside the bag.
2. To avoid this problem, construct a barrier to keep mud, rain, and soil out. Many producers have constructed simple fences or barriers to prevent erosion from damaging the biodigester bag. These may be constructed from wooden stakes and slats of wood. Any mud that washes onto the bag must be cleaned off daily.

There appears to be gas in the bag, but there is no gas coming out of the digester.

1. Check to be sure that the gas valve is open.
2. Crack pipes can cause a leak in the gas line.
3. Regularly inspect your gas lines for damage.
4. Seal any damaged lines securely with glue and rubber ties.

Animals are damaging the digester bag.

1. Animals can quickly cause permanent damage to your biodigesters.
2. Be sure that your biodigester is well protected from animals.

Requirements for a successful system:

1. acceptance by potential users;
2. ability to use the gas when produced;
3. sufficient demand for gas;
4. availability of sufficient raw materials to meet the production requirements; and
5. adequate maintenance and operational control.

Advantages of biogas technology:

1. The BGPs digest, treat, and convert biomass or any other biodegradable materials into two useful end products: (1) inflammable gas as fuel and (2) enriched organic manure. So provides a sanitary way for disposal of human and animal wastes.
2. Weed seeds are destroyed, and pathogens are either destroyed or greatly reduced in number.
3. Rodents and flies are not attracted to the end product of the process. Access of pests and vermin to wastes is limited.
4. Reduces organic content of waste materials by 30%–50% and produces a stabilized liquid effluent. Produces free flowing, almost clear liquid odorless effluent. The effluent is a good liquid fertilizer.

5. The manure obtained from cattle dung using BGP has a higher nutritive value than conventional farmyard manure produced from the same dung.
6. Produces large amounts of methane gas, which can be stored at ambient temperature. Biogas provides a smokeless, high-efficiency fuel for domestic purpose (cooking and lighting), as well as heating and power generation at the village level.
7. Biogas is a clean fuel and keeps kitchen, household, and the surroundings clean.
8. Helps to conserve on imported energy sources.
9. Biogas production technology is an environmentally sound and eco-friendly technology and also a carbon neutral system. Whatever carbon is produced while burning biogas for energy purpose, at least the same amount (quantity) if not more is offset, directly or indirectly, for example, indirectly by carbon offset due to reduction in deforestation (by replacement of firewood) thus reducing the greenhouse gases to the atmosphere as well as directly through use of biogas digested slurry (biomanure) for biomass production, which again absorbs carbon from the atmosphere.
10. Controls environmental pollution, promotes public health through preventing flies and mosquitoes that breed in the fresh dung heaps, staked near the rural house and streets, especially during rainy seasons, and prevents foul odors due to stopping of decomposition in open areas.
11. Digested slurry if applied directly along with the irrigation water to the crops and tree plantation, then less nutrient will be lost from the slurry.
12. Digested slurry is good for backyard horticulture and kitchen garden, undertaken for supply of nutrition from fresh fruits and vegetables to the rural families, and would give additional income to them from the sale of surplus slurry in dried or composted form.
13. BGPs save time in cooking, cleaning utensils, and removing drudgery to women and girl children in the Indian villages.
14. Biogas is a very safe fuel in village home as it cannot explode easily due to 35%–40% of CO_2 (carbon dioxide) in the biogas mixture.
15. Prevents eye and lung diseases in women and children who are normally in the kitchen when food is cooked on firewood and dung cake in traditional stoves.
16. Manure prepared from digested biogas slurry has humus apart from all the nutrients and trace elements that enrich builds and regenerates the soil, thus contributing to better and sustainable crop yield.
17. Application of manure from BGP also increases the water-holding capacity of the soil, which makes it easily available to plants.
18. The application of biogas manure changes texture and structure of the soil and makes it porous for better aeration, thus contributing to better crop yields.

19. Biogas slurry (effluent) can be used for seed treatment, which is found to give better seed germination.
20. Biogas slurry can be used in the intensive composite pisciculture to give better returns to the farmers.
21. The dried slurry can be used as feed for poultry and pigs.

Disadvantages of biogas technology:

1. Liquid sludge presents a potential water pollution problem if handled incorrectly.
2. Proper operating conditions must be maintained in the digester for maximum gas production.
3. Other demerits are as discussed below.
   a. *Investment*:
      i. Except for the small plant, additional money has to be spent for the biogas, despite the subsidy provided by Biogas Support Program India.
      ii. The amount to be spent is huge for the poor people.
   b. *High interest*:
      i. KVIC India has instructed different commercial and development banks to provide loan for the biogas installation. Users have to pay higher interest rate on biogas loan, which is a problem.
   c. *No direct income*:
      i. Although biogas has several benefits, it does not generate income. Farmers prefer direct income to enable them to pay back the principal and the interest on loan.
      ii. They have very little income generating opportunity to utilize the time saved from the installation of biogas and, therefore, hesitate to invest the loan money.
   d. *Land for mortgage and space for biogas construction*:
      i. Mortgage is necessary to obtain loan from the bank. Usually, the farmers have to take or might have already taken loan for the seed, fertilizer, and pesticide.
      ii. Some farmers do not have enough land even for farming, and others are even landless. Furthermore, they prefer the essential and short-term farming loan rather than long-term biogas loan.
      iii. Many farmers, therefore, face mortgage problem for the loan. Furthermore, installation of biogas requires space close to the house. Many families do not have land near their house, and some do not consider good to utilize the nearby space for biogas construction.
   e. *Daily feeding*:
      i. Biogas requires daily feeding of cow dung mixed with water for the smooth operation.
      ii. Users consider it as a burden as it is new extra work for them.

f. *More water to collect*:
   i. Feeding of biogas requires mixing of dung and water in equal proportion. Larger the biogas, greater the amount of dung as well as the water required. Collecting water will be a problem if it is not available nearby.
   ii. Because of this, biogas installation is not recommended if the source of water is farther than 20 m.

g. *Maintenance problem*:
   i. Biogas is being used mostly in kitchen by housewives. Most of them are poorly educated.
   ii. Therefore they could grasp a little of the instruction they are given about maintenance.
   iii. Since the technology is new, the helping hand is not easy to find in the village.
   iv. Biogas Support Program had initiated "Women Users Training." Still, the maintenance problem is not completely solved.
   v. They have to depend on the biogas company. In addition, the spare parts like mantle and glass lamp, stopcock, and water drain are not available in the local market. Visit to the manufacturing companies for spare parts and maintenance is a problem, as most of the companies are located far away.

3.7 Finishing works and instructions to users

Checking of gastightness and watertightness:

After the completion of the construction of structural components and installation of pipes and appliances, and before feeding with mix of dung and water, BGPs should be checked for watertightness (the digester) as well as gastightness (the gasholder—dome and conveyance system—pipes, and appliances). If the plant is not watertight, there will be the risk of leaking of nutrients from the slurry as well as the alteration of water–dung ratio, which affects the HRT adversely. A leaking BGP hence produces inferior quality biofertilizer. Likewise, if the gasholder, pipes, and appliances are not airtight, the produced gas will escape into the atmosphere, resulting in less gas available for application (at the microlevel) and detrimental consequences to environment (at the macrolevel). In other words, the efficiency and effectiveness of BGP highly depend on the gastightness of the gas storage tank as well as the pipes and appliances, and watertightness of the digester. Small biogas production units used are in part financed by CO_2-reduction credits. The units, however, produce methane, CH_4, which is a much stronger greenhouse gas. Therefore the gas permeability (leakage of methane) of the units is crucial from the point of view of not only production efficiency and safety, but also climate. Most of the units, except in Nepal, consist of a masonry dome, partly constructed in the soil, in which methane is produced. There are different methodologies in practice to check the watertightness and gastightness of structures like BGPs. However, the testing of BGPs should be as simple as possible so that it can be performed at the local level with little time and efforts. The simplest techniques to perform these tests are described below.

Checking of watertightness:

After the completion of the required finishing works inside the digester, it should be checked thoroughly for any minor visible cracks in the walls and the floor. If cracks are seen, these should be repaired with chiseling and plastering. If there are no cracks, the following steps should be followed to check the watertightness.

1. Fill the digester with water till the water level reaches at the slurry overflow level in the OT. Leave it for 3–4 hours to allow the newly constructed walls to absorb water.
2. Mark the level of water or slurry in the outlet wall once the water level becomes somewhat stable.
3. Leave it for 24 hours and check the level of water again.
4. Observe the change in water level after 24 hours. Measure the difference of water levels. If the drop is less than 3 cm in smaller plants (4 and 6 cum) and less than 4 cm in bigger plants (8 and 10 cum), the digester is watertight. However, if the drop in water level is more than 4 cm in 24 hours, the digester is not watertight.
5. If water level drops down gradually, wait till the level become somewhat static. If the drop stops at certain height, it indicates that the leakage is occurring above that level. If the level continuously drops down to the floor level, the leakage should be either in the bottom of the wall or on the floor.
6. A thin layer (5–7 mm) of plaster (1:3) with waterproofing compounds should be applied in the digester walls to avoid leakage.

Checking of gastightness:

a. *Gasholder*:

   To check the gastightness of the gasholder, the following steps should be followed.

   1. Ensure that the digester and the OT are watertight.
   2. From the already filled plant (for checking watertightness), take out some water so that the water level in the outlet drops to 15 cm below the overflow level.
   3. Open the main gas valve placed in the top of the dome.
   4. Pump air through the piping system (preferably by detaching the connection of stove and rubber hosepipe) with the help of small hand/foot pump similar to bicycle air pump till the level of water reaches to the overflow level in the outlet. Alternatively, the pressure could be noted in the pressure gauge fitted in the gas pipeline.

5. Close main gas valve. Check any leakage in main gas valve and ensure that there is no leakage in it.
6. Mark the level of water in the OT. In addition, note the pressure reading in the pressure meter fixed in the gas pipeline.
7. Wait for more than 4 hours.
8. After 4 hours, check the water level in the outlet as well as the pressure reading in the pressure gauge.
9. If the drop of water level in the OT is less than 2 cm, the gasholder is gastight. Alternatively, if the pressure readings in 4 hours do not differ by more than 2-cm water column, the gasholder is airtight. If the drop is more than 2 cm, the dome should once again be treated for gastightness.

Alternatively, the airtightness of the gasholder could be checked with smoke test. For this test, smoke-producing substances, such as sulfur, partially wet sawdust or rice husk, could be kept in a container that floats in the water inside the digester to produce smoke, or smoke can be injected to the plant through gas pipeline. If there is any leakage in the gasholder, smoke will come out, which will be easily visible.

b. *Conveyance system (pipes and appliances)*:

To check any leakage from pipes and appliances, the following steps should be followed.

1. Ensure that there is no leakage from the main gas valve.
2. Close the main gas valve, gas taps, water traps, and any other valves in the pipeline.
3. Pump air in the conveyance system through the rubber hosepipe that connects stove and gas pipelines till the pressure reading in the gas pressure gauge increases by 20-cm water column.
4. Wait for 2 hours.
5. After 2 hours, note the pressure reading in the pressure gauge.
6. If the pressure reading decreases by more than 2-cm water column, then there is gas leakage in the conveyance system.
7. To find out the exact point of leakage, apply soap water solution in the joints and appliances.
8. The bubbles of soap water solution will either shake very fast or burst if there is leakage.
9. Alternately, smoke could be injected in the pipeline to check any leakage through it.
10. Repair the leakage once it is detected.

Precautionary measures with respect constructional aspects of BGPs:

Some of the precautionary measures with respect to BGPs that relate to common constructional aspects, etc. are covered in this section.

- For making cement mortar mixture, use wooden boxes for correctly measuring the volume of cement, fine sand, and coarse sand for making proper ratio of mixture.

- Use the following cement mortar, cement paste and cement, and damp proof cement (DPC) powder mixtures for construction of different components and subcomponents of the BGPs, especially to GBP.
  - **a.** The foundations of Ring Beam, MUP, OC (OT + ODC), and MT are to be cast by using concrete mixture prepared in ratio of either 1:4:8 (1 cement: 4 coarse sand: 8 brick ballast) or 1:3:6 (1 cement: 3 coarse sand: 6 stone chips) by volume.
  - **b.** For the casting as well as the first coat of rough plaster of the Ring Beam, use the cement mortar ratio of 1:3 (1 cement: 3 coarse sand) by volume.
  - **c.** For the second coat of smooth plaster on the Ring Beam, use the cement mortar ratio of 1:4 (1 cement: 2 fine sand + 2 coarse sand) by volume.
  - **d.** For the casting as well as the first coat of rough plaster on the outer surface of MUP, use the cement mortar ratio of 1:3 (1 cement: 3 coarse sand) by volume.
  - **e.** For the second coat of smooth plaster on the outer surface of MUP, use the cement mortar ratio of 1:4 (1 cement: 2 fine sand + 2 coarse sand) by volume.
  - **f.** For the casting as well as the first coat of rough plaster on the inner surface of the top segment of MUP, use the mortar ratio of 1:3 (1 cement: 3 coarse sand) by volume.
  - **g.** For the second coat of smooth plaster on the inner surface of the top segment of MUP, use the mortar ratio of 1:4 (1 cement: 2 fine sand + 2 coarse sand) by volume.
  - **h.** For the casting as well as the first coat of rough plaster on the bottom segment of MUP, use the cement mortar ratio of 1:3 (1 cement: 3 coarse sand) by volume.
  - **i.** For the second coat of smooth plaster on the bottom segment of MUP, use the cement mortar ratio of 1:4 (1 cement: 2 fine sand + 2 coarse sand) by volume.
  - **j.** For the casting as well as the first coat of rough plaster both on the outer and inner surfaces of the rectangular-shaped OT, use the cement mortar ratio of 1:4 (1 cement: 4 course sand) by volume.
  - **k.** For the second coat of smooth plaster both on the outer and inner surfaces of the rectangular-shaped OT, use the cement mortar ratio of 1:4 (1 cement: 2 coarse sand + 2 fine sand) by volume.
  - **l.** For the casting as well as the first coat of rough plaster both on the outer and inner surfaces of the hemispherical-shaped ODC, use the cement mortar ratio of 1:4 (1 cement: 4 course sand) by volume.
  - **m.** For the second coat of smooth plaster both on the outer and inner surfaces of the hemispherical-shaped ODC, use the cement mortar ratio of 1:4 (1 cement: 2 coarse sand + 2 fine sand) by volume.

n. For the casting as well as the first coat of rough plaster both on the outer and inner surfaces of MT, use the cement mortar ratio of 1:4 (1 cement: 2 fine sand + 2 coarse sand) by volume.
o. For the second coat of smooth plaster both on the outer and inner surfaces of MT, use the cement mortar ratio of 1:4 (1 cement: 2 fine sand + 2 coarse sand) by volume.
p. For doing the second coat of smooth plaster for the MUP, the OT and the ODC, add DPC powder in the cement bag at the rate of 1 kg (DPC) to 50-kg cement.
q. For doing the second coat of smooth plaster for the MT, the short-inlet channel, and the MhC, add DPC powder in the cement bag at the rate of 1 kg (DPC) to 50-kg cement.
r. For cement polishing and finishing (only on the inside surface, on top of the second coat of smooth plaster for all the components and subcomponents of sustainable biomass program (SBP)-I), use cement paste in the ratio of 1:1 (1-kg cement: 1-L water).
s. For casting and plastering the BRCM walls of digested slurry (or compost) pits, for both the first coat of rough plaster and the second coat of smooth plaster, use the cement mortar ratio of 1:5 (1 cement: 2 fine sand + 3 coarse sand) by volume.
t. For cement polishing and finishing on top of the second coat of smooth plaster (only on the inside surface) of the walls of the digested slurry (or compost) pits, use cement paste in the ratio of 1:3 (1-kg cement: 3-L water).
u. Before starting construction or plastering of any components and subcomponents of SBP-I model BGP, use cement water in the ratio of 1:5 (1-kg cement: 5-L water) by volume. This cement water has to be poured slowly, using a mug, etc.
v. Before starting construction or plastering of walls of the digested slurry (or compost) pits, use cement water in the ratio of 1:5 (1-kg cement: 5-L water) by volume. This cement water has to be poured slowly, using a mug.

Instructions to users:

Once the construction works are completed, the sites should be cleaned and cleared properly. The remains of construction materials have to be dumped properly in disposal areas. The top of the dome has to be filled with soil, which acts as an insulation to protect the plant. The outside portion of outlet walls and the base of the inlet should be filled with soil and compacted. Proper drainage system should be constructed to avoid rain water entering into the biodigester. After the completion of the entire construction work, the mason has to provide proper orientation to the users on plant operation and minor maintenance. Importance of daily feeding as per required quantity, operation of different appliances, major points to be remembered while operating the plants, etc. should be explained to the users before leaving the construction site. Information on the following aspects of operational activities has to be given to the users:

1. initial and daily feeding of plant;
2. use of main valve;
3. checking leakages;
4. use of water drain;
5. cleaning of outlet;
6. composting/maintaining compost pits;
7. oiling of gas tap;
8. cleaning of gas stove;
9. cleaning of gas lamp;
10. breaking of scum layer; and
11. reading of pressure gauge and adjusting of gas flow as per the reading.

Measurements and instrumentation (principles and techniques):

Measurement principles rely on a number of physical, chemical, or biological techniques, or combinations of these. Some measurement principles can be used to assess more than one parameter, while some parameters can be measured by using various principles. The choice of which principle to use for monitoring may be made based on costs, accuracy, time required for analysis, possible interferences, and requirements for sample preparation.

Compositional analysis of biogas:

For the determination of the constituent gases in biogas, that is, H_2S, CO_2, and CH_4, from the total biogas produced in each digester, parallel sets containing the same and equal quantities of slurry were made. The digesters were separately connected to the inlet of a Buckner filter flask containing some quantity of silica gel that dries the biogas. The outlet of the flask was connected to another Buckner flask containing lead acetate solution in 3-M ethanoic acid. Lead acetate absorbs the H_2S gas from the biogas passing through it, forming black precipitates of lead sulfide, as shown in the following equation:

$$H_2S + Pb(COOCH_3)2 = PbS + 2CH_3COOH.$$

From the Buckner filter flask containing lead acetate, the remaining gas after the absorption of H_2S gas passed into another Buckner flask containing 10% NaOH solution, which absorbs carbon (IV) oxide in the biogas, as shown in the following equation:

$$CO_2 + 2NaOH = Na_2CO_3 + H_2O.$$

The remaining gas coming out from the second Buckner flask is mainly methane (CH_4) gas collected by the downward displacement of water. The average weights of CO_2 and H_2S absorbed in each setup were, respectively, determined and converted to volumes using the Vander Waals Eq. (2.9). The weight of the absorbed gas was calculated by subtracting the weight of the absorbent before absorption from its weigh after the absorption process ended

$$\left(P + an^2\right)(V - nb)/v^2 = n\text{RT}$$

where P is the pressure of the gas, V is the volume of the gas, n is the number of moles of the gas, R is the universal gas constant (8.314 J/mol/K), and "a" and "b" are the constants related to the gas. The efficiency of biogas generation depends on the following factors.

a. Acid formers and methane fomenters must remain in a state of dynamic equilibrium, which can be achieved by proper design of digester.
b. Anaerobic fermentation of raw cow dung can take place at any temperature between 8°C and 55°C. The value of 35°C is taken as optimum. The rate of biogas formation is very slow at 8°C. For anaerobic digestion, temperature variation should not be more than 2°C–3°C. Methane bacteria work best in the temperature range of 35°C–38°C.
c. A pH value between 6.8 and 7.8 must be maintained for the best fermentation and normal gas production. The pH value of above 8.5 should not be used as it is difficult for the bacteria to survive above this pH.
d. A specific ratio of C/N must be maintained between 25:1 and 30:1 depending on the raw material used. The ratio of 30:1 is taken as optimum.
e. The water content should be around 90% of the weight of the total contents. Anaerobic fermentation of cow dung proceeds well if the slurry contains 8%–9% solid organic matter.
f. The slurry should be agitated to improve the gas yield.

Loading rate should be optimum. If the digester is loaded with too much raw material, acids will accumulate and fermentation will be affected.

Instruments used for various measurements:

Gravimetric: Simple method to quantify compounds based on mass (that may in some cases be combined with pretreatment(s), for example, heat to drive off moisture for TS characterization.

Chromatography:

1. Separation of substances by their different affinity between a mobile phase and a stationary phase (based on relative solubility, adsorption, size, or charge).
2. Can be used for liquids or gases and can be used to measure individual volatile fatty acids (VFAs) and gas composition.
3. Techniques are divided in gas chromatography, headspace gas chromatography, and high-performance liquid chromatography.

Electrochemistry:

1. Based on the measurement of electrical potential, current, or resistance using electrodes.

2. Can be used for liquid samples to measure pH, redox, conductivity, and a number of ionic species such as ammonium, calcium, various heavy metals, carbonate, and sulfide. It has also been used to measure the dissolved.

Titrimetric:

1. Measurement of the amount of reagent that reacts with the component to be assessed.
2. Can be used to measure alkalinity and be used as a surrogate method for "*n*" measuring total VFAs.

Electronic noses for gas measurements:

The use of arrays of electronic gas sensors, so-called "electronic noses" or volatile compound mappers have been used to measure metabolic activity indirectly. This type of sensor may have a promising potential in anaerobic digestion as they are noninvasive; however, the liquid–gas phase equilibrium is limited in anaerobic systems, and more research still needs to be performed if electronic noses are to be used in this field.

Eudiometer: The eudiometer is a laboratory glass gas tube, closed on one side, with scale measuring the volume of biogases in the chemical reaction.

Eudiometer construction: It consists of the eudiometer tube (B) of 300–400-mL volume graduated from the top down (the scale for the division of the value is 5 mL) and placed onto an upright bottle with a ground joint-fermenter (A) of 500-mL volume. A connecting tube (C), allowing the entry of the putrefied gas in the upright bottle into the measuring tube, runs through the eudiometer bottom. The connecting tube remains in fixed position due to glass sticks fitted on four sides (E). The glass fermenter from which a sufficiently long-dosed connecting pipe (F) runs to the leveling bulb (G) made of glass or synthetic material (at least 750-mL volume) is located at the eudiometer lower end. A tap cock (H) is provided at the eudiometer top end for gas sampling and for adjustments of the zero point. Fig. 3.27 shows the schematic diagram of eudiometer.

Microbiology and molecular tools used for monitoring anaerobic digestion and biomethane plants:

Monitoring of anaerobic digestion and biogas processing plants relies on a number of analytical methods and techniques that have been developed and are applied in many other biotechnologies, and chemical and engineering processes. However, in some cases, a specific methodology has been devised so that widely applied measuring principles could also be used specifically for biogas systems. In many cases, for example, sample preparation has been required due the biofouling and the high suspended solid content of the anaerobic digestion-related samples. Some of the most significant measurement methods that have been applied to anaerobic digestion systems comprise the following:

1. Techniques that can be used for enumeration of microbes or for DNA/RNA-related analysis; these include microscopy, fluorescence in situ

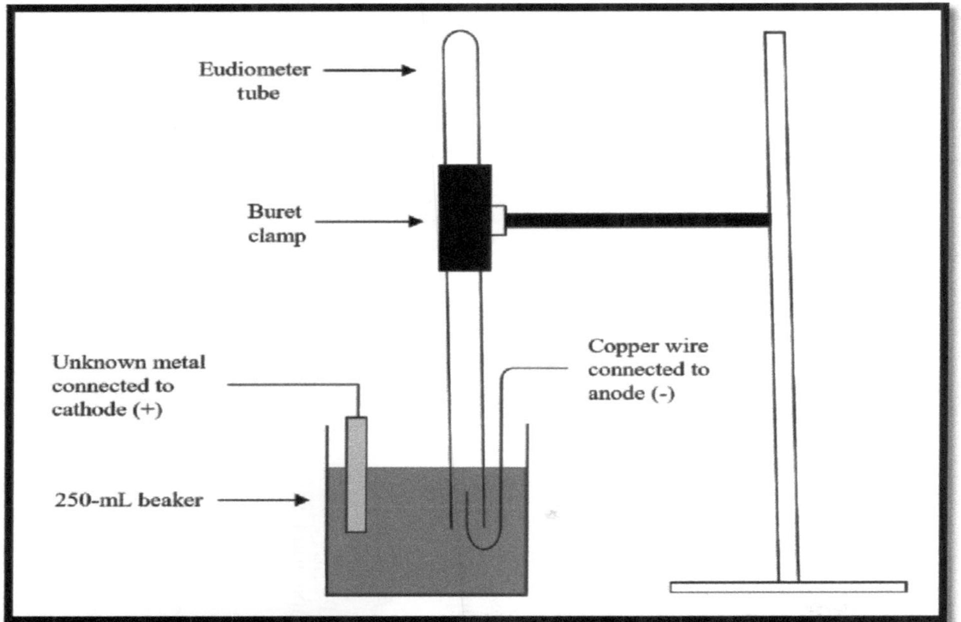

FIGURE 3.27 Schematic diagram of eudiometer.

hybridization, denaturing gradient gel electrophoresis, real-time polymerase chain reaction, and DNA sequencers.

2. These techniques have made significant progress in the last few years, and their application is likely to become more widespread in anaerobic digestion systems in the future.

Spectrometric:

1. Measure the absorbance, transmission, diffusion, or fluorescence of radiation in the UV, visible, and infrared range.
2. Molecular spectroscopy measures liquids, while atomic spectroscopy measures components in the gas phase.
3. Depending on the analysis performed photometrically, measured concentrations, for example, Chemical cxygen demand (COD), $NH_4{-}N$, and VFAs, can suffer from interferences from particulate matter and inherent coloration of the sample.
4. Significant applications for anaerobic digestion of these techniques have been researched in the last decade.

Biosensors:

1. Combine the selectivity of biological substances with microelectronics and optoelectronics.
2. Can be used to measure Biological oxygen demand (BOD), and more recently ammonia and total VFAs.

3.8 MATLAB simulation of biogas power plant

Simulink is part of MATLAB, and MATLAB is a programming tool for numeric computation and data visualization. MATLAB is mostly used for linear system analysis. The main purpose of this simulation (Fig. 3.28) is to provide a process control and instrumentation for the biomass model.

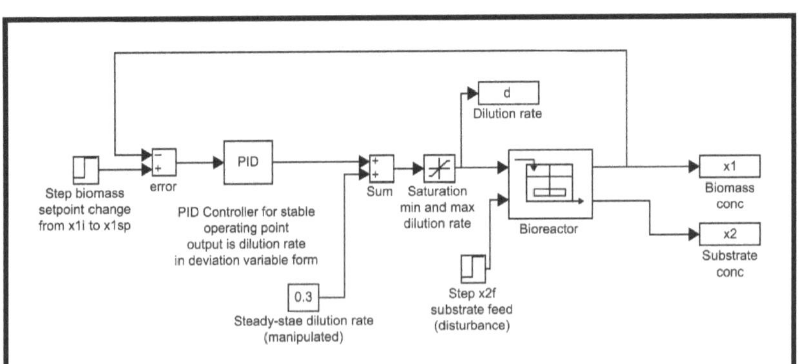

FIGURE 3.28 Simulink model of biomass energy system.

The main objective of this topic intern is to design and tune the biomass process and automate the operation of the biomass process.

Expanding biomass energy to a scale capable of impacting the global emissions of greenhouse gases will require improvements in the growth of feedstock as well as the efficiency of conversion pathways. The majority of the losses in current biomass energy systems are due to the relatively inefficient photosynthesis process, the high-energy requirements of plant processes that support the growth and development of plants, and industrial energy inputs during cultivation and processing. Engineering plants to more efficiently produce photosynthesis or microorganisms to directly produce other energy carriers, such as hydrogen, could relax many of the barriers associated with land, water, and nutrient requirements. Biological conversion processes promise efficiencies higher than thermochemical conversion, but require research to improve microorganisms and molecular-level biological processes. Advances in understanding of genetics and biological conversion processes at the molecular level will allow more control over the efficiency and economics of these pathways. Biomass energy has the potential to make a significant contribution to a carbon-constrained energy future, but technological advances will be required to overcome the low-energy densities and conversion efficiencies, characterizing the present and historical utilization.

3.9 Design of biogas power plant by HOMER software

HOMER, the micropower optimization model, simplifies the task of evaluating designs of both off-grid and grid-connected power systems for a variety of applications. When you design a power system, you must make many decisions about the configuration of the system: What components does it make sense to include in the system design? How many and what size of each component should you use? The large number of technology options and the variation in technology costs and availability of energy resources make these decisions difficult. HOMER's optimization and sensitivity analysis algorithms make it easier to evaluate the many possible system configurations. Fig. 3.29 shows the architecture of HOMER software.

The biomass module allows you to model biomass gasification and biogas fueled or cofired generators. It adds the biomass resource, the biogas fuel, and the biogas-fueled or biogas cofired generator. The biomass module can support users who model systems running on most types of biomass feedstock's and gasification process.

3.9.1 Modeling of biomass–solar energy through HOMER software

Renewable resource inputs: In the present considered system, there are 15 families. Assuming an average of four cows per family, the total

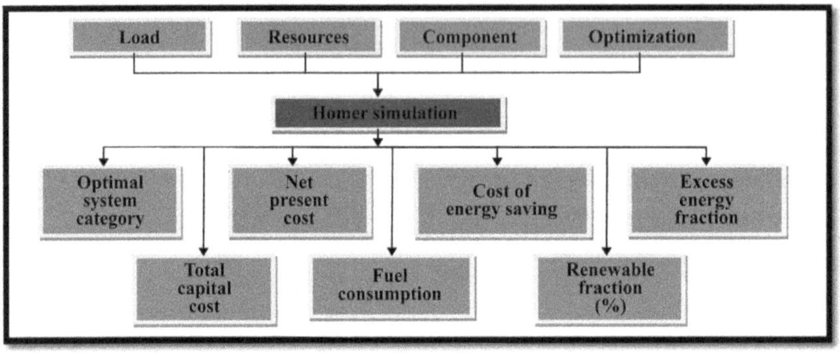

FIGURE 3.29 Architecture of HOMER software.

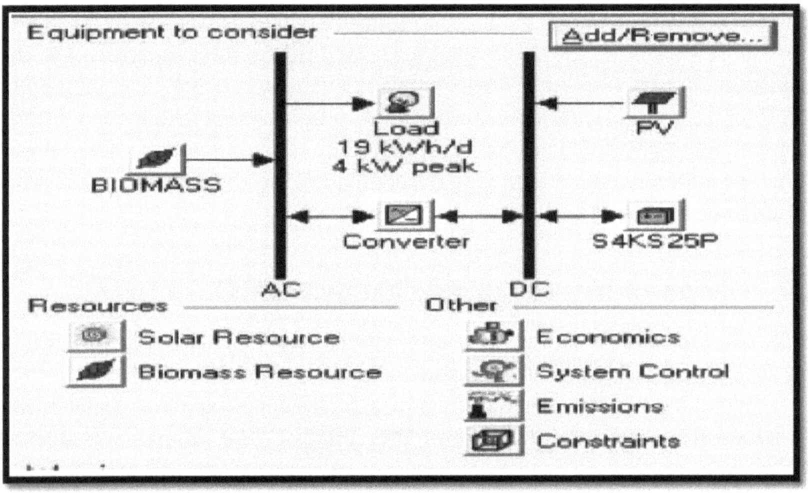

FIGURE 3.30 Biomass simulation model through HOMER.

number of cows becomes 60. One cow produces cow manure of 15 kg; 60 cows = 60 × 15 = 900 kg/day. Hundred men produce human waste of 100 kg/day. Total biogas fuel produced per day = 1 tonne. From one cow's manure, 0.036-m^3 biogas is produced; again, 0.025-m^3 biogas is produced from one human waste. To produce 1 kWh of power, 2–3-m^3 biogas is needed. This costs around $500 in India. The resource of biomass is the source of biogas for generator fuel. Users can give monthly input in tonnes/day manner; HOMER then scales the baseline data up or down to the scaled annual average value. Figs. 3.30 and 3.31 show biomass simulation model through HOMER and HOMER resource window, respectively.

The location of a particular city in India is latitude 20.3° North and 85.8° East; IST (GMT + 05:30). HOMER evaluates Photo-voltaic (PV) array

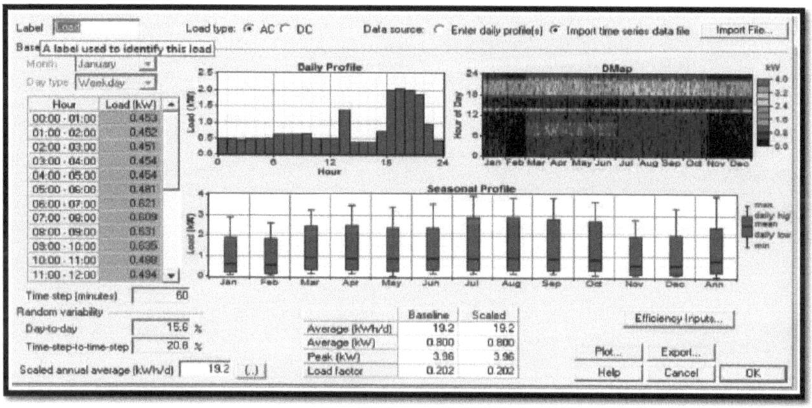

FIGURE 3.31 HOMER resource window.

power for the year on an hourly basis. Again, HOMER uses latitude values to calculate the average daily radiation from the clearness index and vice versa. Users can directly input data to the solar resource, or HOMER acquires solar resource data from NASA Atmospheric Science Data Centre for the provided coordinates.

Primary load: The electrical demand is termed as load. It is designed in HOMER to meet the need for a specific time. The HOMER plots the month-wise hourly load to create a better understanding of the seasonal demand profile. This is the first step in configuring system architecture. Scaled annual average of the system is 19.2 kWh/day, and the peak load is 3.96 kWh. The load factor is 0.202.

Battery: The batteries are used for the purpose of backup and maintain a constant voltage in peak load or shortfall in the generation. HOMER implements a battery bank/string consisting 24 numbers of batteries in a series–parallel connection. The battery chosen for this study is Surrette 4KS25P from the battery types provided by HOMER.

Converter: A converter is an electronic device used in hybrid power generation to maintain the continuity of energy among AC and DC electrical components. It consists of an inverter and a rectifier to perform the conversion from AC to DC and vice versa. Table 3.15 shows the monthly data of biomass waste material.

3.9.2 Result and discussion

Economic issues and constraints are the main factors in the HOMER simulation tool. In this proposed system, we consider 6% annual real interest rate, maximum annual capacity shortage is 1%, and 6.5% is an operating reserve as a percentage of hourly load. Here, the total capacity shortage divided by

TABLE 3.15 Available biomass/day.

Month	Available biomass tonnes/day
January	1.00
February	1.00
March	0.9
April	1.00
May	0.9
June	0.9
July	1.00
August	1.00
September	0.9
October	1.00
November	1.00
December	1.00

the total electrical load is known as the maximum annual capacity shortage. This proposed stand-alone system works properly almost 25 years, in other word say that project lifetime to be 25 years. From the simulation result, the total production of electrical energy is the combination of energy produced by solar system (33%), biomass system (67%), and diesel generator (1%). The net present cost of the hybrid renewable energy system is ₹2,088,660, operating cost is ₹55,620/year, and levelized cost of energy is ₹26.4/kWh. In every time step, HOMER calculates the renewable penetration, and it is given by the ratio of the total renewable electrical power output at a given time step (kilowatt) to the total electrical load served in this time step (kilowatt). Solar system and biomass system produce 4745 and 9667 kWh/year, respectively, which shows that the percentage contributions of solar and biomass systems are 33% and 67% of the total electricity production. Excess electricity is surplus electrical energy that must be dumped, because it cannot be used to serve a load or charge the batteries. Excess electricity occurs when there is a surplus of power being produced (either from a renewable source or by the generator when its minimum output exceeds the load), and the batteries are unable to absorb it all. The capacity shortage fraction is equal to the total capacity shortage divided by the total electrical demand. HOMER considers a system feasible (or acceptable) only if the capacity shortage fraction is less than or equal to the maximum annual capacity shortage.

Exercise

1. Explain the general description and types of biogas power plant.
2. Write the different types and shapes of biogas power plant.
3. Explain the function of different components of biogas power plant.
4. What is the utilization of gasholder or gas storage chamber in the biogas power plant?
5. What is the utilization of gas outlet pipe and valve in the biogas power plant?
6. What is the utilization of inlet chamber in the biogas power plant?
7. Write the classification of biogas power plant according to the geometrical shapes.
8. Write the classification of biogas power plant according to the gas storage.
9. Write the classification of biogas power plant according to the inlet and the outlet.
10. Write the classification of biogas power plant according to the buried position.
11. What are the advantages of fixed-dome biogas power plant?
12. Write the types of fixed-dome biogas power plant.
13. Explain the operation of Janata model of biogas power plant.
14. Explain the operation of Deenbandhu model of biogas power plant.
15. Explain the operation of floating-drum-type biogas power plant.
16. Explain the operation of water-jacket floating-drum type of biogas power plant.
17. What are the advantages of balloon-digester biogas power plant?
18. What types of wastes produce through biogas?
19. How biogas energy system developed through HOMER software?
20. How biogas energy system developed through MATLAB software?

Chapter 4

Control system of biomass power plant

Chapter outline

Objectives

- To provide knowledge about control system of biogas power plant
- To provide knowledge about automatic control system of biogas power plant
- To provide knowledge about process control of biogas power plant

4.1 Automatic control of biomass power plant

A control system is a system that provides the desired response by controlling the output, and in other manner, the control system of biomass energy systems is a combination of different elements that worked together to get desired output, and in the case of biomass energy system, the desired output fulfills the electricity demand of the consumer, so that for such type of requirement, it is necessary to develop the system that generates required amount of electricity. Fig. 4.1 shows the general block diagram of control system.

Here, the control system is represented by a single block. Since the output is controlled by the varying input parameter and it is also necessary to output according to the requirement, the control system is classified into two

Design and Optimization of Biogas Energy Systems. DOI: https://doi.org/10.1016/B978-0-12-822718-3.00004-6

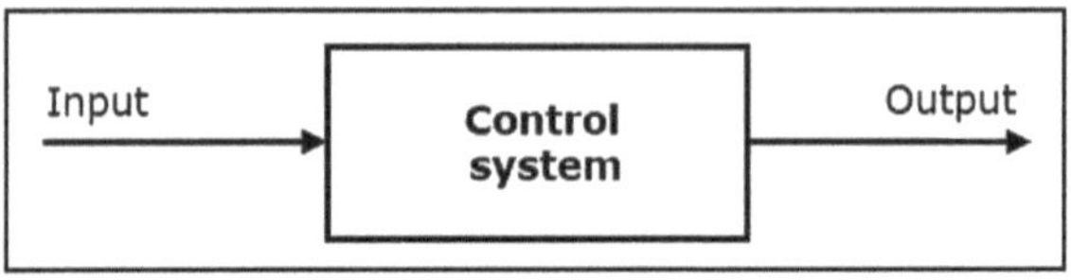

FIGURE 4.1 General block diagram of control system.

categories: an open-loop system and closed-loop system. In biomass power plant if generation of electricity is according to the requirement than system is control by the open loop control system but if generated electricity doesn't fulfill the consumer's requirement than is necessary to change some input parameter of biomass power plant and in this case path is provided from output to input and such type of control system is called closed loop control system.

The traffic light control system is an example of control system and it comes under the category of open-loop system as well as a closed-loop control system. Here, a sequence of input signal is applied to this control system and the output is one of the three lights that will be on for some duration of time. During this time, the other two lights will be off. Based on the traffic study at a particular junction, the on and off times of the lights can be determined. Accordingly, the input signal controls the output. Therefore the traffic light control system operates on a time basis, but timing of red and green light does not depend on the density of vehicle, and at this, it is a part of open-loop system, because we cannot change the output according to the input. On the other hand, it is possible to develop a sensor that changes the time duration of red and green light according to the density of vehicle than traffic control system that becomes a closed-loop control system.

Rather than open-loop and closed-loop control systems, the control system is also classified as follows:

1. continuous-time control system;
2. discrete-time control system;
3. single-input and single-output control system;
4. multiinput input and multioutput control system; and
5. multiinput and single-output system.

Continuous control is the one in which the variables and parameters are continuous and analog, and discrete control is that in which the variables and parameters are discrete, mostly binary discrete. In a single-input single-output system, only one output parameter is found out with the help of single input, but in the multiinput multioutput system, "*n*" represents the number of input and the output parameter is used for controlling the overall system. Industrial control systems used in the process industries have tended to emphasize the control of continuous variables and parameters. By contrast, the manufacturing industries produce discrete parts and products, and the

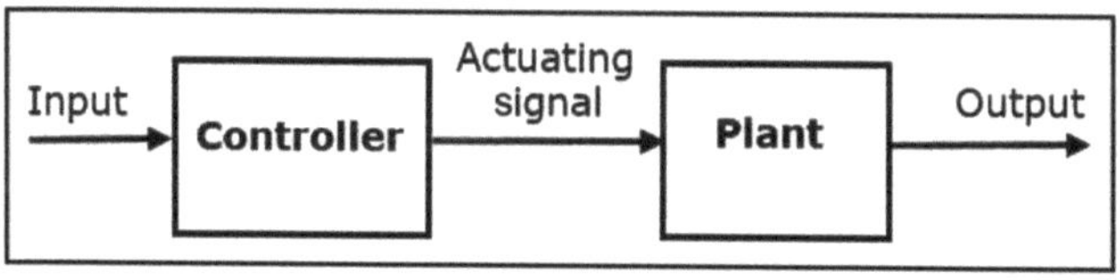

FIGURE 4.2 Forward path of control system.

TABLE 4.1 Comparison between continuous and discrete control systems of biomass power plant.

Comparison	Continuous control in biomass power plant	Discrete control in biomass power plant
Typical measurement of product output	Site assessment and quantity of waste	Undesirable gas and number of components
Typical quality measure	Consistency and absence of aerobic digestion	Dimensions and solid concentration
Typical variables and parameter	Aerobic and anaerobic digestion, voltage, and current	Retention time and nutrient concentration
Typical equipment	Anaerobic digester and mixing tank	Switches and wires
Typical process time constant	Tonne (kWh)	Less than a second

controllers used here have tended to emphasize discrete variables and parameters. Fig. 4.2 shows the forward path of control system.

Here, an input is applied to a controller and it produces an actuating signal or controlling signal. This signal is given as an input to a plant or process that is to be controlled. Therefore the plant produces an output, which is controlled. The traffic light control system, which we discussed earlier, is an example of an open-loop control system. In closed-loop control systems, output is fed back to the input. So the control action is dependent on the desired output. Table 4.1 shows the comparison between the continuous and discrete control systems of biomass power plant.

Fig. 4.3 shows the block diagram of the negative-feedback closed-loop control system.

The error detector produces an error signal, which is the difference between the input and the feedback signal. This feedback signal is obtained from the block (feedback elements) by considering the output of the overall system as an input to this block. Instead of the direct input, the error signal is applied as an input to a controller. So the controller produces an actuating signal that controls the plant. In this combination, the output of the control

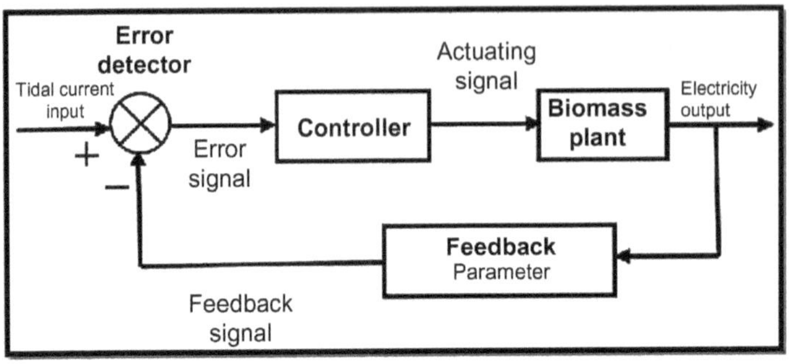

FIGURE 4.3 Negative-feedback closed-loop control.

system is adjusted automatically until we get the desired response. Hence, the closed-loop control systems are also called the automatic control systems. The traffic light control system having a sensor at the input is an example of a closed-loop control system.

The need for automatic control of biomass power plant is as follows:

1. Installation cost and control with protection equipment is costly in a biomass power plant but automatic control will provide all-time protection in relatively chipper cost.
2. Biomass plants can start and stop more frequently comparing with other power plant.
3. Provide more effective and very smooth operation.
4. Generally, biomass plants are situated in a rural area, so manual control is very difficult in rural area; in this situation, automation is a very good option.
5. The operating cost will decrease very significantly.

Based on the foregoing analysis, the control of the biomass power generation system may be organized in a cascaded manner, because in the biomass power plant, output of the biomass conversion process is working as the input of the mixing tank. The biomass power plant consists of two control loops. The inner loop controls the direct and thermochemical conversion process and the outer power control loop regulates the anaerobic digestion system that depends on the value of temperature, loading rate, and solid concentration. And this varying or fluctuating condition is controlled by using different types of differential, distance relay, switches, contractors, analog and digital timers, and isolators such as a single-break isolator, double-break isolator, bus isolator, and line isolator and different logic components.

4.2 Control strategies of biogas conversion system

The primary control and automation system at a biomass power plant are coupled with the start and stop mechanisms of the single- and multiunit and optimum running control of active, reactive, and real power, maximum voltage, and fundamental frequency, which is also composed of a number of harmonics. Supervisory control and data acquisition and retrieval are used to cover some deficiency related to the instantaneous system efficiency, monthly plant factor, and demand factor, and it also sends the essential technical information to the operators and managers. The number of control systems and also different paths of control to be applied to a biomass power plant are affected by some factors such as number, rating, and type of mixing tank and anaerobic digester.

The robust control system for a biomass power plant includes essential control circuit/logic, number of control devices, indicating and recording instrument, proper protection, and annunciation on the main circuit board and on the single unit control board for a biomass electricity generation, conversion, and transmission operation including grid-interconnected operation of biomass power stations. These some special features are necessary to provide operators with the facilities required for the organization and management of the biomass power station's major and auxiliary equipment. In the design of these, feature considerations must be given to the size and importance of the biomass station with respect to other stations in the conventional and nonconventional power systems, the distance between the main control room with respect to the equipment to be controlled, and all other station features that influence the control system.

4.2.1 Conventional control system with relay logic

Practically, various methods of control system are broadly classified under three main headings—manual, automatic, and supervisory depending on the method of operation of biomass power plant. In the manual control of biomass power plant operation of mixing tank, anaerobic digester is performed or controlled manually, whereas in automatic control of anaerobic, a sequence of controlled operation is performed automatically, but the initiation of the mixing tank and generation of operations may be performed manually or automatically.

Manual control: Here, each component of biomass power plant in the string of the prestarting checks of the component, synchronization between the aerobic digester and the anaerobic digester, loading and the stopping sequence of biomass plant is selected and performed in turn by hand whether mechanically or by the push buttons of each component.

Semiautomatic control: Here, from a solitary manual beginning drive, a unit might be conveyed to the preparations to synchronize condition by the programmed choice, execution, and giving of an arrangement of biomass component controls. Moreover, a comparative ceasing motivation totally closes down the unit. Synchronizing and stacking and running control stay manual capacities from the nearby and remote control focuses.

Fully automatic control: Here, means are provided for running up, automatically synchronizing, and loading up to a predetermined quantity on receipt of a single starting impulse. Subsequent manual variations of loading and excitation may be provided as a remote control function. The corresponding stopping impulse will cause the load to be reduced, the unit to be disconnected from the bus bars, and the turbine to be shut down completely.

Off-site supervisory control: Starting, stopping, switch closing, or opening and other functions are initiated from a remote point, together with indications of successful operations of voltage and load control and of the repetition of alarm conditions at the remote control point. The equipment is ancillary to either semiautomatic or fully automatic unit control.

4.2.2 Control of unit operation

The control of the unit activity of biomass power plant is for the most part as taking over and this kind of unit has begun from the single unit control board situated close to the biomass unit or senator board; however, synchronization and stacking of the general creation of biomass power plant are performed from the focal control room that is close to the age framework. In general, biomass unit might have been begun, blended, and stacked from the focal control room in the incorporated control framework. In view of control of unit activity and kind of control plans of prebegin checks, beginning, synchronizing, stacking, and preventing from a focal control room are made. Beginning of the unit might be performed by methods for a succession ace controller switch introduced on the control board of every unit. The ace controller switch in the initial step of its arrangement for the most part opens the principle channel valve and begins unit helpers. In the second step, the turbine is started and raised to speed no heap and field breaker is shut. In the third step, the paralleling of the unit is done and unit is synchronized with the generator transport by an end generator breaker. In the last advance, the stacking of the unit to a preset esteem is completed. Master controller switch comparably is utilized for controlling activity close down. Beginning, synchronizing, and stacking naturally on receipt of single beginning motivation are given at programmed tidal stations.

The unit control board is designed to perform the following functions of the biomass power plant:

1. information receipt and monitoring of temperature of the substrate and loading rate;
2. start/stop control sequencing of mixing tank and anaerobic digestion;
3. annunciation of alarm conditions during lower pH value of waste material;
4. temperature information monitoring of rural area;

5. metering and instrumentation signals display of electricity generation through the biomass power plant;
6. event recording, when required;
7. synchronizing and connecting the anaerobic digester to the system; and
8. control of real/reactive power of the biomass energy system.

4.2.3 Information and control signals

Basically, there are four types of signals that may be provided between the control board and any particular component of biomass power plant.

1. analog inputs to transmit variable signals from the instrument transformer such as current and potential transformer, resistance temperature detectors, and thermocouples for temperature measurement;
2. digital inputs to provide status or digitized values of variable quantities from the equipment;
3. digital outputs to send command signals (ON and OFF) from the control board to the biomass component; and
4. analog outputs to transmit variable signals from the control board to equipment such as the governor, voltage regulator, etc.

The correspondence connects between the control board and the hardware ought to be sufficient to transmit data and control signals. Data signals are the signs sent to the control board. Control signals are the yields leaving the control load up to a different gear. Information signals to the control board come from the following:

1. mixing tank and their terminal component; and
2. aerobic and anaerobic digester component.

Information and control signals are needed between the control board and each of the following:

1. step-up transformer;
2. SF6 and other circuit breaker and operating switches;
3. homogenizing tank; and
4. inlet and outlet valve.

4.2.4 Biomass equipment control

Controlling the operation of mixing tank is the most essential task of the biomass power plant, because the speed and load control of mixing tank and homogenizing tank are the main controller performance in which the system regulates the mixing of waste material through the mixing tank to balance the input power of biomass power plant with the load. In the case of small biomass power plant, the load control is also used, where the excess load is sidetracked to model load to maintain a constant speed of mixing tank. With an isolated biomass energy system, the governor controls the frequency of the overall plant and in interconnected system; the governor may be used to regulate the unit load and may contribute to the system frequency control.

Waste material injection: The injection of waste material to the mixing tank in accordance with electricity demand is controlled by the needle, and pressure is adjusted by nozzle and guide vanes. All these forms of biomass control mechanism employ oil servomechanisms. Gradual fluctuations in the quantity of waste material may cause disastrous waste material fractures, and minimum governor closing times in mixing tank are fixed and gradual closure/opening is provided to avoid hammer pressure rise or drop limits.

4.2.5 Load frequency control

In an isolated biomass system consisting of a generator and load, the varying demand of the load can be satisfied entirely by the governor action. The governor of the unit is set to maintain the frequency at 50 Hz by setting the speed droop indicator to zero. The machine speed will be maintained exactly at 50 Hz with varying load demands provided the amount of load is not greater than the unit's ability to carry it.

When a unit is operating in a large interconnected system, it is not and is indeed virtually impossible to set all governors to respond isochronously to maintain constant frequency. In such cases, unit speed droop is set at 3%–5% depending on the system's load sharing requirement. A governor set at 5% speed droop will cause its generator to accept 100% of its capacity when there is a frequency droop of 5%. Depending on its regulating ability, the unit can be adjusted to help maintain system frequency, which is the exact indication of the balance between supply and demand. The operator in the control room, on the receipt of orders from the central load dispatch office, adjusts the speed level of the unit to assist the system to maintain the frequency at 50 Hz. In the case of units fitted with automatic load-frequency control device, the speed level is adjusted by the load frequency control equipment and even by the load dispatch office itself.

4.2.5.1 Transmission line protection

In any power framework, the larger part of deficiencies happens to overhead transmission lines; these being the most uncovered components of the framework. Defensive transferring for transmission lines is, consequently, generally critical. The greater part of flaws on high-voltage lines starts with the flashover of the protection at a certain point, that is, an L-G blame because of lightning or superfluous questions on lines, for example, trees, kites, and so on. Nonetheless, L-L-G deficiencies are very normal because of lightning and every so often synchronous L-L-L-G faults, because lightning will happen. The L-G and L-L-G shortcomings will create a remaining current on a grounded nonpartisan framework, yet the L-L-L-G fault does not for the most part does as such. Deficiencies of the L-L and L-L-L write will happen with the wind or slush storms because of conductors swinging together. These deficiencies do not create leftover streams. Line assurance must, accordingly, cover stage-to-stage shortcomings that are free of the ground and in addition stage to ground issues. Nitty gritty short-out investigations are required to be conveyed before advancing the assurance for the transmission lines.

Lines may be divided into the following three classes:

a. two-terminal lines;
b. parallel lines; and
c. radial lines.

4.2.5.2 Two-terminal lines

Transmission lines, unlike apparatuses or buses, have their terminals some distance apart. For this reason, the zone differential that has been found to be the ideal protection for station zones cannot be used in its usual form on lines.

The general requirements for ideal line protection are as follows:

a. The relays must operate instantaneously for transmission system of biomass power plant.
b. The relay scheme must be inherently selective according to the kilometer range of transmission line of biomass power plant.
c. The relays at both ends of the line must operate simultaneously for all line faults during the operation of biomass energy system.
d. The relays must not respond to surging between the generating sources as long as the generators do not fall out of step, in which case the relays should operate.
e. The protection must cover all phase and ground faults of biomass power plant.

The protective schemes in general uses for phase and ground protection of two terminal lines are as follows:

a. overcurrent—with or without direction;
b. distance protection—usually with direction;
c. pilot wire for short lines; and
d. carrier current for long lines.

4.2.5.3 Overcurrent protection

The overcurrent relay plot in either the rapid or planned enlistment write is not generally attractive for transmission lines due fundamentally to varieties in the extent of the overcurrent under a different framework working conditions and to the relative estimation of the base overcurrent and the full load current of the line.

Since rapid overcurrent transfers must be set securely over the most extreme through short out of the line area, their zone of safe particular task is normally constrained to just a little segment of the line. The coordinated acceptance overcurrent handoff used to cover the rest of the line must be set securely over the most extreme load current with adequate planning to be particular or outside flaws. Where least estimations of fault current are close to the full heap of the line, overcurrent transfers are not relevant. In any case, where they can be utilized, they have the upside of straightforwardness.

4.2.5.4 Phase-to-ground faults

On a framework with grounded neutrals, any present, which leaves a line conductor and comes back to the neutrals by means of ground as in a blame to ground, is a leftover current and can be estimated in the impartial (normal) heap of the star-associated secondarys of an arrangement of current transformers in the three periods of that line. This current by and large skewering speaks to a blame current very autonomous of load. It can, in this way, be utilized as a part of a present transfer of delicate setting.

Where ground security fast requires the directional hand-off can be gotten in a rapid short for either present or voltage polarization. The above kind of ground security, be that as it may, is off for use on a framework, which is grounded through a high protection or reactance. For this situation, the estimation of the nonpartisan current is controlled by the impartial condition and the area of the blame has little effect to the estimation of the unbiased current. In such a framework, some kind of separation hand-off would need to be utilized and directed by the above sort of directional hand-off.

4.2.5.5 Distance protection

The distance relay of the impedance and reactance and more type are used.

4.2.5.6 Phase-to-phase faults

4.2.5.6.1 The impedance type

In this type, phase-to-phase voltage is mechanically balanced in the relay against the current of the proper phase, with the current acting to close the contacts. For ratio E/I greater than a certain value, which is adjustable and is the distance setting of the relay, the contacts will be kept open, whereas for lower values, the contacts will be closed. This ratio represents an impedance, where neglecting resistance of a fault itself is a constant for any given length of line, that is, an impedance relay can be set to cover any desirable length of line. For all types of faults involving more than one-phase wire of a line, the impedance relay has the following desirable features:

a. It gives accurate and dependable high-speed protection for these faults over approximately 80% of the line. It gives timed protection over the whole line with a backup effect for terminal high voltage (H.V.) buses. Terminal station zones and tandem line sections must be clearly sufficiently fast to be selective with this timing.

b. It is to a very large extent independent of connected generating capacity and configuration of the system, both for its accuracy and speed and will even operate at fault currents under full load. It will operate accurately on any current that will give 5% drop over the section protection. It can be designed to operate in 1/60 second or less for all faults except those close to the boundary conditions of fault current or distance.

c. It can be used on any line sufficiently long that the fault impedance (arc resistance) does not add a sufficient amount to the line impedance to prevent operation. The arc resistance is proportional roughly to the length and usually about 300–500 V/ft. A high-speed relay will cause tripping before an arc starts to expand. Fig. 4.4 shows different control strategies of biogas power plant.

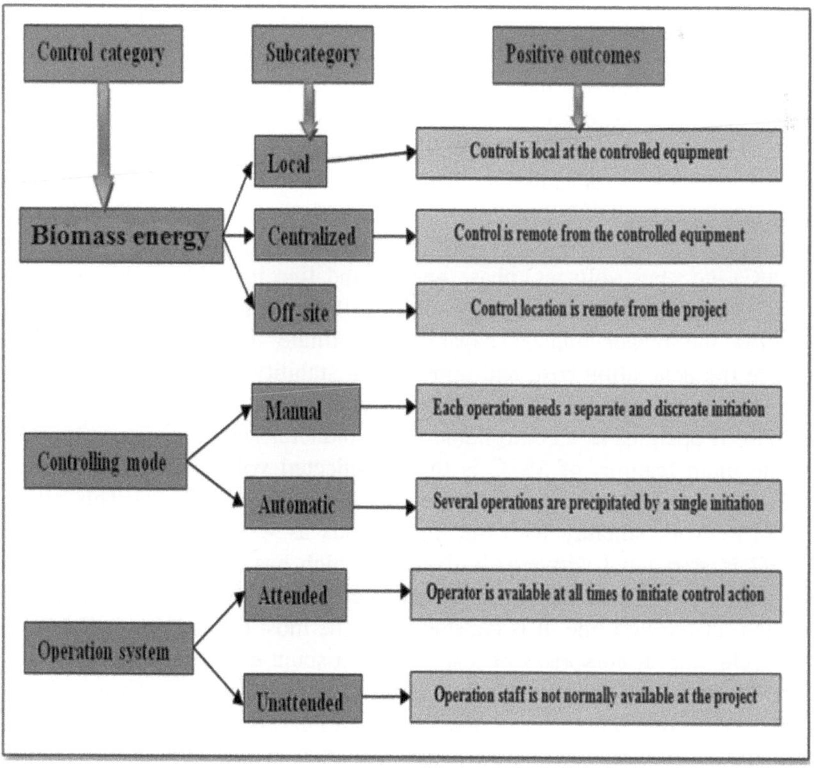

FIGURE 4.4 Control strategies of biogas power plant.

d. The impedance relay is relatively simple in construction.

4.3 Reactive power control of biogas system

Reactive power can be stated as the amount of "unused" power that is developed by reactive components, such as inductors or capacitors in an alternating current (AC) circuits. For improving the performance of an AC power system, the reactive is to be managed in an efficient way; this is known as reactive power compensation. There are two aspects to the problem of reactive power compensation: load compensation and voltage support. Load compensation comprises of the improvement of power factor, balancing of the real power drawn from the supply, better voltage regulation, etc. of large fluctuating loads. Voltage support consists of reduction in voltage fluctuation at a given terminal of the transmission line. There are two types of compensation that can be used: series and shunt compensations. These modified the parameters of the system to give improved volt ampere reactive (VAR) compensation. These quite suitably do the job of absorbing or generating reactive power with a faster time response and which come under flexible AC transmission systems (FACTS). This will allow increasing the transfer of apparent power through a transmission line and much better stability by the adjustment of parameters that govern the power system, that is, current, voltage, phase angle, frequency, and impedance. In order to control the power flow in the system, it is necessary to control the reactive power in the transmission line. There are many different FACT devices that are used to compensate the reactive power in transmission line. The various FACT methods are static var compensator (SVC), static synchronous series compensator (SSSC), static synchronous compensator (STATCOM), and unified power flow controller (UPFC) etc.; bus voltages, phase angles, and line impedances in the power system can be regulated rapidly and flexibly. Hence, FACT devices can develop the power transfer capability, facilitate the power flow control, reduce the generation cost, and improve the stability and security of the system. SSSC is an FACT device, which is connected in series with the power system. It operates as a controllable series inductor and series capacitor. One of the main features of SSSC is that its injected voltage is managed separately and is not in any case related to the intensity of line. This allows SSSC to work suitably with the lower loads as well as higher loads. The UPFC is associated with FACT devices, which has striking features. It has the ability to manage the entire the parameters that affect the flow of power in the transmission line. It is considered as the most refined power flow control technique. It comprises of a series and a shunt converter connected by a dc link capacitor, which can all together complete the role of transmission line real/reactive power flow control. The UPFC bus voltage and shunt reactive power are controlled by the shunt converter. While the transmission line real and active power is controlled by series converter by injecting a series voltage of changeable magnitude and phase angle. The parallel part that is a STATCOM injects a sinusoidal current of variable magnitude. SVC is a part of FACT device, which regulates the voltage, harmonics, and power factor, and stabilizes the system. SVC is an automatic impedance matching machine that is used to design to take the system closer to unity power factor. There are two main situations where SVCs are used, when connected to the power system for regulating the transmission voltage and when connected close to the large industrial loads for the improvement of power quality. SVC uses a thyristor-controlled reactor (TCR) to lower the voltage and consume reactive power from the system in the case of capacitive (leading). In inductive case (lagging), a capacitor bank automatically switches in and provides a higher voltage. The result is constantly variable, leading or lagging by connecting the TCR along with the capacitor bank. An STATCOM is a modifiable device that is used on an AC transmission network. It works as a source or sink of reactive AC power. STATCOM is also called a static synchronous condenser. It based on the voltage source converter. There is a new approach in which solid-state synchronous voltage sources are engaged for the actual time control of power transfer in transmission systems and the dynamic compensation. The accomplishment of the synchronous voltage source is made possible by a multipulse inverter with gate turn-off thyristors. With this, it generates the reactive power that is essential for network compensation and also interfaces with a suitable energy storage device to settle real power trade with the AC system. This creates a widespread management of power flow control for series compensation, shunt compensation, and phase angle control. To improve the system safety, FACT devices are being used to control the power flow device, which provides the chance to control voltages and power flows. Since the previous years, STATCOM plays an important role in regulation of voltage within AC transmission systems. In the three-phase AC transmission lines, an analysis of voltage reregulation problem frequently occurs and its solution by taking forward the development of a voltage regulation system using an FACT device STATCOM has been discussed.

Control algorithm for reactive power: Optimal reactive power dispatch is a nonlinear and mixed-integer optimization problem, which includes both discrete and continuous control variables. To find the situation of the control variables such as generator voltages, tap position of the tap changing transformer, and the amount of reactive compensation devices, the planned algorithm is used to optimize a certain purpose. The power transmission loss, voltage profile, and voltage stability are optimized discretely. It means that the "pure" passive filters provide inadequate performance in terms of harmonic filtering. The source current becomes nearly sinusoidal when the active filters were started and the active filter will improve the filtering performance of the passive filter (De La Roza et al., 2002). In both the cases, the result proves that the performance of the shunt active power filter with a hybrid-fuzzy controller is better than that with the conventional P-I controller. The transient response of the power system network has been improved greatly and the dynamic response becomes faster by using a hybrid fuzzy controller. Nowadays, for compensation of reactive power, the fixed capacitor-thyristor-controlled reactor (FCTCR) is used, which is controlled by the neural network. The algorithm that is used to instruct the neural networks is back-propagation. In the lagging power factor, the TCR gives constantly controllable reactive power. A fixed-capacitor bank is connected in shunt with the TCR to expand the dynamic controllable range to leading power factor. By selecting a suitable quantity of inductive/capacitive, the reactive power can compensated. The rapid and dynamic balancing of the system is possible by having a control circuit that makes use of computer-based neural network rather than traditional discrete load switching. The neural network-based FCTCR can give quick response to the reactive power of the system. In long transmission line, reactive power plays an important role in voltage stability and power transfer capability in power system. For controlling the reactive power in long transmission line, the shunt-connected compensators are usually used. Under light load condition, the TCR is used for the controlling of reactive power. The reactive power in the transmission lines can be controlled by controlling the firing angle of the thyristors. In the system, the TCR introduces the harmonic current. The harmonics in current can be brought within the particular limit by managing the reactive power injection. A Fuzzy Logic Controller is implemented to acquire best possible management of reactive power of the compensator to uphold voltage and harmonic in current within the prescribed limits. An algorithm that optimizes the firing angle in each fuzzy subset is estimated for the construction of rules in Fuzzy Logic Controller. The uniqueness of the algorithm is that it uses an easy error formula for the control of the rank of the possible firing angles in each fuzzy subset. Proportional–integral–derivative (PID) controllers is the simplest and the most widely used control method, but the main problem occurs from tuning the PID parameters to meet the desired specifications for a wide range of operating conditions. For different loading conditions, the self-tuning artificial neural network (ANN)-PID controller is compared with the ANN controller. It is observed that the ANN-based PID controller is faster, with less peak overshoot and settling time than the ANN controllers.

4.4 Power system stability of biogas power plant

Biogas plants are designed to produce methane (CH_4) and carbon dioxide (CO_2) from organic material in the absence of oxygen. This conversion is called anaerobic digestion and its end product is biogas. While there are big differences between the scale and operation of industrial and agricultural biogas plants, the basic plant design and components are essentially the same in most cases. Each biogas plant consists of one or several storage tanks for organic material, a fermentation tank, and a final storage tank for fully digested sludge. The fermentation tank has two phases: a gas and a liquid phase, where organic

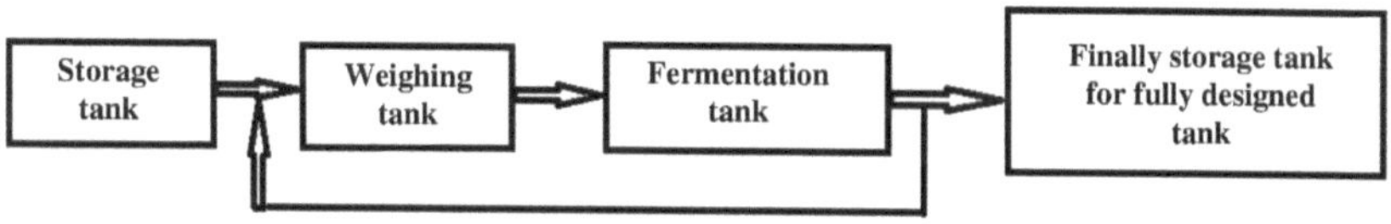

FIGURE 4.5 Block diagram of biogas plant.

material is digested by anaerobic bacteria in a relatively complex biochemical process. Fig. 4.5 shows the block diagram of biogas power plant.

There are four processes involved in biogas production.

1. Hydrolysis breaks complex organic structures open to make them accessible to the following processes.
2. Acidogenesis produces organic acids as well as hydrogen, CO_2, different alcohols, and a small amount of acetic acid out of organic material.
3. Acetogenesis uses organic acids, hydrogen, and CO_2 to produce acetic acid.
4. Methanogenesis produces CH_4 from acetic acid and to a lesser extent from hydrogen and CO_2.

4.4.1 Common control and optimization strategies

Better control and optimization of the anaerobic digestion process is one of the most effective ways of improving the efficiency of biogas power plants, but other approaches can also be used to improve plant operation. These include control of the cogeneration units with respect to variations in biogas amount and quality and the optimization of stirring intervals for the agitators inside the fermentation tank to maximize biogas production. Stirring of the contents of the fermentation tank is very important, because it improves the contact between the anaerobic bacteria and the available substrate. Moreover, stirring on a regular basis helps to reduce sedimentation and improves homogeneity inside the fermentation tank. Nevertheless, continuous strong stirring can also have a negative effect on the speed of anaerobic digestion as increasing shearing forces aggravate the contact between the bacteria and the substrate. Another way to monitor and improve plant operation is to perform laboratory analysis of the fermentation sludge and substrate feed on a regular basis. Organic acids, ammonium, and heavy metal concentrations are the most important parameters that need to be examined. Knowledge of these parameters allows efficient process operating conditions to be determined. However, performing the analysis and interpreting the results require detailed knowledge about the fermentation process, and access to such expertise is generally only cost-effective for the largest biogas production facilities. Furthermore, while the high level of detail obtained by a laboratory analysis allows a very precise assessment of the state of the process, the analysis of samples generally takes a few days, so that the information is not available in a timely fashion. Clearly, this is not satisfactory for the detection of critical process states requiring immediate attention.

These basic control and optimization strategies using proportional–integral or PID controllers for temperature and combustion control as well as regularly laboratory analysis have proven to be efficient and have made a significant contribution to improved biogas plant operation. Nevertheless, new technologies and optimization strategies that take into account new online measurements offer new possibilities for faster and more efficient reactions to varying process states. Some developments in this area have already been demonstrated in lab-scale applications and simulation case studies by Genovesi and Alferes.

4.4.1.1 Online-measurements

The strong progress and decreasing costs in the Automation and IT sector make it possible to broaden the application of online-measurement equipment on biogas plants. This is of considerable benefit to operators, as the provision of up-to-date information on process states allows them to make better decisions based on more information, and hence increases the likelihood that plants are operated efficiently. In addition, available measurement data can be used to develop computer-based simulation and optimization models, which allow a further increase in productivity with minimum effort for the operator. In the following sections, the most common and most interesting parameters that are currently measured on biogas plants are described.

4.4.1.2 Common online measurements

Every biogas plant has a certain number of online-measurement devices that are used to monitor the most critical process parameters. However, the number and quality of the equipment used depend on planned investment volume and regular maintenance. Monitoring of the fermentation temperature is crucial, as CH_4-forming bacteria only survive in relatively narrow temperature bands. Thus it is necessary to have a reliable value of this parameter at all time. This also applies for redox potential, which is used to monitor the anaerobic environment necessary for biogas production (around −500 mV). Very small amounts of oxygen directly result in an increase in redox potential. To control effectively cogeneration units, input biogas flow and composition have to be monitored closely. The biogas composition, in particular, is very important. The concentration of CH_4 in biogas has to be above 50% most of the time to guarantee continuous efficient operation of cogeneration units. Furthermore, high concentrations of H_2S in the biogas can inhibit anaerobic digestion processes as well as cause severe emission and corrosion problems in cogeneration units. However, no precise limit values can be specified for H_2S concentrations, because the inhibition depends on the adaptation of the process to H_2S.

4.4.1.3 Innovative online-measurements

Anaerobic digestion processes are still considered as black boxes. Many important process parameters can only be measured using complex and expensive laboratory equipment. Nevertheless, online-measurement sensors have recently come on the market, which can indirectly measure parameters such as organic acids and the amount of total solids. These new sensors are very promising but still need to be validated in long-term operation. In particular, the calibration of these sensors is very complicated and requires expert knowledge. The measurement of total and organic solids is used to determine the quality of the substrate feed, its potential for biogas production, and, if measured inside the fermentation tank, digester load. The more the digestible biomass a substrate contains, the higher the amount of organic solids. In agricultural biogas plants that primarily use renewable energy crops, there is only a small difference between the total solids and the organic solids. With total and organic solids, it is common to use a drying closet to evaporate water in a sample. By weighing before and after drying, the percentage of dry material can be measured. To get the amount of organic solids, the dried sample is then put into a muffle furnace where all the organic material is burned up. Again by comparing the weight before and after, the amount of organic solids can be deduced.

However, this method has two main disadvantages:

i. The process is very time-consuming, as the drying and burning take more than 24 hours for one sample.
ii. The energy consumption during these processes is enormous. New online-measurement probes can be used to measure these parameters directly in the process instead of taking samples. For total solids, there are already systems available for field use. For example, an ultrasound measurement unit is increasingly being used in modern biogas plants. This can measure the amount of solids in a substrate as it is pumped through the feed line.

The online measurement of organic solids is more complicated. One promising method in this context is the use of Near-infrared spectroscopy (NIRS). The measurement equipment basically consists of an NIR-light source and a detector. The light source generates pulses of NIR light, which are used to illuminate the sample material. The detector then measures the reflected light spectrum (800–1600 nm) and compares it with the source spectrum.

4.4.2 Programmable logic controller-based biogas plant parameters automatic control

In modern coal-, oil-, or gas-fired power stations and waste incineration plants, maximum availability is critical. Whether dealing with seasonal or daily peak loads or average loads, availability must be guaranteed. To achieve this, plant safety is a major focus. The environment is also a key issue. That is why reliable treatment of waste gases and safe handling of soot and ash also play a central role for large power stations. Other factors to consider include high pressures and temperatures, harsh ambient conditions, changing ambient temperatures and humidity, as well as corrosive and, in some cases, potentially explosive atmospheres. Some products and system solutions are designed to meet these requirements, and therefore contribute to the efficient operation and safety of the plants. In this topic, the objective was to make a programmable logic controller (PLC)-based biogas plant's parameter control system.

Anaerobic digestion can occur within a wide range of environmental conditions. However, really specific conditions are needed to reach optimum production. The key parameters intervening in the stability and efficiency control are temperature, pH, and hydraulic retention time (RT), organic loading rate, inhibitor concentration, concentrations of total volatile fatty acid, and substrate composition.

The most important of these parameters is temperature. Anaerobic digestion can occur in a very large range of temperature, going from 10°C to 71°C. However, two specific regimes have been noted at optimal temperatures: the so-called mesospheric regime, at around 35°C, and the thermopiles regime, at around 53°C. The thermopile regime implies a higher biogas production rate, but it needs a greater energy supply, because anaerobic bacteria do not generate sufficient heat. It may also be important to note that biogas production falls off significantly between 39.4°C and 51°C and above 55°C.

Concerning the relationship between the amount of carbon and nitrogen present in the organic material (C/N ratio), microorganisms need a 20–30:1 ratio of C:N. However, agricultural residues contain low nitrogen levels and have a ratio around 60–90. That is why nitrogen needs to be supplemented, for example, with ammonia (inorganic form) or with manure, urea, or food wastes (organic form). However, this issue can also favor the use of fertilizer for crops intended for biogas production.

Although biogas is composed of two greenhouse gases, its use is quite neutral from a greenhouse effect point of view, because its production enters in the carbon cycle: the use of biogas only releases in the atmosphere the carbon that has already been taken out of the atmosphere by vegetal plants before. Anaerobic digestion can be compared with respect to a number of environmental effects and sustainability criteria including energy balance, nutrient recycling, global warming mitigation potential, and so on.

A PLC is a microprocessor-based controller with multiple inputs and outputs. It uses a programmable memory to store instructions and carry out functions to control machines and processes.

The PLC performs the logic functions of relays, timers, counters, and sequencers.

It has the following advantages:

- low cost;
- reliability; and
- reprogramability.

Fig. 4.6 shows the model of a PLC.

Inputs:

- waste-level detector used to monitor the quantity of waste in mixing tank;
- water-level detector to monitor the quantity of water, which will be mixed with the waste to have the slurry;
- pressure sensor to monitor the pressure in the anaerobic digestion; and
- temperature sensor to control the temperature inside the digester so that it remains in the range.

Outputs:

- An alarm system containing buzzers is used to indicate temperature and pressure drop in their respective levels.
- Solenoid valves are used for liquid flow in the entire process.
- A stirrer is used to mix waste and water in order to obtain the slurry.
- Water pump is used for water injection through the inlet valve upon sensing a low level.
- The controlling device to be used in this is shown in Fig. 4.7, PLC and the control system IOs are also shown.

The step-by-step controller operations are as follows:

1. Scanning, sensing, scaling, and categorizing of biogas power plant data.

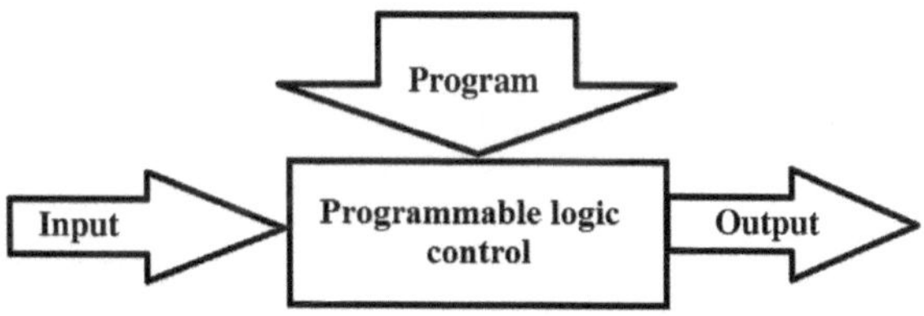

FIGURE 4.6 Model of a PLC.

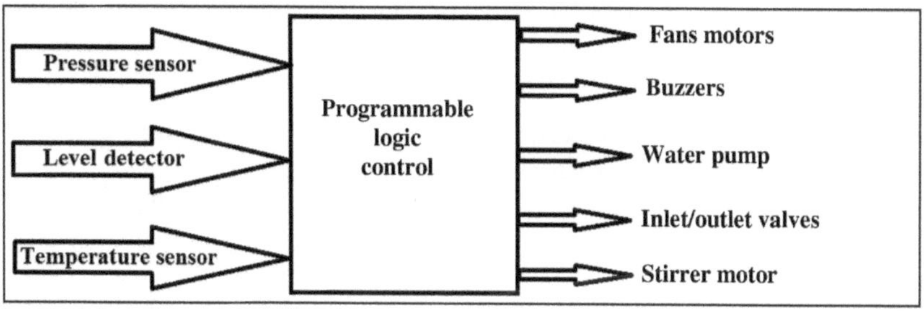

FIGURE 4.7 Biogas plant control parameter.

2. Check conditions (less, equal, or more) then execute to the specific process (increase pressure, keep constant, or decrease pressure) individually.
3. The third phase loops to the first step. The controller measures the values from sensors, classifies them, processes, and then gives a value to actuators.

The controller receives all values from pressure and volume sensors on digester and postdigester. The values are processed in the controller. The controller decides the right actions for fans based on logic conditions. If the actual value is not in the allowable range, then adjusted pressure value is used as set point.

4.4.2.1 Digester system design

All extension-service advice concerning biogas plants must begin with an estimation of the quantitative and qualitative energy requirements of the interested party. Then, the biogas generating potential must be calculated on the basis of the given biomass incidence and compared with the energy demand. Both the energy demand and the gas-generating potential, however, are variables that cannot be very accurately determined in the planning phase. Since determination of the biogas production volume depends in part on the size of the biogas plant, that aspect is included in this section.

4.4.2.2 Determining the energy demand

The energy demand of any given farm is equal to the sum of all present and future consumption situations, that is, cooking, lighting, cooling, power generation, etc. The following alternative modes of calculation are useful:

- determining biogas demand on the basis of present consumption;
- calculating biogas demand via comparable use data; and
- estimating biogas demand by way of appliance consumption data and assumed periods of use.

4.4.2.3 Determining the biogas production

The quantity, quality, and type of biomass available for use in the biogas plant constitute the basic factor of biogas generation. The biogas incidence can and should also be calculated according to different methods applied in parallel.

These methods are listed below:

- measuring the biomass incidence (quantities of excrement and green substrate);
- determining the biomass supply via pertinent literature data; and
- determining the biomass incidence via regional reference data.

4.4.2.4 Sizing the plant

The size of the biogas plant depends on the quantity, quality, and kind of available biomass and on the digesting temperature.

4.4.2.4.1 Sizing the digester

The size of the digester, that is, the digester volume (V_d), is determined on the basis of the chosen RT and the daily substrate input quantity (Sd).

$$\text{Digester volume } (m^3) = \text{Retention time} \times \text{Daily substrate input quantity}$$

The RT, in turn, is determined by the chosen/given digesting temperature. For an unheated biogas plant, the temperature prevailing in the digester can be assumed as 1°C–2°C above the soil temperature. Seasonal variation must be given due consideration, however, that is, the digester must be sized for the least favorable season of the year. For a plant of simple design, the RT should amount to at least 40 days. Practical experience shows that RTs of 60–80 days, or even 100 days or more, is no rarity when there is a shortage of substrate. On the other hand, extra-long RTs can increase the gas yield by as much as 40%. The substrate input depends on how much water has to be added to the substrate in order to arrive at a solids content of 4%–8%.

$$\text{Substrate input} = \text{Biomass} + \text{Water}$$

In most agricultural biogas plants, the mixing ratio for dung (cattle and/or pigs) and water (B:W) amounts to between 1:3 and 2:1.

4.4.2.4.2 Calculating the daily gas production (G)

The amount of biogas generated each day (G m^3 gas/day) is calculated on the basis of the specific gas yield (Gy) of the substrate and the daily substrate input (Sd).

The calculation can be based on the following.

4.4.2.4.3 Sizing the gasholder

The size of the gasholder, that is, the gasholder volume (V_g), depends on the relative rates of gas generation and gas consumption. The gasholder must be designed to:

a. cover the peak consumption rate (V_{g1}); and
b. hold the gas produced during the longest zero-consumption period (V_{g2}).

Let us have the following:

gc, max = maximum hourly gas consumption (m^3/h);
tc, max = time of maximum consumption(h);
vc. max = maximum gas consumption (m^3);
G = gas production (m^3/h); and
tz, max = maximum zero-consumption time (h).

We can now calculate

$$V_{g1} = gc_{max} \times tc_{max}$$

$$V_{g2} = G \times tz_{max}$$

The larger V_g-value (V_{g1} or V_{g2}) determines the size of the gasholder.

A safety margin of 10%–20% should be added. Practical experience shows that 40%–60% of the daily gas production normally has to be stored.

The ratio V_d:V_g is a major factor with regard to the basic design of the biogas plant. For a typical agricultural biogas plant, the V_d/V_g-ratio amounts to somewhere between 3:1 and 10:1, with 5:1–6:1 occurring most frequently.

As described in Section 3.1, the biogas energy production (P) must be greater than the energy demand (D). This central requirement of biogas utilization frequently leads to problems, because small farms with only a few head of livestock usually suffer from a shortage of biomass. In the case of a negative balance, the planner must check both sides' production and demand against the following criteria:

energy demand (D)

investigate the following possibilities:

- shorter use of gas-fueled appliances, for example, burning time of lamps; and
- omitting certain appliances, for example, radiant heater and second lamp.

4.4.2.5 Design in MATLAB

The following section will describe (left column) the MATLAB program (right column). The program was designed for sizing a plant based on what inputs the operator will planning on using as feed. In this way, the service becomes more universal, and plants can produce biomass with more ease by manufacturing. The MATLAB program was developed to help in plant design. My aim was to decrease the time that had used to calculate some parameters in for biogas plant, and as a result, this time has been reduced to the remarkable level (approximately 90%). The task will only be to collect some data and put them in the program, and after being executed and manipulated, all needed values to design and implement the biogas plant will be displayed by MATLAB toolbox. The parameters that will be displayed are listed below:

> Daily water to add to system (L), Volume Digester (m^3), Diameter Digester (m), Volume Gasholder (m^3), Height of Digester (m), daily gas production (m^3), daily solid fertilizer production (kg), and maximum pressure at which to set the estimated recovery value (ERV) (kPa).

Compared with relay control, PLCs are more flexible, more powerful, smaller, cheaper, easier to troubleshoot, faster to implement, easier to document, and have features (like communication) that cannot be accomplished

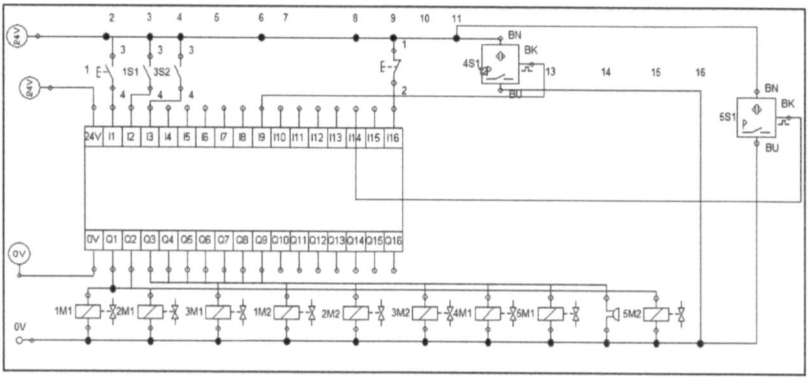

FIGURE 4.8 Simulation using FluidSim software.

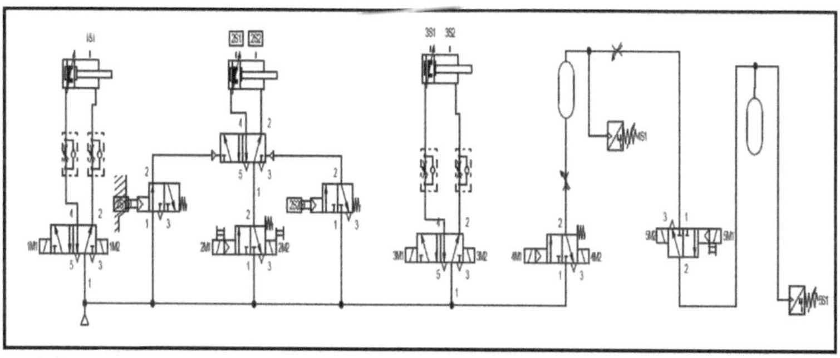

FIGURE 4.9 Equipment connection and control diagram.

by hard wiring. The control system for biogas plant using PLC was designed and simulated using FluidSim software and is shown in Fig. 4.8.

The circuits shown in Figs. 4.9 and 4.10 are for hardware connection, where six inputs and nine outputs are identified. This shows the importance of PLC-based control, because the configuration of relay is performed internally.

This configuration was performed for the purpose of simulation, but it can be used for hardware implementation in the case we have the EASYPORT CONNECTION. For my case, the ladder logic program will be developed for the use of SIMANTIC software, because the interfacing device is available from SIEMENS.

4.4.2.6 Case study 1: visualization and control of the processes at the Pacov Biogas Power Plant

In 2011 a new biogas power plant (Fig. 4.11) with the output of 750 kW was put into operation in the town of Pacov. The power plant is used to process biological waste—slurry from the breeding of livestock. During the process

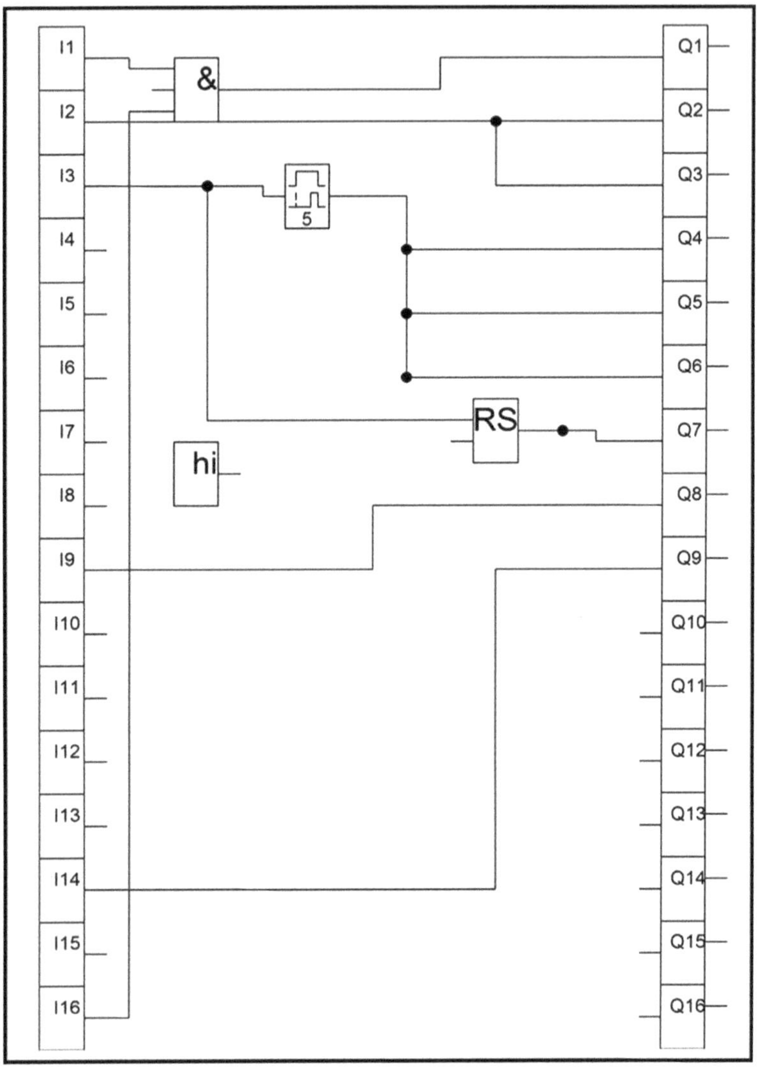

FIGURE 4.10 Controller configuration.

of anaerobic digestion, biogas—CH_4—is produced. It is collected in storage tanks and then used for the simultaneous generation of electricity and heat (cogeneration). Thanks to the connection of the cogeneration unit to the high-voltage grid, the produced electricity is delivered to the power grid. The cogeneration unit's cooling system is the source of heat energy. Heat energy is then used for the heating of the biogas power plant itself and the adjacent poultry house. The biogas power plant control and visualization

FIGURE 4.11 Pacov Biogas Power Plant.

TABLE 4.2 Specification of Pacov Biogas Power Plant.

Country	Czechia
Title	Visualization and control of the processes at the Pacov Biogas Power Plant
Implemented by	ELTOS Hradek
Year	2010–11
Control system	5 × Unitronics V280
SCADA software	1 × Reliance 4 control server with 500 data points Reliance 4 Web Client

systems were implemented by the South Moravian company ELTOS Hradek (Table 4.2).

4.4.2.6.1 Control system

Prior to the installation of the control and visualization systems, all processes had been analyzed. Subsequently, a control system by Unitronics and the Reliance SCADA/HMI system were installed. Reliance is responsible for the visualization of the biogas power plant processes. Five Unitronics V280 PLCs are used to control the processes. The primary PLC receives information from the four secondary PLCs and simultaneously provides operating data to Reliance. The entire system is interconnected by a computer network and its parts communicate with one another via the TCP/IP protocol. The Reliance system is installed on the operator room's PC. For the sake of clarity, the biogas power plant processes are visualized on five visualization windows that help the operating personnel to monitor or control:

- the overview screen displaying real-time values from the biogas power plant;
- the substrate treatment process (grinding slurry—the substrate is used as the primary material for the production of CH_4);
- the process of biowaste pasteurization;
- the process of anaerobic digestion;
- the collection of CH_4 in storage tanks;
- the process of storing digestate (the residual material remaining after anaerobic digestion);
- the process of heating the biogas power plant; and
- the process of heating the poultry house.

To track down errors from the operation of the power plant, a report and an alarm database were created in the Reliance system. Error messages and alarms can be sent to recipients as email or text messages. The operating and managing personnel can also access the visualization project remotely via Reliance 4 Web Client.

4.4.2.7 Case study 2: analytical control of fermentation processes in biogas plants (Fig. 4.12): Lellbach Biogas Plant, 1.2 MWel. (Source: EnerCess)

Against the background of finite fossil fuels and the contentious use of nuclear energy, renewable sources of energy are steadily gaining importance. Leading the way is the use of CH_4 gas obtained from fermentation processes in biogas plants. Plant operators in Germany receive a high return for feeding electricity generated from renewable raw materials into the national grid—up to €0.18/kWh. All plant operators, therefore, have a vested interest in running their biogas plant as efficiently as possible. If excessive amounts of biomass are fed into such plants however, this may have drastic economic consequences and may even inactivate the biomass, necessitating a cost-intensive restart. On the other hand, long-term underloading of a plant also has financial consequences, as less electricity and heat are generated and revenue is, therefore, lost. A precise and reliable analysis of the fermentation processes using photometric cuvette tests, easy-to-use titrators, and process measurement technology for online monitoring ensures stable process management cost effectively. In a biogas plant, natural fermentation and decomposition processes produce biogas, which is used to generate electricity as efficiently as possible. In the first phase, the substrate is made available, stored, and treated in accordance with requirements and fed into the bioreactor (Fig. 1). In the second phase, anaerobic fermentation processes take place in the digester, producing biogas. In the third phase, the gas is treated, stored, and utilized. Finally, in

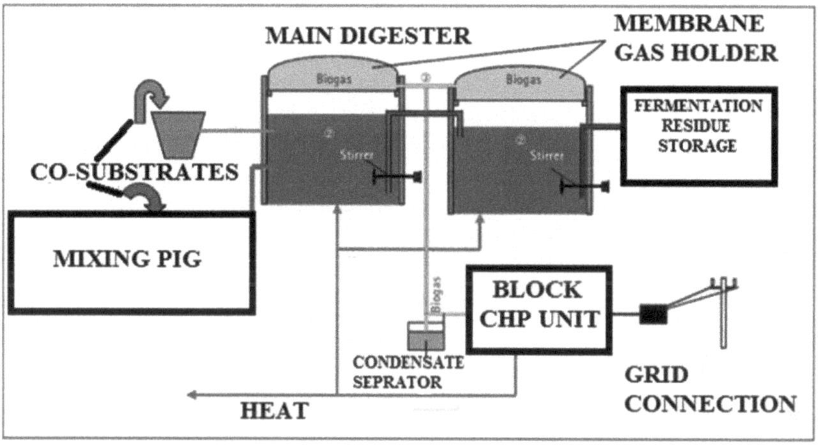

FIGURE 4.12 Analytical control of fermentation processes in biogas plants.

the fourth phase, the fermentation residues are utilized (e.g., as fertilizer in the agricultural sector). The main component of biogas is CH_4 (about 50%–70%). Biogas of 1 m^3 contains about 6 kWh of available energy and is equivalent to about 0.6 L of fuel oil in terms of its average calorific value. The heat generated during its combustion is fed into the fermentation process as process heat or used to heat on-site living and working quarters and livestock buildings or sold to external customers (e.g., operators of local heat networks).

There are basically two types of plants in which wet fermentation processes take place: plants that use renewable raw materials and cofermentation plants. The former use renewable raw materials such as maize, grass, complete cereal plants, and grains, sometimes together with manure slurry. In cofermentation plants, substrates of nonrenewable raw materials are used, such as residues from fat separators, food residues, flotation oil, industrial waste products (glycerol or oil sludge), and domestic organic waste. Manure slurry is also usually used. Biogas is produced by a highly sensitive and complex process. Without instrumentation, biogas plants are often underloaded, that is, the biomass feed rate is too low, so that electricity generation is not cost-effective. This can result in substantial financial shortfalls. The effects of overloading are especially drastic. Such unintentional "overfeeding" slows down or stops the biological fermentation process and may cause a total system breakdown. A cost-intensive plant restart is then necessary.

Control parameters for a cost-effective process: To achieve optimal control of the degradation process in a biogas plant, a detailed knowledge of the key chemical and physical parameters is necessary. *Temperature*: Temperature plays a crucial role. Biogas plants are usually mesophilic or thermophilic. The former function most efficiently in the temperature range from 35°C to 41°C, while the latter prefers 57°C. Methanogenic bacteria in particular are extremely sensitive to temperature fluctuations. The temperature of the fermentation process should, therefore, be kept constant to within a maximum of ± 1°C.

Total solids content/total organic solids content content: The total solids content (TS) or total organic solids content (TOS) is used to estimate the volumetric loading of the digester for the purpose of managing the solid streams. Wet digesters are usually run with a TS content of 8%–10%, while special digesters may be operated with a TS of up to 20%. The TOS is very important for the operation of the plant. If it is too high [e.g., >3 kg oTS/(m^3/day)], there is a danger of overloading the digester. In this case, the substrate input must be immediately reduced.

Redox potential: The redox potential of a digester is a measure of the oxidizability or reducibility of its content. Biogas production only proceeds efficiently in an anaerobic environment, that is, the redox potential must be less than 330 mV. In general, the use of oxidation promoting substrates, that is, substrates that contain oxygen, sulfate, or nitrate groups, may significantly change the redox potential and thus cause a shift in the pH. Such a negative development for the fermentation process can be triggered by, for example, a change in the substrate.

pH: pH just like the temperature, there is more than one optimum pH value. During hydrolysis and acidification, the best pH is between 4.5 and 6.3. The optimal pH range for CH_4 formation is the narrow window between 7.0 and 7.7. Continuous pH metering gives an early indication of any acute disruption of the process. However, the plant cannot be controlled reliably simply on the basis of the current pH. This is especially true of plants whose digester has a high buffer capability, as an unintentionally large input of organic acids does not necessarily result in a drop in pH.

Monitoring digesters with photometric tests: The most important parameters in the fermentation process, which can be monitored using chemical or photometric methods, are the organic acids formed as intermediates, the Chemical oxygen demand (COD), and the ammonium concentration. Until now, the necessary measurements have been carried out in external service laboratories, with the associated high costs and sometimes considerable delays between the sampling and availability of the results. A late response to negative processes in the digester may enable considerable inhibition of the biogas productivity to occur, even to the extent that the biomass may be inactivated, bringing the total plant to a halt. For this reason, it is advisable to monitor the digester directly and as quickly as possible with photometric cuvette tests, which have been regarded as the "state of the art" in wastewater monitoring for several years. The suitability of photometric cuvette tests for monitoring organic acids, ammonium, and COD was tested on numerous samples from a cofermentation biogas plant (1.5 MWel.) in Lower Saxony. Reference analyses were carried out by the contract laboratory of a nearby mechanical biological waste treatment plant. A decisive factor for the assessment was the comparability of the results to those obtained using corresponding standard methods and the consistency of the results obtained from diluted and spiked samples. Samples taken from a digester may remain biologically active. The possibility of postsampling formation of acetic acid in the samples cannot, therefore, be excluded.

Exercise

1. Define the control system.
2. What is automatic control system?
3. Write the different types of control system.
4. Explain the control system of biogas power plant.
5. What is the different parameter, which is used for control mechanism of biogas power plant?
6. What is the cost control parameter of biogas power plant?
7. Explain reactive power control of biogas power plant.
8. What are the different control strategies of biogas power plant?
9. Explain control mechanism of biogas power plant through programmable logic control.
10. Explain analytical control of fermentation processes in biogas power plant.

Reference

De La Roza, C., Laca, A., Garcia, L.A., Diaz, M., 2002. Stirring and mixing effects at different cider fermentation scales. Food Bioprod. Process. Trans. Inst. Chem. Eng. Part C. 80 (2), 129–134.

Chapter 5

Reliability assessment of biogas power plant

Chapter Outline

5.1 Maintainability and availability function of biogas power plant

Reliability is defined as the probability of a biogas device or system performing its purpose adequately for the intended operating period of time for generation of electricity. Biogas energy system reliability is defined as the ability of a biogas electrical power system to supply the system load with reasonable continuity and quality of supply. Major subdivisions of biogas or biogas power system reliability are "system adequacy" and "system security." The term "adequacy" is related to the existence of sufficient facilities within the system to satisfy the consumer load demand who take the electricity from biogas energy system and system operational constraints. Figs. 5.1 and 5.2 explore the basic theme, step, and process of reliability analysis, respectively. Fig. 5.1 shows the reliability analysis; first, direct and indirect information are collected, and then system security and system adequacy are determined through direct and indirect acquisition, respectively. In the biogas

Design and Optimization of Biogas Energy Systems. DOI: https://doi.org/10.1016/B978-0-12-822718-3.00005-8

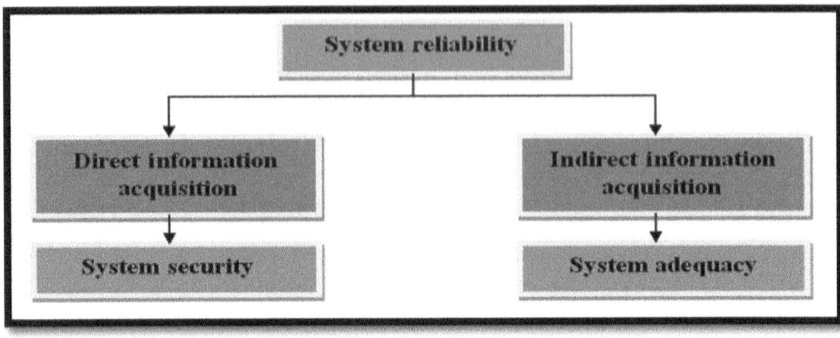

FIGURE 5.1 Step of reliability analysis.

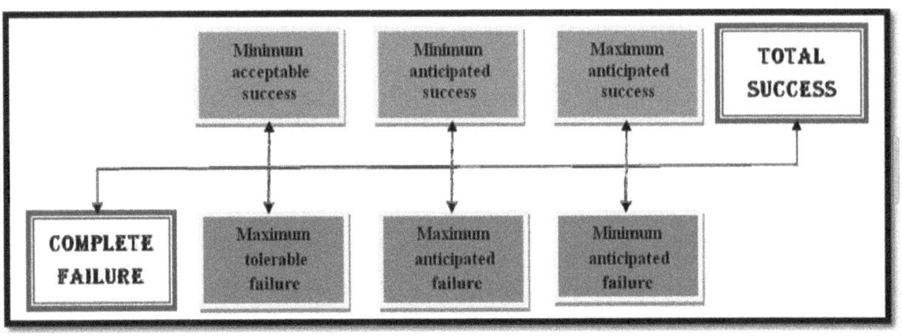

FIGURE 5.2 Process of reliability analysis.

energy system, information related to the parameter of prefeasibility analysis, technical parameter, design parameter, and cost of the plant is coming under the categories of direct and indirect information. Fig. 5.2 shows the success and failure rates of biogas energy system, which is determined through the reliability analysis. The total success of biogas energy system always depends on minimum and maximum anticipated successes of different components of a biogas energy system.

The definition brings into focus four important factors, namely:

- The reliability of biogas equipment is expressed as a probability.
- The biogas power plant is required to give adequate performance.
- The duration of adequate performance of biogas power plant is specified.
- The environment or operating conditions for biogas power plants are prescribed.

Reliability is defined as the ability of an item to perform a required function under stated conditions for a certain period of time, which is often measured by probability of survival and failure rates. The reliability of biogas power plant depends on the system architecture, lower-level component, assemblies, environmental and operational stresses, and human factor. Field experiences reveal that power electronic converter is usually one of the most critical assemblies in terms of failure rate, lifetime, and operating and maintenance cost in renewable energy.

Reliability is a measure of the continuous delivery of correctional service. The objectives of reliability engineering on the order of priority are:

1. to apply engineering knowledge and specialist techniques to prevent or to reduce the likelihood or frequency of failure of biogas power plant;
2. to identify and correct the causes of failure of biogas power plant that does occur despite the effort to prevent them;
3. to determine ways of coping with failure that does occur if their cases have not been corrected; and
4. to apply methods for estimating the likely reliability of new designs and for analyzing reliability data.

To get the real profit of biogas energy system and confidence by continual working of the energy system to supply the load, we require calculating the system reliability. Several reliability indices have been used:

1. loss of load expected;
2. loss of energy expected;
3. loss of power supply probability (LPSP); and
4. equivalent loss factor.

LPSP is defined as the probability that an insufficient power supply results when the biogas energy system is unable to satisfy the load demand. It is a feasible measure of the system performance for an assumed or known load distribution.

5.1.1 Maintainability

In a reliability and probability theory, if any biogas energy system and their different components or elements are repaired in a specific period of time, then that is known as maintainability of biogas energy system. Maintainability is the most important part of any specific condition, specific component, and specific physical phenomenon of biogas power plant. Maintainability is an apex part of engineering; for example, in biogas plant engineering, it is necessary to maintain power system in a proper way, and then it generates large amounts of electrical power for the consumer.

Maintainability is a very specific property of modeling of a technical or engineering framework of biogas energy system. It is always related to the concept of economy, replacement, and safety measures, which is very helpful to run any biogas energy system. Maintainability of any biogas energy system is classified into the following categories:

- component activity of biogas plant;
- component repair time of biogas plant;
- biogas system repair time;
- biogas system downtime; and
- biogas verification time.

5.1.1.1 Biogas power plant component activity

A specific task performed by individual component of biogas power plant of any specific physical arrangement is known as component activity. Maintainability of individual component of biogas power plant is directly effects the maintainability of the overall electricity generation system. The component activity is also related to the manner in which individual components are connected. If individual component is connected in series, then the performance of the overall system is affected at large level through the performance of individual components of biogas power plant. For example, anaerobic digester and mixing tank are connected in series, and if faults occur in anaerobic digester, then mixing tank is also not worked properly. If possible, in biogas power plant, it is necessary to connect a large number of components in parallel to maintain the framework of the complete system in a proper way.

5.1.1.2 Component repair time

A time required to minimize the failure of individual components of biogas power plant is known as component repair time. Component repair time is directly proportional to the maintainability of the overall biogas power plant.

5.1.1.3 System repair time

A time required to minimize the failure of the overall biogas system is known as the system repair time. System repair time is directly proportional to the maintainability of the overall biogas energy system.

5.1.1.4 System downtime

A time required for the overall biogas energy system, when it reaches from ON condition to OFF condition.

5.1.1.5 Verification time

A time required to testing of any individual component and overall biogas energy system. Table 5.1 shows the comparison between three parameters of plant.

TABLE 5.1 Comparison between three parameters of plant.

Reliability of biogas plant	Maintainability of biogas plant	Availability of biogas plant
Constant	Decreases	Decreases
Constant	Increases	Increases
Decreases	Constant	Increases
Decreases	Constant	Decreases

5.1.1.6 Availability

It deals with the duration of uptime for operations of biogas plant and is a measure of how often the system is alive and well. It is often expressed as:

$$\text{Availability} = \frac{\text{Uptime of Biogas Plant}}{\text{Uptime of Biogas Plant} + \text{Downtime of Biogas Plant}}$$

$$\text{Availability} = \frac{\text{MTBF}}{\text{MTBF} + \text{MTR} + \text{MTWS}}$$

where MTBF = mean time between failures of biogas plantMTR = mean time to repair of biogas plantMTWS = mean time to waiting for spares, reflecting supply.

In reliability, one is worried about outlining a thing to keep going, to the extent that this would be possible without disappointment; in maintainability, the accentuation is on planning a thing, so a disappointment can be redressed as fast as would be prudent. A case of a discrete maintainability parameter is the quantity of support activities finished in some time *t*, while a case of a nonstop practicality parameter is an ideal opportunity to finish an upkeep activity.

If we consider performance of any component so that concept of maintainability in terms of functions that are analogous to the terms of reliability, they may be derived in a way identical to that done for reliability, by merely substituting *t* (time-to-restore) for *t* (time-to-failure), μ (repair rate) for λ (failure rate), and M(t) probability of successfully completing a repair action in time *t*, or $P(T \leq t)$ for F(t) probability of failing by age *t*. In other words, the following correspondences prevail in maintainability and reliability engineering functions.

- The time-to-failure probability density function (PDF) in reliability corresponds with the time-to-maintain PDF in maintainability.
- The failure rate function in reliability corresponds with the repair rate function in maintainability. Repair rate is the rate with which a repair

action is performed and is expressed in terms of the number of repair actions performed and successfully completed per hour.

- The probability of system failure, or system unreliability, corresponds with the probability of successful system maintenance, or system maintainability.

These and other analogous functions that are utilized to find out reliability of biogas power plant are summarized in the following way in Table 5.2:

As illustrated in Fig. 5.3, maintainability of biogas power plant can be expressed either as a measure of the time (T) required for repairing a given percentage ($P\%$) of all system failures, or as a probability (P) of restoring the system to operational status within a period of time (T) following a failure.

TABLE 5.2 Comparison of reliability and maintainability functions.

Reliability	Maintainability
Time to failure $f(t)$	Time to repair $r(t)$
Reliability $R(t) = \int_t^{\infty} f(t)dt$	Maintainability $M_a(t) = \int_o^t r(t)dt$
Failure rate $\lambda(t) = \frac{f(t)}{R(t)}$	$\gamma(t) = \frac{r(t)}{1 - M_a(t)}$
Mean time to failure $\text{MTTF} = \int_{-\infty}^{\infty} f(t)dt = \int_0^{\infty} R(t)dt$	Mean time to repair $\text{MTTF} = \int_{-\infty}^{\infty} r(t)dt$
PDF of time to failure $f(t) = \lambda(t)R(t) = \lambda(t)e^{-\int_0^t \lambda(t)}dt$	PDF of time to repair $r(t) = \gamma(t)(1 - M_a(t))$

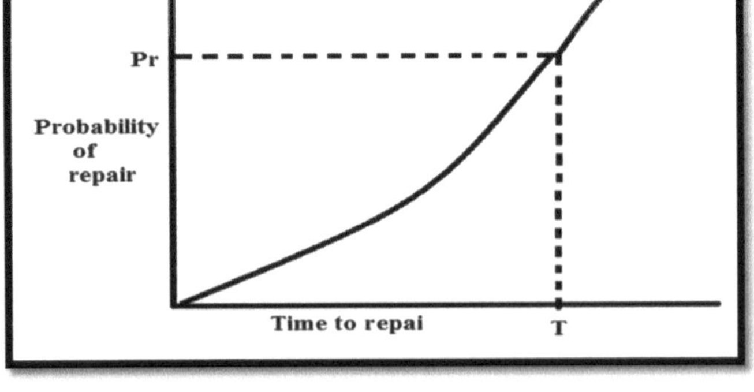

FIGURE 5.3 Probability of repair.

There are three categories of availability:

1. Inherent availability of biogas plant:

$$A_I = \frac{\text{MTBF}}{\text{MTBF} + \text{MTTR}}$$

2. Achieved availability of biogas plant:

$$A_A = \frac{\text{MTMA}}{\text{MTMA} + \text{MMT}}$$

3. Operational availability of biogas power plant:

$$A_O = \frac{\text{MTMA}}{\text{MTMA} + \text{MDT}}$$

where MTBF = mean time between failuresMTTR = mean time to repair

There are six components of maintainability of biogas power plant:

1. Element activity: There are simple maintenance actions of short duration and relatively small variance that do not vary appreciably from one system to another. Operation of gas holder and gas pipe of biogas power plant is an example of elemental activity.
2. Malfunction active repair time consists of:
   (i) preparation time of waste material;
   (ii) repair time of digester and mixing pit, and probability of repair is shown in Fig. 5.3;
   (iii) malfunction verification time of slurry;
   (iv) final malfunction test time of electrical energy; and
   (v) fault location time of gas holder and gas outlet.
3. Malfunction repair time:
   (i) active repair time of technical component of biogas power plant; and
   (ii) administrative time of plant.
4. System repair time = malfunction repair time × number of malfunctions
5. System downtime:
   (i) logistic;
   (ii) repair time; and
   (iii) system final test time.

Maintainability is an inherent characteristic of the design of a biogas power plant. It relates to the ease, economy, and safety maintenance of individual components of a biogas power plant. Most engineered objects require maintenance throughout their life cycle. Maintainability deals with duration of maintenance outages or how long it takes to achieve the maintenance

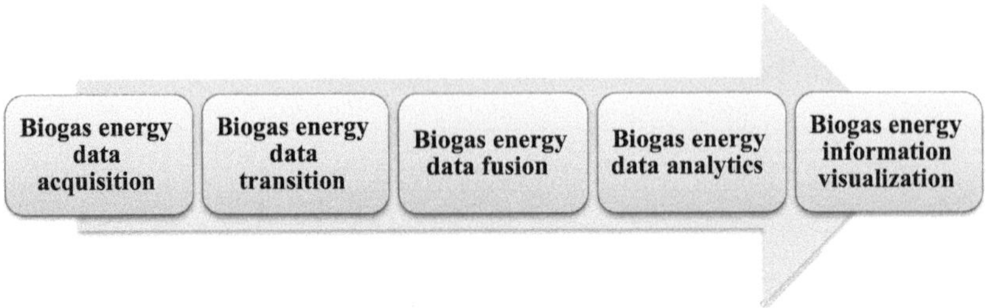

FIGURE 5.4 Path of maintainability.

actions compared with a datum. The key figure of maintainability is mean time to repair (MTTR)

$$M(t) = 1 - \exp\left(\frac{t}{\text{MTTR}}\right) = 1 - \exp(\mu t)$$

where μ = constant maintenance rate.

In the manufacturing company, productivity is the last stage of certain events, and this event always depends on the availability and operability of certain elements. In any manufacturing company, availability is always related to the quantity of raw material, and operability is depending on the number of employees of that company. Fig. 5.4 shows the path of the maintainability, which is the completion of data acquisition, data transition, data fusion, data analytics, and information visualization.

5.2 System reliability and redundancy technique of biogas power plant

5.2.1 Components in series in biogas power plant

A number of components in biogas power plant connected together to form an overall system are said to be in series configuration if the working of all the components causes the system to function properly; that is, failure of a component may lead to failure of the biogas power plant. The series system is also called chain system, as each link of the chain is crucial for successful operation of the chain. In Chinese biogas plant, slurry inlet, gas outlet, and slurry outlet are connected in series; if one of the elements is damaged, then the performance of the overall system is also deteriorating. Fig. 5.5 shows the series reliability of a system having "n" components. This block diagram is known as a reliability block diagram (RBD).

It is observed that such block diagrams do not show how the components are interconnected physically, yet indicate how they should be treated from a reliability perspective. There is a probability that the two components to be

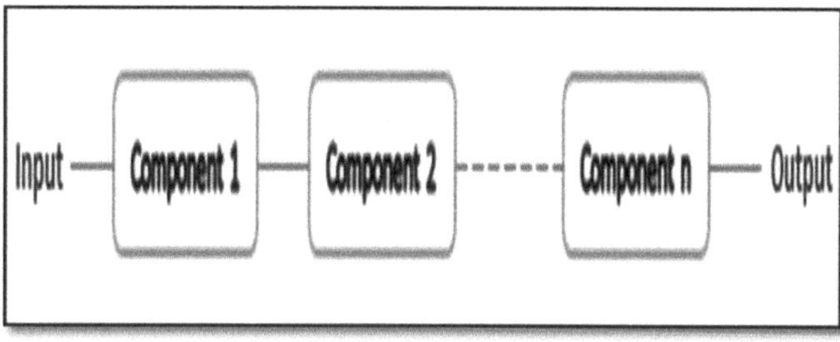

FIGURE 5.5 Reliability block diagram for a series system.

physically associated in parallel yet to be in series reliability (insofar as failure of either causes system failure). Similarly, two components might be physically in the series, however in parallel from a reliability perspective.

Let E_n be the condition that the nth component of biogas power plant is working properly and $\bar{E}_n$ be the condition that the nth component is not working. The reliability of the system is given by:

$$\begin{aligned} R(t) &= P(E_1 \cap E_2 \cap \ldots\ldots E_n) \\ &= P(E_1)P(E_2/E_1)P(E_3/E_1E_2)\ldots\ldots \end{aligned} \tag{5.1}$$

where $P(E_2/E_1)$ is the probability of event E_2 when E_1 has already occurred. If the probabilities of failure of "n" components of biogas power plant are given and assumed to be independent, then the reliability of the system will be the product of reliabilities of "n" components. It can be expressed as:

$$R(t) = R_1(t).R_2(t)\ldots\ldots R_n(t) \tag{5.2}$$

If independent components are assumed, then

$$R(t) = P(E_1 \cap E_2 \cap \ldots\ldots E_n) = P(E_1)P(E_2)\ldots\ldots P(E_n) \tag{5.3}$$

It is clear from Eq. (5.3) that the generalized formula for system reliability of "n" series-connected components of biogas power plant is:

$$R(t) = \prod_{k=1}^{n} \mathbf{P}_k(\mathbf{t}) \tag{5.4}$$

where $\mathbf{P}_k(\mathbf{t})$ is the probability of kth component to be good at time t. This can be expressed in terms of failure events as:

$$R(t) = 1 - (\text{probability of the system failure})$$

If one of the components fails to operate, then the system will fail

$$R(t) = 1 - P(\overline{E}_1 \cup \overline{E}_2 \cup \ldots\ldots\ldots \overline{E}_n) \tag{5.5}$$

From Eqs. (5.3) and (5.5), it can be seen that

$$P(E_n) + P(\overline{E}_n) = 1 \tag{5.6}$$

5.2.1.1 Failure rate of series system

Suppose that the time to failure of components is exponentially distributed, then

$$P_k(t) = \exp(-\lambda_k t)$$
$$R(t) = \prod_{k=1}^{n} \exp(-\lambda_k t) = \exp\left(-t\sum_{k=1}^{n}\lambda_k\right) \tag{5.7}$$

Therefore the system reliability can be expressed as:

$$R(t) = \exp(-\lambda t)$$

where $\lambda = \sum_{k=1}^{n} \lambda_k$ is the system failure rate. Therefore the system failure rate is equal to the sum of the failure rates of individual components of biogas power plant. As the failure rate of the biogas component is constant, then the failure rate of the system will also be constant.

Mean time to failure of the system is:

$$\text{MTTF} = \int_0^\infty R(t)dt = \frac{1}{\sum_{k=1}^{n}\lambda_k} = \frac{1}{\sum_{k=1}^{n}\frac{1}{\text{MTTF}_k}} \tag{5.8}$$

From Eq. (5.8), it can be seen that system MTTF and component MTTF are related by:

$$\frac{1}{\text{MTTF}} = \frac{1}{\text{MTTF}_1} + \frac{1}{\text{MTTF}_2} + \ldots\ldots\ldots + \frac{1}{\text{MTTF}_n} \tag{5.9}$$

Q. 5.1: Consider a four-component gas holder, slurry inlet, gas outlet, and slurry outlet of biogas energy system of which the components are exponentially distributed with the constant failure rate of 0.075×10^{-3}, 0.20×10^{-3}, 0.80×10^{-3}, and 0.925×10^{-3} per hour, respectively. Evaluate the failure rate and MTTF of the system. Also calculate the reliability of each component of biogas power plant and the whole system at 500 hours.

Solution:

$$\lambda_1 = 0.075 \times 10^{-3} \text{ per hour}, \quad \lambda_2 = 0.20 \times 10^{-3} \text{ per hour}$$
$$\lambda_3 = 0.80 \times 10^{-3} \text{ per hour} \quad \lambda_4 = 0.925 \times 10^{-3} \text{ per hour}$$

Reliability of each component:

$$R_1(t) = \exp\left(-0.075 \times 10^{-3} \times 500\right) = 0.9632$$
$$R_2(t) = \exp\left(-0.20 \times 10^{-3} \times 500\right) = 0.9048$$
$$R_3(t) = \exp\left(-0.80 \times 10^{-3} \times 500\right) = 0.6703$$
$$R_4(t) = \exp\left(-0.925 \times 10^{-3} \times 500\right) = 0.6297$$

System failure rate:

$$\lambda = \lambda_1 + \lambda_2 + \lambda_3 + \lambda_4 = (0.075 + 0.20 + 0.80 + 0.925) \times 10^{-3}$$
$$= 2 \times 10^{-3} \text{ per hour}$$

System reliability:

$$R(t) = \exp(-\lambda t) = \exp\left(2 \times 10^{-3} \times 500\right) = 0.3679$$

$$\text{or } R(t) = R_1(t)R_2(t)R_3(t)R_4(t) = 0.9632 \times 0.9048 \times 0.6703 \times 0.6297 = 0.3679$$

MTTF of the system:

$$\text{MTTF} = \frac{1}{\lambda} = \frac{1}{2 \times 10^{-3}} = 500 \text{ hours}$$

So that if biogas components are connected in series, then system reliability is 36.79%, because mean time to failure is 500 hours out of 8760 hours in a year.

Q. 5.2: Consider a five-component biogas energy system of which the components are exponentially distributed with an identical constant failure rate. If $R(200) = 0.96$ is the system reliability, find the individual component MTTF.

Solution:

$$R(200) = \exp(-\lambda t) = \exp(-\lambda \times 200) = 0.96$$

$$\text{System failure rate,} \quad \lambda = 2.041 \times 10^{-4} \text{ per hour}$$

$$\text{Components failure rate,} \quad \lambda_k = 4.082 \times 10^{-5} \text{ per hour}$$

$$\text{MTTF of component,} \quad \text{MTTF} = 1/\lambda_k = 1/4.082 \times 10^{-5} = 24496.6 \text{ hours}$$

$$\text{MTTF of system,} \quad \text{MTTF} = 1/\lambda = 1/2.041 \times 10^{-4} = 4899.3 \text{ hours}$$

5.2.2 Effect of component reliability in series system

In series configuration, the biogas component with lower reliability has the greatest impact on the reliability of overall biogas energy system. As the component with least reliability will fail first due to which the whole system will collapse and this weakest component rules the system reliability. Thus the reliability of the series system will always be less than the reliability of least reliable component.

Consider a biogas energy system of three components in the series having reliabilities $R_1 = 60\%$, $R_2 = 70\%$, and $R_3 = 80\%$. The first row of Table 5.3 shows the reliability of each component and the corresponding overall system reliability. In the second row, the reliability of the first component is increased by 10%, while the reliabilities of the other two components remain constant. Similarly, in the third row, the reliability of the second component is increased by 10%, and in the fourth row, the reliability of the third component is increased by 10%. It is observed that the reliability of each component affects the overall system reliability. It can be seen that the reliability of

TABLE 5.3 Effect of each component's reliability on overall system reliability.

Component 1	Component 2	Component 3	Biogas system
Biogas digester	Mixing tank	Biogas generator	
0.6	0.7	0.8	0.3360
0.7	0.7	0.8	0.3920
0.6	0.8	0.8	0.3840
0.6	0.7	0.9	0.3780

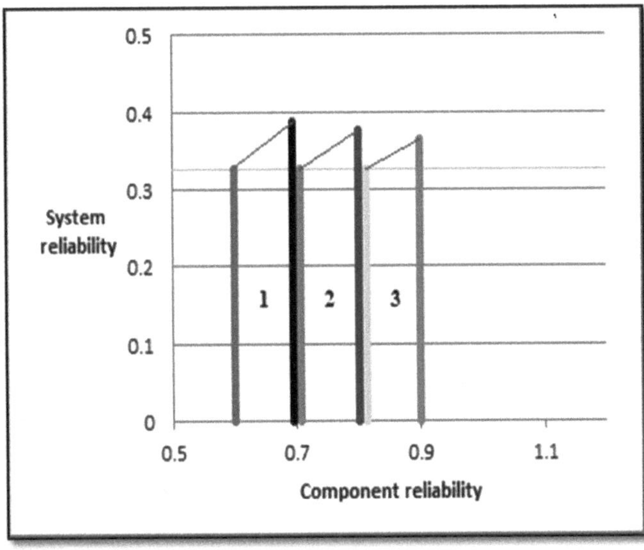

FIGURE 5.6 Rate of change of system reliability of biogas power plant.

the system is highest, when the reliability of the first component is increased (component with lower reliability initially). Fig. 5.6 shows the rate of change of system reliability of biogas power plant.

Table 5.3 shows the effect of each component's reliability on overall system reliability.

The rate of change of system reliability of biogas energy system with respect to change in reliability of each component is shown in Fig. 5.6. It can be seen that the slope of the first component is steeper than the other two. This indicates that the first component has higher reliability importance as an increase in the reliability of the first component will cause the highest increase in system reliability.

5.2.3 Effect of number of components in series system

The number of components in a series system also affects the system reliability considerably. As the number of components connected reliability-wise in series increases, the system reliability decreases. Consider two systems that consist of only one component having reliabilities of 94% and 92%. Thus the system reliabilities will be 94% and 92%, respectively. Table 5.4 shows the system reliability, if components are successively increasing in the series having the same reliability.

Fig. 5.7 illustrates the impact that the quantity of components has on the reliability of the system, especially when the component having low reliability. Therefore the reliability of the component should be high to achieve high system reliability, particularly for the systems with numerous components organized reliability-wise in series.

TABLE 5.4 Effect of number of components on overall biogas system reliability.

Number of components of biogas power plant	System 1 reliability ($R = 0.94$)	System 2 reliability ($R = 0.92$)
1	0.9400	0.9200
2	0.8836	0.8464
4	0.7807	0.7164
6	0.6899	0.6064
10	0.5386	0.4344
15	0.3953	0.2863

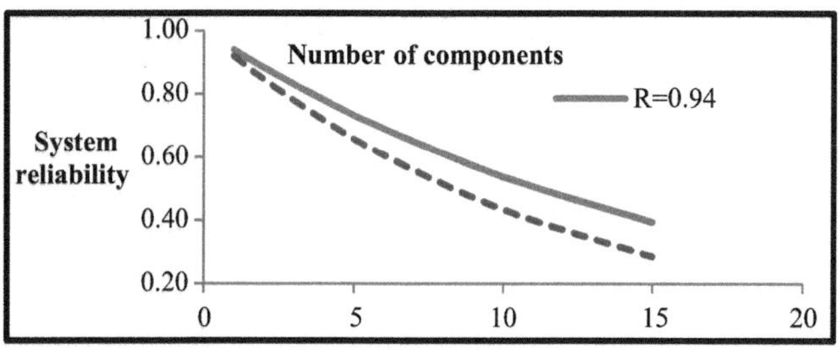

FIGURE 5.7 Change in system reliability with number of components.

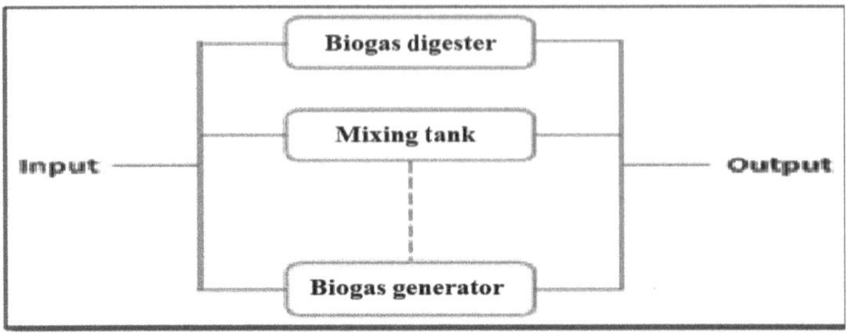

FIGURE 5.8 Reliability block diagram for a parallel system.

5.2.4 Components in parallel

A number of components connected together to form a biogas system are said to be in parallel configuration if the working on any one of the "n" components causes the system to function properly; that is, failure of all components of a biogas power system may lead to failure of the whole biogas energy system. Fig. 5.8 shows the RBD of a system having "n" components connected reliability-wise in parallel. The RBD of the parallel configuration is having "n" components, which are connected across each other, and there are "n" parallel paths between input and output terminals.

System reliability can be expressed in terms of successful events E_n and failure events $\bar{E}_n$ of the nth components as $R(t) = 1 -$ (probability of all "n" components fails, i.e., probability that at least one component does not fail)

$$R(t) = 1 - P(E_1 \cup E_2 \cup \ldots\ldots E_n) = 1 - P(\overline{E}_1 \cap \overline{E}_2 \cap \ldots\ldots\ldots \overline{E}_n) \tag{5.10}$$

Assume that all components are independent

$$\begin{aligned} R(t) &= 1 - P(\overline{E}_1)P(\overline{E}_2)\ldots\ldots P(\overline{E}_n) \\ &= 1 - (1 - P_1)(1 - P_2)\ldots\ldots\ldots(1 - P_n) \end{aligned} \tag{5.11}$$

where $P_1 = P(E_1)$, and reliability in general as time function is expressed as:

$$R(t) = 1 - \prod_{k=1}^{n}(1 - \mathrm{P}_k(t)$$

If all the "n" components in a system have the same reliability, then

$$R(t) = 1 - (1 - P(t))^n$$

5.2.4.1 *Failure rate of parallel system*

Suppose that the time to failure of components is exponentially distributed with a constant failure rate λ_k, then

$$R(t) = 1 - \prod_{k=1}^{n}(1 - \exp(-\lambda_k t))) \tag{5.12}$$

The system reliability for "n" parallel components with an identical failure rate λ is given by:

$$R(t) = \mathbf{1} - (\mathbf{1} - \mathbf{exp}(-\boldsymbol{\lambda} t))^n \tag{5.13}$$

The mean time to failure of the system is:

$$\text{MTTF} = \int_0^\infty R(t)dt = \int_0^\infty (1 - (1 - \exp(-\lambda t))^n)dt \tag{5.14}$$

Let $(1 - \exp(-\lambda t)) = x$; from Eq. (5.16), MTTF can be expressed as:

$$\begin{aligned} \text{MTTF} &= \frac{1}{\lambda}\int_0^1 \left\{\frac{1 - x^n}{1 - x}\right\}dx \\ &= \frac{1}{\lambda}\int_0^1 \left(1 + x + x^2 + \ldots.. + x^{n-1}\right)dx \\ &= \frac{1}{\lambda}\left(1 + \frac{1}{2} + \frac{1}{3} + \ldots\ldots\ldots + \frac{1}{n}\right) \\ &= \frac{1}{\lambda}\sum\nolimits_{k=1}^{n} \frac{1}{k} \end{aligned} \tag{5.15}$$

The mean time to failure having an unequal failure rate for all the components in parallel can be calculated as:

$$\begin{aligned} \boldsymbol{MTTF} &= \int_0^\infty \boldsymbol{R(t)dt} = \int_0^\infty \left(\mathbf{1} - \prod\nolimits_{k=1}^{n}(\mathbf{1} - \mathbf{exp}(-\boldsymbol{\lambda}_k \mathbf{t}))\right)\boldsymbol{dt} \\ &= \sum\nolimits_k \frac{1}{\lambda_k} - \sum\nolimits_{k<l} \frac{1}{\lambda_k + \lambda_l} + \sum\nolimits_{k<l<m} \frac{1}{\lambda_k + \lambda_l + \lambda_m} - \ldots\ldots\ldots \\ &\quad \ldots\ldots(-1)^{n-1}\frac{1}{\lambda_1 + \lambda_2 + \ldots. + \lambda_n} \end{aligned} \tag{5.16}$$

$$\text{MTTF} = \frac{1}{\lambda_1} + \frac{1}{\lambda_2} - \frac{1}{\lambda_1 + \lambda_2} \tag{5.17}$$

If both components are having the same failure rate, then

$$\text{MTTF} = \frac{2}{\lambda} - \frac{1}{2\lambda} = \frac{3}{2\lambda}. \tag{5.18}$$

A repeated formula for calculating MTTF can be obtained as:

$$\begin{aligned} \text{MTTF}_n &= \frac{1}{\lambda}\sum\nolimits_{k=1}^{n}\frac{1}{k} = \frac{1}{\lambda}\sum\nolimits_{k=1}^{n-1}\frac{1}{k} + \frac{1}{n\lambda} \\ \text{MTTF}_1 &= \frac{1}{\lambda} \end{aligned} \tag{5.19}$$

Therefore

$$\mathrm{MTTF}_n = \mathrm{MTTF}_{n-1} + \frac{1}{n\lambda} \quad \text{for } n \geq 2 \tag{5.20}$$

Now assume that the system reliability is given as:

$$R(t) = 2e^{-\lambda t} - e^{-2\lambda t}$$

Then, the hazard rate of two independent and identical component systems can be evaluated as:

$$\begin{aligned} h(t) &= \frac{f(t)}{R(t)} = -\frac{R'(t)}{R(t)} \\ R'(t) &= -2\lambda e^{-\lambda t} + 2\lambda e^{-2\lambda t} \\ h(t) &= 2\lambda\left[\frac{1-e^{-\lambda t}}{2-e^{-\lambda t}}\right] \\ &= 2\lambda\left[\frac{(1-P(t))}{(2-P(t))}\right] \end{aligned} \tag{5.21}$$

Eq. (5.22) shows that the failure rate of the system is time-dependent even when the failure rates of the components are constant.

Q. 5.3: Consider a four-component biogas energy system of which the components are exponentially distributed with constant failure rates of 0.075×10^{-3}, 0.20×10^{-3}, 0.80×10^{-3}, and 0.925×10^{-3} per hour. Evaluate the failure rate and MTTF of the system. Also calculate the reliability of each component and the whole system at 500 hours.

Solution:

$$\begin{aligned} &\lambda_1 = 0.075 \times 10^{-3} \text{ per hour}, \quad \lambda_2 = 0.20 \times 10^{-3} \text{per hour} \\ &\lambda_3 = 0.80 \times 10^{-3} \text{ per hour} \quad \lambda_4 = 0.925 \times 10^{-3} \text{ per hour} \end{aligned}$$

Reliability of each component:

$$\begin{aligned} R_1(t) &= \exp\left(-0.075 \times 10^{-3} \times 500\right) = 0.9632 \\ R_2(t) &= \exp\left(-0.20 \times 10^{-3} \times 500\right) = 0.9048 \\ R_3(t) &= \exp\left(-0.80 \times 10^{-3} \times 500\right) = 0.6703 \\ R_4(t) &= \exp\left(-0.925 \times 10^{-3} \times 500\right) = 0.6297 \end{aligned}$$

System reliability:

$$R(t) = 1 - (1 - R_1(t))(1 - R_2(t))(1 - R_3(t))(1 - R_4(t))$$
$$R(t) = 1 - (1 - 0.9632)(1 - 0.9048)(1 - 0.6703)(1 - 0.6297) = 0.9996$$

System failure rate:

$$MTTF = \frac{1}{\lambda 1} + \frac{1}{\lambda 2} + \frac{1}{\lambda 3} + \frac{1}{\lambda 4} - \frac{1}{\lambda 1 + \lambda 2} - \frac{1}{\lambda 2 + \lambda 3} - \frac{1}{\lambda 3 + \lambda 1}$$
$$- \frac{1}{\lambda 1 + \lambda 4} - \frac{1}{\lambda 2 + \lambda 4} - \frac{1}{\lambda 3 + \lambda 4} + \frac{1}{\lambda 1 + \lambda 2 + \lambda 3}$$

$$+ \frac{1}{\lambda 1 + \lambda 2 + \lambda 4} + \frac{1}{\lambda 1 + \lambda 3 + \lambda 4} + \frac{1}{\lambda 2 + \lambda 3 + \lambda 4}$$

$$- \frac{1}{\lambda 1 + \lambda 2 + \lambda 3 + \lambda 4} = 14,753.6 \text{ hoursSystem failure}$$

$$\text{rate } \lambda = \frac{1}{MTTF} = \frac{1}{14,753.6} = 6.778 \times 10^{-5} \text{ per hour}$$

Q. 5.4: Consider a five-component biogas energy system of which the components are exponentially distributed with an identical constant failure rate. If $R(200) = 0.96$ is the system reliability, find the individual component MTTF.

Solution:

$$R(200) = 1 - (1 - \exp(-\lambda t))^n = 1 - (1 - \exp(-\lambda \times 200))^5 = 0.96$$
$$\text{Components failure rate,} \quad \lambda = 0.0037 \text{ per hour}$$
$$\text{MTTF of component,} \quad MTTF = 1/\lambda = \frac{1}{0.0037} = 268.43 \text{ hours}$$
$$\text{MTTF of system,} \quad MTTF = \mathbf{1/\lambda\sum\_(k=1)\hat{n}} \times 1/\boldsymbol{k} = 1/\lambda$$

$$\left(1 + \mathbf{1/2} + \mathbf{1/3}\right) = 11/6\lambda = 495.5 \text{ hoursSystem failure rate,}$$

$$\lambda_S = \frac{1}{MTTF} = 0.002 \text{ per hour}$$

5.2.5 Effect of component reliability of biogas power plant in parallel system

When series configuration of biogas system is analyzed, it is found that the minimum reliability component has the greatest impact on the reliability of the system. However, the component with the maximum reliability in a parallel configuration of the biogas energy system has the greatest impact on the system reliability, since the most reliable component is the one that will fail last. This is an essential property of the parallel configuration, explicitly in the design and enhancement of the system.

TABLE 5.5 Effect of each component's reliability on overall system reliability.

Component 1	Component 2	Component 3	Biogas system
Biogas digester	Mixing tank	Biogas generator	
0.5	0.6	0.7	0.94
0.6	0.6	0.7	0.952
0.5	0.7	0.7	0.955
0.5	0.6	0.8	0.960

Consider a system of three components of biogas power plant in parallel having reliabilities $R_1 = 50\%$, $R_2 = 60\%$, and $R_3 = 70\%$. In Table 5.5, the effect on each component's reliability on the overall system reliability is analyzed. The first row of the table shows the reliability of each component and the corresponding overall system reliability. In the second row, the reliability of the first component is increased by 10%, while the reliabilities of the other two components remain constant. Similarly, in the third row, the reliability of the second component is increased by 10%, and in the fourth row, the reliability of the third component is increased by 10%. It is observed that the reliability of each component affects the overall system reliability. It can be seen that the reliability of the system is highest, when the reliability of components, which is more reliable, is increased (component with the highest reliability initially). The rate of change of system reliability with respect to change in reliability of each component is shown in Fig. 5.9. It can be seen that the slope of the third component is steeper than the other two. This indicates that the third component has higher reliability importance as an increase in the reliability of this component will cause the highest increase in system reliability.

5.3 Biogas plant component failure and failure mode

5.3.1 Failure distribution model of biogas energy system

The failure rate is the frequency with which a biogas system or biogas component fails, expressed in failures per unit of time. It is often denoted by the Greek letter λ (lambda) and is highly used in reliability engineering. The failure rate of a biogas system usually depends on time, with the rate varying over the life cycle of the system. For example, in biogas power plant, failure rate in its fifth year of service may be many times greater than its failure rate during its first year of service. One does not expect to replace an exhaust pipe, overhaul the digester, or have major transmission problems in a new system.

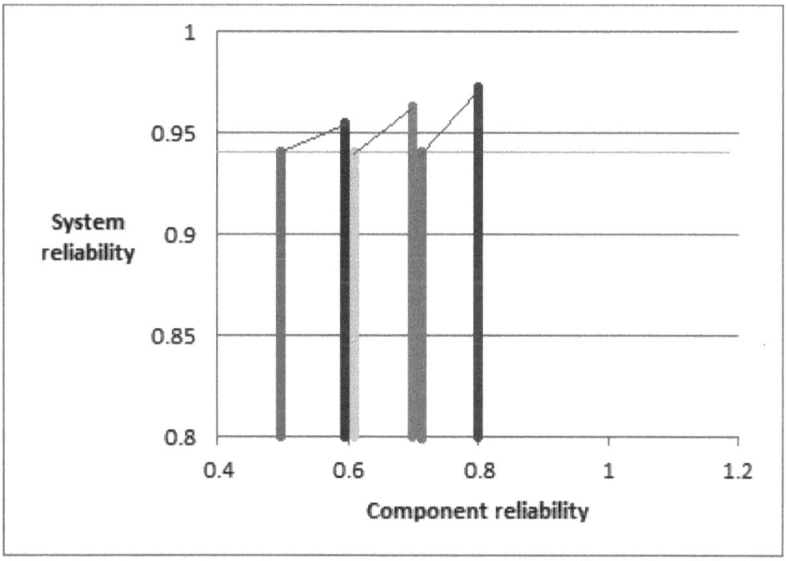

FIGURE 5.9 Rate of change of system reliability.

A mistake that is often made when calculating reliability of biogas power plant metrics is trying to use the failure rate function instead of the probability of failure function. These two functions, along with the PDF and the reliability function, make up the four functions that are commonly used to describe reliability data of biogas power plant. The failure rate is the frequency with which a biogas system or component fails, expressed, for example, in failures per hour. In practice, the closely related mean time between failures is more commonly expressed and used for high quality biogas components or systems. The failure rate is usually time-dependent, and an intuitive corollary is that the rate changes over time versus the expected life cycle of a system. For example, as a digester older, the failure rate in its fifth year of service may be many times greater than its failure rate during its first year of service. One simply does not expect to replace an exhaust pipe, overhaul the plant, or have major transmission problems in a biogas energy system.

5.3.1.1 The Jelinski–Moranda model

The Jelinski–Moranda model is a time between failure models. This model makes the following assumptions about the fault detection and correction process in biogas power system.

- The rate of fault detection is proportional to the current fault content of the biogas energy system.

- All failures are equally likely to occur and are independent of each other in biogas power plant.
- Each failure is of the same order of severity as any other failure of any component of biogas power plant.
- The failure rate remains constant over the interval between failure occurrences, when biogas energy system is in operation.
- During the test, the software is operated in a similar manner as the expected operational usage.
- The faults are corrected instantaneously without introduction of new faults into the biogas energy system.

5.3.1.2 Frequency of failure

The number of failure events of biogas energy system that occur is divided by the total elapsed calendar time during which those events occur or by the total number of consumer demands, as applicable. The frequency of failure of biogas energy system is directly related to the component failure and component failure mode of biogas energy system.

5.3.1.3 Failure density

This is the ratio of the number of failures during a given unit interval of time to the total number of components of the primary stage of the test.

Q.: In the performance testing of 100 small and big components of a biogas energy system, the number of failures during each time interval of 20 hours is shown in Table 5.6. Estimate the mean time to failure for these specimens.

TABLE 5.6 Time interval and number of failures.

Time interval hours	Number of failures during the interval
$T<1000$	0
$1000<T<1010$	30
$1010<T<1030$	35
$1030<T<1050$	15
$1050<T<1070$	15
$1070<T<1100$	5

Solution:

$$\text{Mean Time to Failure} = \frac{1}{100[30(1010) + 35(1030) + 15(1050) + 15(1070) + 5(1100)]}$$
$$= \frac{1}{100[30,300 + 36,050 + 15,750 + 16,050 + 5500]}$$
$$= \frac{1}{100[103,650]}$$
$$= 1036 \text{ hours}$$

Q.: In the performance testing of 200 components of a biogas power plant, the number of failures during each time interval of 20 hours is shown in Table 5.7. Estimate the mean time to failure for these specimens.

Solution:

$$\text{Mean Time to Failure} = \frac{1}{100[60(1010) + 70(1030) + 30(1050) + 30(1070) + 10(1100)]}$$
$$= \frac{1}{200[60,600 + 72,100 + 31,500 + 32,100 + 11,000]}$$
$$= \frac{1}{200[207,300]}$$
$$= 1036 \text{ hours}$$

5.3.1.4 Biogas energy repairable system

In order to find out the assessment of repairable system of biogas energy system, we already discussed above the function of availability, where availability is the probability that the particular biogas energy system, component, and element is operating and working satisfactorily at time "*t*." On the other hand, reliability is the probability that the given system has operated

TABLE 5.7 Time interval and number of failures.

Time interval hours	Number of failures during the interval
$T < 1000$	0
$1000 < T < 1010$	60
$1010 < T < 1030$	70
$1030 < T < 1050$	30
$1050 < T < 1070$	30
$1070 < T < 1100$	10

satisfactorily over the time interval of 0 to t. Repairable biogas energy system always increases the system life span and the performance of the given system. To appropriately manage repairable frameworks, we have to first see how components in these biogas energy systems are reestablished. By and large, upkeep is characterized as any activity that reestablishes fizzled units to an operational condition or holds a nonfizzled unit in an operational state. For repairable frameworks, support assumes a fundamental part in the life of a framework. It influences the framework's general dependability, accessibility, downtime, the cost of the task, and so forth. A repairable system is a system that can be restored to an operating condition of a failure. Questions of interest in repairable systems' analysis include:

- How many failures in biogas energy plant will occur over a fixed-time interval?
- What is the probability of a failure of biogas power plant in the next time interval?
- What is the availability of the biogas energy system?
- How many spare parts should be purchased for biogas energy system?
- What is the cost of maintaining the biogas energy system?
- What is the optimum overhaul time in biogas energy system?

To find out instantaneous repair rate of any biogas energy system is considered in which 1500 technical devices are undergoing repairs simultaneously. Let the number of equipment repaired during every interval of 1 hour. Let all devices be in the repair state at the end of $T = 75$ hours. For example, at the end of 2 hours, a total number of 23 units were repaired. Hence, the probability that a device will get repaired within 2 hours is given by:

$$\text{Maintainability}(\mathbf{t} = \mathbf{2}) = \frac{23}{1500} = 0.015334$$

Similarly, at the end of "t" hours, the number of units that got repaired was 1270. Hence, the maintainability for "t" hours is given by:

$$M(t) = \frac{1270}{1500} = 0.8467$$

During the first unit interval, eight devices got repaired, and the average population on the repair bench was ½ (1500 + 1492) = 1496.

Hence, the repair rate during the first unit time interval is:

$$r(t) = r(1) = \frac{8}{\frac{(1500 + 1492)}{2}} = \frac{8}{1496} = 0.00534$$

We can now generalize the calculations of $M_a(t)$ and $r(t)$. Let there be a total number of undergoing repairs under identical conditions.

Let the number of units repaired during each unit interval of time be $n1$, $n2, \ldots, nT$. Then

$$M_a(t) = \frac{n1 + n2 + \ldots\ldots + nT}{N} = \frac{1}{N}\left(\sum^{t} 1 \times ni\right)$$

$$r(t) = \frac{(n1 + n2 + \ldots\ldots + nt) - (n1 + n2 + \ldots\ldots + n(t-1))}{\frac{1}{2}\left[\{N - (n1 + n2 + \ldots\ldots + nt)\} + \{N - (n1 + n2 + \ldots\ldots + n(t-1))\}\right]}$$

Dividing the numerator and the denominator by N, we get:

$$r(t) = \frac{(M_a(t) - M_a(t-1))}{\left[\left(1 - \frac{1}{2[M_a(t) - M_a(t-1)]}\right]}$$

If the time interval, instead of being unity, is t, then the repair rate is:

$$r(t) = \frac{\frac{1}{2(M_a(t) - M_a(t-1))}}{\left[\left(1 - \frac{1}{2[M_a(t) - M_a(t-1)]}\right]}$$

In the limit t---0

$$r(t) = \frac{1}{(1 - M_a(t))} \cdot \frac{d}{dt[M_a(t)]}$$

This is also known as the instantaneous repair rate of biogas energy system.

5.3.1.5 Mean time to repair of biogas energy system

MTTR is a basic measure of the maintainability of repairable items. It represents the average time required to repair a failed component or device of a biogas energy system. Expressed mathematically, it is the total corrective maintenance time for failures divided by the total number of corrective maintenance actions for failures during a given period of time, when biogas energy system is in operation. It generally does not include lead time for parts not readily available or other administrative or logistic downtime. The equation for MTTR will be obtained in two ways: MTTR is the average time required to troubleshoot and repair failed equipment and return it to normal operating conditions of biogas energy system. It is a basic technical measure of the maintainability of equipment and repairable parts. Maintenance time is defined as the time between the start of the operation of biogas energy system and the moment the system is returned to production. This includes notification time, diagnostic time, fix time, wait time (cool down), reassembly, alignment, calibration, test time, back to production of electricity, and so on. It generally does not take into account lead time for parts. MTTR ultimately reflects how well an organization can respond to a problem and repair it.

Expressed mathematically, the MTTR calculation is the total maintenance time divided by the total number of maintenance actions over a specific period

$$\text{MTTR} = \frac{\text{(Total Maintenance Time)}}{\text{(No. of Repairs)}}$$

Let there be N components of biogas energy system undergoing repairs simultaneously under identical conditions. Let t be the time intervals when we note down the number of units that have been repaired. Let $n1$ be the number of units repaired during the first t interval and so on nk be the number of units repaired during the kth t interval. Let the last t interval when all the N units have been repaired be the lth t interval

$$n1 + n2 + \ldots\ldots\ldots\ldots nk + \ldots\ldots\ldots nl = \sum (i=1)^l \times ni = N$$

Let $kt = t$. Then, the number of units repaired within time t is:

$$n1 + n2 + \ldots\ldots\ldots\ldots nk = \sum (i=1)^k \times ni$$

Hence, the maintainability $M_a(t = k\Delta t) = \frac{n1 + n2 + \ldots\ldots\ldots nk}{N} = \frac{1}{N}\sum (i=1)^k \times n\_i$

The number of items repaired during the kth t interval was n_k. Its value is obviously

$$nk = N[M_a(t = k\Delta t) - M(t = k\Delta t - \Delta t)]$$

$$\text{MTTR} = \frac{1}{N}[n1\ \Delta t + n2\ 2\Delta t + \ldots\ldots nk\ k\Delta t + \ldots\ldots + nl\ l\Delta t]$$

$$= \frac{1}{N}\sum (i=1)^l \times n\_k(k\Delta t)$$

5.3.1.6 *Practical model of biogas repairable system*

When we develop any complex system such as biogas energy system and after the failure of those systems, there are two possibilities, they are repaired and may be replaced, but some complex systems, such as automobiles, data communication, computer peripherals, traction systems, and so on, are repaired and are not replaced when they fail. The reliability of biogas energy repairable system is different from a reliability of actual systems. The reliability of repairable system depends on the number of failures during the warranty period of each equipment of biogas energy system, which is mentioned by the shopkeeper, the total lifetime of the product, and the replacement cost of the product of biogas energy system. The most powerful theory to determine the reliability of repairable system is a power law model. The power law model is easy to understand, and it is used to find out the frequency of failures of biogas energy system. In statistics, a power law is a functional relationship between two quantities, where a relative change in one quantity results in a proportional relative change in the other quantity, independent of the initial size of those quantities: one quantity varies as a power of another. If power law is applied in reliability assessment, then performance of individual component of biogas energy system is directly related to the performance of the overall biogas energy system.

5.3.2 Confidence level of biogas repairable system

In the case of biogas repairable system, it is necessary to find out at what level fault is minimized, and the system is repaired at the desired level, where efficiency of system is increased. The assessment of a repairable system was found out by the concept of confidence level, the confidence level is the probability of a biogas energy system at a standard parameter, and this parameter lies between lower and upper limits. The confidence level of a repairable system also depends on the desired output and reference output of electricity in biogas energy system. The upper limits of the interval are known as confidence limit of the overall biogas energy system. The confidence level of the biogas energy system is done in two manners:

1. confidence level of individual component of biogas energy system; and
2. confidence level of overall biogas energy system.

There are three possible cases developed to a confidence level of individual and overall biogas energy systems.

1. The confidence level is greater than the actual reliability of biogas energy system.
2. The confidence level is equal to the actual reliability of biogas energy system.
3. The confidence level is less than the actual reliability of biogas energy system.

Fig. 5.10 shows the confidence level of reliability. If the confidence level of biogas energy system is greater than the actual reliability, then faulty system of biogas energy system is repaired more than the desired level and it increases the lifetime of the overall biogas energy system. If the confidence level of biogas energy system is less than the actual reliability, then it is necessary to replace the individual component or overall system with a new one. The ideal condition is found out when the confidence level is equal to the actual reliability of biogas energy system.

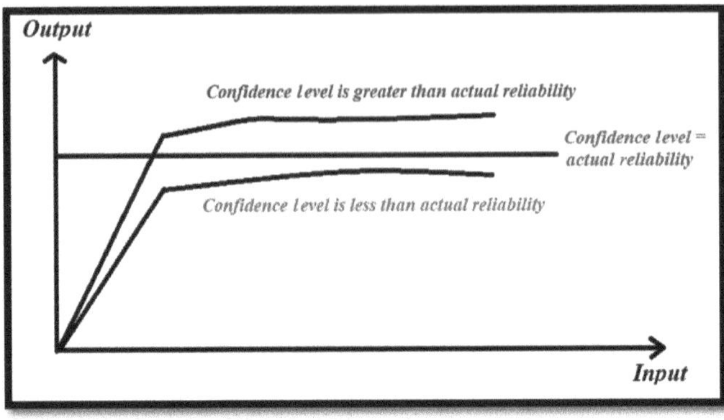

FIGURE 5.10 Confidence of reliability of biogas energy system.

The probability of biogas energy repairable system is a ratio of the number of times successful maintenance (M_S) is to be done to repair the system to the total number of times maintenance (M_T) is to be done

$$\text{Probability of Repairable System}(R_P) = \frac{M_S(t)}{M_T(t)}$$

Rewriting the numerator yields:

$$R_P = \frac{M_T(t) - M_F(t)}{M_T(t)} = 1 - \frac{M_F(t)}{M_T(t)} = 1 - Q(t)$$

where $M_F(t)$ = number of missions of "t" duration of each unsuccessful maintenance$Q_{(t)}$ = probability of unsuccessful maintenance.

This applies when one unit undertakes $M_T(t)$ missions out of which $M_s(t)$ mission succeed, and when the unit fails, it is restored to the same condition it was started on the first mission. If all maintenance of repairable system is failing, then $M_s(t) = 0$.

If $M_s(t)$ missions are completed successfully, the definition of probability of repairable system as $\frac{M_S(t)}{M_T(t)}$ is again conditional to the fact that all mission objectives and designed for performance objectives are attained. Their design adequacy is 100%. A better way of seeing the complete picture of repairable system or system effectiveness is defined as:

$$\text{Effectiveness of System} = O \times R \times D = \frac{M_{AC}}{M_T}$$

where O is the probability that the actual system is either available at the beginning of the project, R is the probability that has completed the successful maintenance without failure, and D is the probability of after the maintenance system is performed successfully and this is called design adequacy.

Q.: In a biogas energy system, 150 identical systems are put into operation, and each one's performance is monitored for a predetermined period of 20 hours. The number of units that fail to function satisfactorily during these 10 hours is counted to be 10.

1. What is the estimate of the reliability of biogas energy system for a target of 20 hours?
2. If one such unit undertook 50 such identical targets, each one of 20 hours, it failed in 10 of these targets, and it was restored to its original state after maintenance. What is the reliability estimate of such units?

Solution:

$$R_P = \frac{M_T(t) - M_F(t)}{M_T(t)} = 1 - \frac{M_F(t)}{M_T(t)} = 1 - \frac{10}{150} = 1 - 0.067 = 0.933 = 93.3\%$$

$$R_P = \frac{150 - 10}{150} = 0.933 = 93.3\%$$

Q.: In a biogas energy system, 200 identical systems are put into operation, and each one's performance is monitored for a predetermined period of 10 hours. The number of units that fail to function satisfactorily during these 10 hours is counted to be 5.

1. What is the estimate of the reliability of biogas energy system for a target of 10 hours?
2. If one such unit undertook 50 such identical targets, each one of 10 hours, it failed in five of this target, and it was restored to its original state after maintenance. What is the reliability estimate of such units?

Solution:

$$R_P = \frac{M_T(t) - M_F(t)}{M_T(t)} = 1 - \frac{M_F(t)}{M_T(t)} = 1 - \frac{5}{200} = 1 - 0.025 = 0.975 = 97.5\%$$

$$R_P = \frac{200 - 5}{200} = 0.975 = 97.5\%$$

Q.: To repair the overall biogas energy system, 100 identical units of energy system were tested for 40 hours. One unit fails before completing 10 hours of operation, three units fail just before completing 15 hours of operation, and two units fail just before completing 40 hours of operation.

1. What is the reliability estimate of these units for a target of 40 hours?
2. What is the reliability estimate for the end of each failure period, including previous failures, when the failed units are replaced with good ones?

3. What is the reliability estimate for each failure period if the failed units are not replaced?

Solution 1:

$$R_P = 1 - \frac{M_F(t)}{M_T(t)} = 1 - \frac{1+3+2}{100} = 1 - \frac{6}{100} = 1 - 0.06 = 0.94 = 94\%$$

Solution 2: First, failure from 0 to 10 hours:

$$R_P = 1 - \frac{M_F(t)}{M_T(t)} = 1 - \frac{1}{100} = 1 - 0.01 = 0.99 = 99\%$$

Second, 0–15 hours:

$$R_P = 1 - \frac{M_F(t)}{M_T(t)} = 1 - \frac{1+3}{100} = 1 - 0.04 = 0.96 = 96\%$$

Third, 0–40 hours:

$$R_P = 1 - \frac{M_F(t)}{M_T(t)} = 1 - \frac{1+3+2}{100} = 1 - 0.06 = 0.94 = 94\%$$

$$R_P(0 - 10 \text{ hours}) = 1 - \frac{M_F(t)}{M_T(t)} = 1 - \frac{1}{100} = 1 - 0.01 = 0.99 = 99\%$$

$$R_P(10 - 15 \text{ hours}) = 1 - \frac{M_F(t)}{M_T(t)} = 1 - \frac{3}{100-1} = 1 - 0.030 = 0.97 = 97\%$$

$$R_P(15 - 40 \text{ hours}) = 1 - \frac{M_F(t)}{M_T(t)} = 1 - \frac{2}{100-1-3} = 1 - 0.020 = 0.98 = 98\%$$

Q.: To repair the overall biogas energy system, 100 identical units of energy system were tested for 40 hours. One unit fails before completing 10 hours of operation, three units fail just before completing 15 hours of operation, and two units fail just before completing 40 hours of operation.

1. What is the reliability estimate of these units for a target of 40 hours?
2. What is the reliability estimate for the end of each failure period, including previous failures, when the failed units are replaced with good ones?
3. What is the reliability estimate for each failure period if the failed units are not replaced?

Solution 1:

$$R_P = 1 - \frac{M_F(t)}{M_T(t)} = 1 - \frac{1+3+2}{100} = 1 - \frac{6}{100} = 1 - 0.06 = 0.94 = 94\%$$

Solution 2: First, failure from 0 to 10 hours:

$$R_P = 1 - \frac{M_F(t)}{M_T(t)} = 1 - \frac{1}{100} = 1 - 0.01 = 0.99 = 99\%$$

Second, 0–15 hours:

$$R_P = 1 - \frac{M_F(t)}{M_T(t)} = 1 - \frac{1+3}{100} = 1 - 0.04 = 0.96 = 96\%$$

Third, 0–40 hours:

$$R_P = 1 - \frac{M_F(t)}{M_T(t)} = 1 - \frac{1+3+2}{100} = 1 - 0.06 = 0.94 = 94\%$$

$$R_P(0 - 10\text{ hours}) = 1 - \frac{M_F(t)}{M_T(t)} = 1 - \frac{1}{100} = 1 - 0.01 = 0.99 = 99\%$$

$$R_P(10 - 15\text{ hours}) = 1 - \frac{M_F(t)}{M_T(t)} = 1 - \frac{3}{100-1} = 1 - 0.030 = 0.97 = 97\%$$

$$R_P(15 - 40\text{ hours}) = 1 - \frac{M_F(t)}{M_T(t)} = 1 - \frac{2}{100-1-3} = 1 - 0.020 = 0.98 = 98\%$$

Q.: A biogas energy system's operational readiness is 85%, target reliability is 90%, and design adequacy is 96%.

1. What is its system effectiveness?
2. If there were $M_T = 150$ such systems in various states of existence, how many will be available to start their target successfully?
3. How many such systems will complete their mission successfully?
4. How many systems worth of mission objectives would be achieved?

Solution:
System effectiveness $= O \times R \times D = 0.85 \times 0.90 \times 0.96 = 0.735 = 73.5\%$
The number of systems that will start a target is obtained from:
$150 \times 0.85 = 128$ systems
The number of systems that will complete their mission successfully is obtained from:
$150 \times 0.85 \times 0.90 = 115$ systems
The number of systems that effectively achieve all designed for mission objectives is:
$150 \times 0.85 \times 0.90 \times 0.96 = 111$ systems
Q.: A biogas energy system's operational readiness is 80%, target reliability is 85%, and design adequacy is 92%.

1. What is its system effectiveness?
2. If there were $M_T = 150$ such systems in various states of existence, how many will be available to start their target successfully?

3. How many such systems will complete their mission successfully?
4. How many systems worth of mission objectives would be achieved?

Solution:

System effectiveness $= O \times R \times D = 0.80 \times 0.85 \times 0.92 = 0.625 = 62.5\%$

The number of systems that will start a target is obtained from:

$150 \times 0.80 = 120$ systems

The number of systems that will complete their mission successfully is obtained from:

$150 \times 0.80 \times 0.85 = 102$ systems

The number of systems that effectively achieve all designed for mission objectives is:

$150 \times 0.80 \times 0.85 \times 0.92 = 94$ systems

Q.: A biogas energy system's operational readiness is 75%, target reliability is 80%, and design adequacy is 86%.

1. What is its system effectiveness?
2. If there were $M_T = 150$ such systems in various states of existence, how many will be available to start their target successfully?
3. How many such systems will complete their mission successfully?
4. How many systems worth of mission objectives would be achieved?

Solution:

System effectiveness $= O \times R \times D = 0.75 \times 0.80 \times 0.86 = 0.516 = 51.6\%$.

The number of systems that will start a target is obtained from:

$150 \times 0.75 = 113$ systems

The number of systems that will complete their mission successfully is obtained from:

$150 \times 0.75 \times 0.80 = 90$ systems

The number of systems that effectively achieve all designed for mission objectives is:

$150 \times 0.75 \times 0.80 \times 0.86 = 78$ systems

5.3.2.1 Constant failure modes of biogas energy conversion system: Bayes' theorem

It is a formula that describes how to update the probabilities of hypotheses when given evidence. It follows simply from the axioms of conditional probability, but can be used to powerfully reason about a wide range of problems involving belief updates. Given a hypothesis *H* and evidence *E*, Bayes' theorem states that the relationship between the probability of the hypothesis before getting the evidence *P*(*H*) and the probability of the hypothesis after getting the evidence *P*(*H*|*E*) is:

$$P(H \uparrow \mathrm{E}) = \frac{(P(E \uparrow H))}{(P(E))P(H)}$$

Many modern machine learning techniques rely on Bayes' theorem. For instance, spam filters use Bayesian updating to determine whether an email is real or spam, given the words in the email. In addition, many specific techniques in statistics, such as calculating *p*-values or interpreting medical results, are best described in terms of how they contribute to updating hypotheses using Bayes' theorem.

Q.: A digester plant with MTTF = 1000 hours and MDT = 5 hours has an average availability

$$A\_av = \frac{\text{MTTF}}{(\text{MTTF} + \text{MDT})} = \frac{1000}{(1000 + 5)} = 0.995$$

On the average, the machine will perform function 99.5% of the time. The average unavailability is 0.5%, which has 44 hours of downtime per year, when the machine is supported to run continuously.

Q.: An energy unit of a biogas energy system consists of three subunits: A, B, and C. For the successful operation of the unit, all the three units of biogas energy system A, B, and C have to work satisfactorily. The probability that the biogas subunit A does not meet the specification is 0.079, the probability that the biogas subunit B does not meet the requirement is 0.049, and the probability that the biogas subunit C does not meet the specification is 0.06. What is the probability that the unit comprising the three biogas subunits does not work satisfactorily?

Solution: For the failure of the system, it is necessary that at least one of the subunits fails. The probability that at least one biogas subunit fail is:

$$1 - (1 - 0.079)(1 - 0.049)(1 - 0.06)$$
$$1 - (0.921)(0.951)(0.94) = 1 - 0.8233 = 0.177$$

This means that, on the average, 17 or 18 out of 100 units are rejected as being defective.

Q.: An energy unit of a biogas energy system consists of three subunits: A, B, and C. For the successful operation of the unit, all the three units of biogas energy system A, B, and C have to work satisfactorily. The probability that the biogas subunit A does not meet the specification is 0.069, the probability that the biogas subunit B does not meet the requirement is 0.059, and the probability that the biogas subunit C does not meet the specification is 0.07. What is the probability that the unit comprising the three biogas subunits does not work satisfactorily?

Solution: For the failure of the system, it is necessary that at least one of the subunits fails. The probability that at least one biogas subunit fail is:

$$1 - (1 - 0.069)(1 - 0.059)(1 - 0.07)$$
$$1 - (0.931)(0.941)(0.93) = 1 - 0.8147 = 0.1853$$

This means that, on the average, 18 or 19 out of 100 units are rejected as being defective.

Q.: The probability of a biogas generator A fulfilling the electricity demand is 0.8 and that of another biogas generator B fulfilling the electricity demand is 0.7. If both biogas generators A and B aim to fulfill the electricity demand simultaneously, what is the probability of either biogas generator A or B to fulfill the electricity demand?

Solution: Here, $P(\text{A and B})$ is equal to $P(\text{A}) \times P(\text{B})$, since the two events are independent. Hence

$$P(\text{A and B}) = 0.8 \times 0.7 = 0.56$$
$$P(\text{A or B}) = P(\text{A}) + P(\text{B}) - P(\text{A and B})$$
$$= 0.8 + 0.7 - 0.56 = 0.94$$

Q.: A manufacturer of a biogas generator buys his requirements of generator from four different suppliers. On an average, firm A_1 supplies 30% of its requirements, firm E_2 supplies 25%, firm A_3 supplies 22%, and firm A_4 supplies 23%. A quality test carried out on the tidal generator supplied by each firm reveals that:

6% of the supplies from A_1 are below standard.
8% of the supplies from A_2 are below standard.
10% of the supplies from A_3 are below standard.
12% of the supplies from A_4 are below standard.

Whenever a substandard sensor is used, the actuator is found to give unsatisfactory performance.

Solution: According to Bayes' theorem:

From the data given

$$P(\text{A1})P(\text{E}|\text{A1}) = (0.3)(0.06) = 0.018P(\text{A2})P(\text{E}|\text{A2}) = (0.25)(0.08)$$
$$= 0.020P(\text{A3})P(\text{E}|\text{A3}) = (0.22)(0.10) = 0.022P(\text{A4})P(\text{E}|\text{A4})$$
$$= (0.23)(0.12) = 0.0276$$

Q.: A biogas energy system is composed of four biogas units connected in series. The failure rates for these units are $\lambda1 = 0.002$, $\lambda2 = 0.003$, $\lambda3 = 0.004$, and $\lambda4 = 0.007$. It is desired that the maximum failure rate for the system is $\lambda5^* = 0.01$. Allocate this among the four units.

Solution: The sum of the unit failure rates is:

$$\lambda s = \lambda1 + \lambda2 + \lambda3 + \lambda4$$
$$= 0.002 + 0.003 + 0.004 + 0.007 = 0.016$$

Hence, the allocated unit failure rates are:

$$\lambda 1^* = \frac{0.01}{0.016} \times 0.002 = 0.0013$$

$$\lambda 2^* = \frac{0.01}{0.016} \times 0.003 = 0.0019$$

$$\lambda 3^* = \frac{0.01}{0.016} \times 0.004 = 0.0025$$

$$\lambda 4^* = \frac{0.01}{0.016} \times 0.007 = 0.0044$$

5.3.2.1.1 Biogas model uncertainty

Biogas model uncertainties consider uncertainties related to physical models of biogas energy system in order to estimate, for example, environmental conditions, loads, or stresses. In the following, examples where modeling uncertainties should be considered for biogas energy conversion system applications are given:

- model of waste collection;
- load calculation models;
- stress estimations based on loads onto the structure (structural analysis); and
- models to assess damage/fatigue.

Model uncertainties are often applying a specific and biogas energy conversion system model type specifically due to the fact that the same resistance and load models are not always considered or the environmental conditions are different. In order to determine modeling uncertainties, validation of the model is necessary. A validation assesses the accuracy of the model.

Structural failure of biogas energy system, which means complete or partial loss of load-carrying capacity, of components or structural systems may occur due to different failure modes. Failure modes are used to describe a certain type of structural failure and are also considered as limit state conditions when designing structural parts of biogas energy system. Commonly, failure modes are divided into the following categories.

Ultimate limit states, which contain structural failures due to excess of maximum load-carrying load capacity:

- loss of structural resistance of biogas energy system (excessive yielding and buckling);
- failure of biogas components due to brittle fracture;

- loss of static equilibrium of the plant structure, or of a part of the structure, considered as a rigid body, for example, overturning or capsizing of biogas energy system;
- failure of critical components of the biogas plant structure caused by exceeding the ultimate resistance (in some cases, reduced by repeated loads) or the ultimate deformation of the biogas components; and
- transformation of the biogas structure into a mechanism (collapse or excessive deformation).

Fatigue limit states, which focus on failure modes due to cyclic loading and structural damage accumulation:

- cumulative damage due to repeated loads of consumer.

Accidental limit states, which are failure modes resulting from accidental loads caused by, for example, collisions, floods, explosions, or fire:

- ultimate resistance of damaged structures of biogas power plant;
- maintain biogas structural integrity after local damage or flooding; and
- loss of station keeping (free drifting).

Serviceability limit states, which occur when excessive vibrations, leakages, deflections, or drainage disable the function for which the structural part was built:

- deflections that may alter the effect of acting forces;
- deformations that may change the distribution of loads between supported rigid objects and the supporting structure; and
- excessive vibrations producing discomfort or affecting nonstructural components.

5.3.3 Reliability analysis of biogas energy system by fault tree analysis

Reliability is defined as the probability of a device or system performing its purpose adequately for the intended operating period of time. It is also defined as the ability of biomass energy system to supply the system load with reasonable continuity and quality of supply. There are different reliability prediction analyses used in a study, such as an RBD, fault tree analysis, and Markov analysis.

Fault tree analysis is a systematic and stylized process in which undesired event is defined. In this analysis, event is resolved into its immediate causes, and the resolution of events continues until basic causes are identified. Fault trees and RBDs are both symbolic analytical logic techniques that can be applied to analyze system reliability and related characteristics. The fault tree of the biogas energy systems is established according to the past performance data. The faults are summarized on the basis of the working process

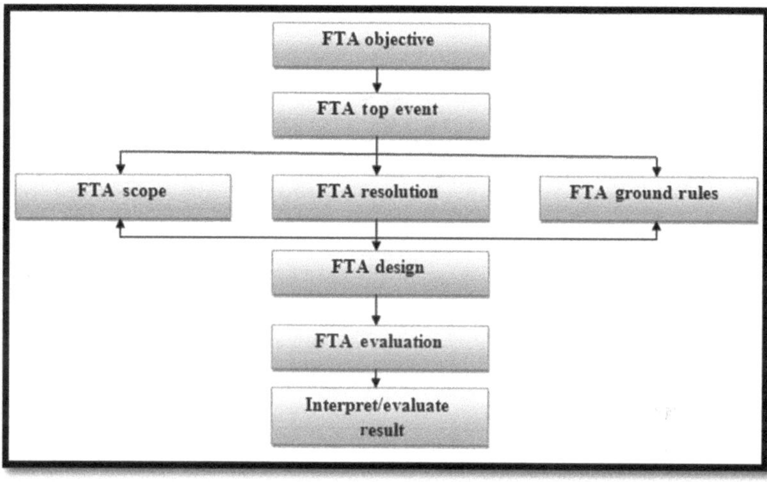

FIGURE 5.11 Flow diagram of fault tree analysis.

of the biogas energy system. Fig. 5.11 shows a flow diagram of fault tree analysis. All of the events are connected by the logical symbols, AND gate, and OR gate. In FTA analysis, Boolean algebra is commonly used to simplify the formula, getting the expression of the top event. Therefore if the fault probability of each unit is acquired, the fault probability of the whole system can be instantly obtained.

The fault tree explicitly shows all the different relationships that are necessary to result in the top event. It is also a tangible record of the systematic analysis of the logic and basic causes, leading to the top event. Fault tree provides a framework for thorough qualitative and quantitative evaluation of the top event. Fig. 5.12 shows the event of fault tree in which the top event is distributed in two intermediate event, they are connected with AND gate function, and the intermediate event takes the basic event with the connectivity of OR gate function. Table 5.8 presents the fault condition and failure rate of biogas energy system. In this table, alphabetical symbols present the condition of fault, and event or failure rate is converted into top event, intermediate event, and basic event. In this fault tree analysis, biogas energy system does not develop sufficient amount of energy worked as top event. Based on this table, develop a fault tree diagram (Fig. 5.13) of biogas energy system that is connected with AND and OR gate functions. In this figure, the faults B, C, and D are connected with the top event A by OR gate function, then an equation is developed $A = B + C + D$, and a similar procedure is followed for all the fault conditions.

The four necessary steps required to begin a fault tree are as follows.

1. Define the undesired event to be analyzed.
2. Define the boundary of the system.

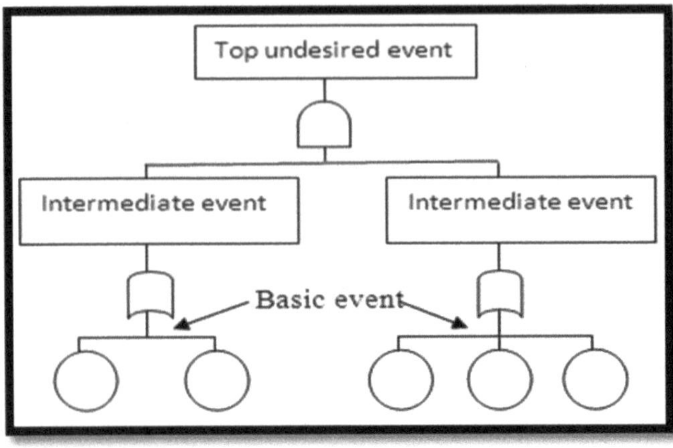

FIGURE 5.12 Fault tree event.

3. Define the basic causal events to be considered.
4. Define the initial state of the system.

5.3.4 Reliability measurement

According to fault tree analysis (Fig. 5.13),

Total failure rate:

$$A = B + C + D$$

From Eqs. (5.1)–(5.8) (see Fig. 5.13), we conclude:

$$(O \times P \times I) + S + T + U + V + (M \times N) + K + L + (L \times L) \times (M \times M) \times (N \times N) + Z \times (A \times A) + (H \times H) \times (I \times I) \times (J \times J) + (F \times F) \times (G \times G) + (B \times B) \times (C \times C)$$

According to Table 5.8, put the values of failure rate in the following equation:

Total failure rate = A =

$$(0.0011 \times .03 \times .025) + (.001) + (.001) + (.035) + (.022) + (0.019 \times .022) + (0.022) + (0.025) + (0.055 \times 0.055 \times 0.055) + (0.0015 \times 0.00155) + (0.0018 \times 0.0018 \times 0.0018) + (0.045 \times 0.045) + (0.025 \times 0.025)$$

Total failure rate = 0.109 × 100 = 10.9

According to the fault tree analysis, the fault rate is 10.9, which shows that, based on all the faults, system is 89.1% reliable.

TABLE 5.8 Fault condition and failure rate of biogas energy system.

Symbol	Condition of fault	Events/failure rate
a	Biogas energy does not develop sufficient amount of energy	Top events
b	The generator does not work properly	Intermediate events
c	The biogas energy system is primary failed	Intermediate events
d	The battery does not work properly	Intermediate events
e	Inappropriate environment condition	Basic events
f	Component fault in biogas energy system	Basic events
g	Unimproved power quality	Basic events
h	Effect of waste material	0.0408
i	Effect of temperature	0.025
j	Loading rate effect	0.022
k	Solid concentration	0.022
l	Hydrogen ion concentration	0.025
m	Retention time	0.019
n	Due to synchronism of anaerobic digester	0.022
o	Variation due to surface of mixing tank	0.0011
p	Variation with toxic substances	0.03
q	Biogas generator performance	Basic events
r	Controller in biogas system	Basic events
s	Deteriorate the performance of biogas generator	0.001
t	Uncontrolled operation of controller	0.001
u	Improper arrangement of converter with biogas energy system	0.035
v	Improper switching regulator with biogas energy system	0.022
w	Bad environment condition	Basic events
x	Component fault	0.045
y	Unimproved power quality	Basic events
z	Effect of variation of pH value	0.0015

(Continued)

TABLE 5.8 (Continued)

Symbol	Condition of fault	Events/failure rate
aa	Effect of variation of temperature	0.00155
bb	Voltage fluctuation	0.025
cc	Current fluctuation	0.025
dd	Fault in biogas system arrangement	0.055
ee	Controller in overall system	Basic events
ff	Improper arrangement of converter with biogas energy system	0.045
gg	Improper switching regulator with biogas energy system	0.045
hh	On the basis of thickness of waste material	0.0018
ii	On the basis of junction structure of biogas system	0.0018
jj	Retention period	0.0018
kk	Unbalance of system component	Basic events
ll	Improper distribution panel of overall biogas energy systems	0.055
mm	Improper wiring and connector of overall biogas energy systems	0.055
nn	Improper arrangement of junction box of overall biogas energy systems	0.055

5.4 Time-dependent hazard model and bathtub curve

Hazard function (also known as failure rate or hazard rate function) is defined as the rate of failure of a biogas power plant component or system, given that the failure has not occurred prior to time t. It is also expressed as the ratio of PDF over reliability function, and is given by:

$$h(t) = \frac{f[t]}{R[t]} = \frac{f[t]}{1 - U[t]} \tag{5.22}$$

The hazard rate function depends on the number of failures in a given small time period $[t]$ and the number of components of biogas energy system subjected to failure. Therefore it is defined as:

$$h(t) = \frac{\text{Number of failures in biogas power plant per unit time}}{\text{Number of biogas plant components subjected to failure}}$$

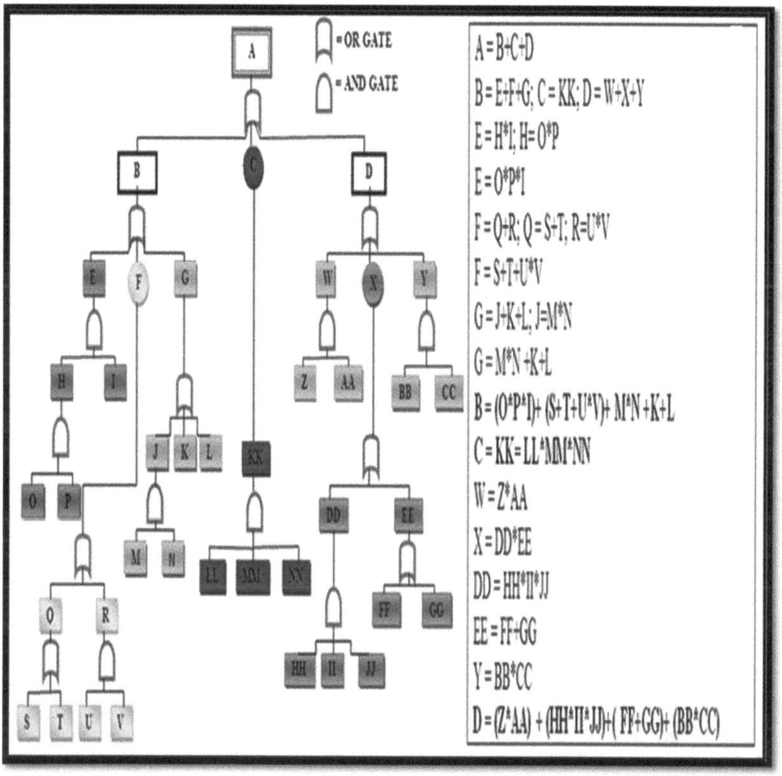

FIGURE 5.13 Fault tree analysis of biogas energy system.

It provides the probability of failure occurring within an interval $[t, t+t]$. From Eq. (5.15)

$$P\{t \leq T \leq t + \Delta t\} = R(t) - R[t + \Delta t] \tag{5.23}$$

Given that the biogas component has survived up to time t, then the conditional probability of a failure in a given time interval $[t, t+t]$ is:

$$P\{t \leq T \leq t + \Delta t | T \geq t\} = \frac{R(t) - R[t + \Delta t]}{R[t]} \tag{5.24}$$

Then, the conditional probability of failure per unit time will be:

$$\frac{R(t) - R[t + \Delta t]}{\Delta t.R[t]} \tag{5.25}$$

The hazard function is defined as the instantaneous failure rate of biogas component, as small time t approaches to zero and is expressed as:

$$\begin{aligned} h(t) &= \lim_{\Delta t \to 0} \frac{R(t) - R[t + \Delta t]}{\Delta t} \cdot \frac{1}{R[t]} \\ &= \lim_{\Delta t \to 0} - \frac{[R(t + \Delta t) - R(t)]}{\Delta t} \cdot \frac{1}{R[t]} \end{aligned} \tag{5.26}$$

$$= -\frac{dR[t]}{dt} \cdot \frac{1}{R[t]} = \frac{f[t]}{R[t]} \tag{5.27}$$

Reliability function can be derived from the known hazard rate function or failure rate function of biogas power plant

$$h(t) = -\frac{dR[t]}{dt} \cdot \frac{1}{R[t]}$$

$$h(t)dt = -\frac{dR[t]}{R[t]}$$

Integrating both sides

$$\int_0^t h(t')dt' = \int_1^{R[t]} -\frac{dR[t]}{R[t]}$$

where R[0] = 1 at $t = 0$.

Then

$$-\int_0^t h(t')dt' = lnR(t)$$

$$R(t) = \exp\left[-\int_0^t h(t')dt'\right] \tag{5.28}$$

Eq. (5.28) can be used to determine reliability function of biogas power plant from hazard rate function of biogas power plant

$$\begin{aligned} F(t) &= 1 - R(t) \\ &= 1 - \exp\left[-\textstyle\int_0^t (t')dt'\right] \end{aligned} \tag{5.29}$$

$$f(t) = \frac{dF[t]}{dt}$$

$$\begin{aligned} &= h[t]\exp\left[-\textstyle\int_0^t (t')dt'\right] \\ &= h(t)R(t) \end{aligned}$$

Many years of understanding of failure information of different components of biogas power plant has demonstrated that failures can be organized in various modes depending upon the type of failure of a particular biogas plant component. At first, an extensive number of failures have been noticed, bringing about high hazard rate. These early failures are known as the initial failures or infant mortality. This initial failure period is called the *debugging* or *burn-in* period as, during this period, all the faulty units are removed from the system. During this initial period, the number of failures in biogas plant that are demonstrated is primarily because of unsatisfactory design, wrong manufacturing and erroneous techniques, incompatible supply of material, and insufficient quality. The hazard rate gradually decreases from there on. After this initial period, the hazard rate becomes constant, which corresponds to the useful life period. Less number of failures that occur during this long period of operation has been observed. There is no specific reason for failures that occur during this period. They happen because of a sharp change in the parameters deciding execution of the units, either because of the difference in the working stress or surrounding conditions. In either case, it is hard to anticipate the amplitude of stress deviation and their occurring period; hence, the failures during this period are frequently called random failures or catastrophic failures. This period is described by constant number of failures per unit time and is the period of normal operation.

Over the long haul on, the units get outworn and start to degrade. A slow change in the estimations of the technical parameters of biogas power plant deciding the execution of the unit results, and when these parameters cross the limits of compatibility, the unit stops operating. This period is known as wear-out period of biogas operation. During this period, hazard rate of biogas plant starts increasing, and a large number of failures have been noticed during the operation of biogas power plant for generation of electricity. This is due to wearing degradation of material and ageing of digester. Adequate maintenance strategies can be utilized for efficiently stretching the period of useful life and procrastinating the beginning of the ageing period. A curve denoting the three modes of failure is shown in Fig. 5.14, and it is known as *bathtub curve* in reliability analysis because of its shape.

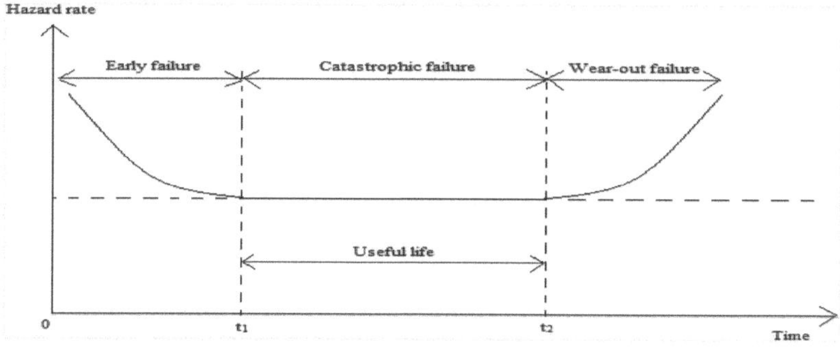

FIGURE 5.14 Bathtub curve.

5.5 Exercise

1. Explain maintainability and availability functions of biogas power plant.
2. What is the significance of reliability in the field of biogas power plant?
3. Explain the steps of reliability analysis of biogas power plant.
4. Write the comparative analysis of maintainability and availability functions.
5. Write the significance of inherent, achieved, and operational availability of biogas power plant.
6. Explain system reliability and redundancy technique of biogas power plant.
7. Explain the function of reliability if components of biogas energy system are connected in series.
8. Explain the function of reliability if components of biogas energy system are connected in parallel.
9. Consider a four-component gas holder, slurry inlet, gas outlet, and slurry outlet of biogas energy system of which the components are exponentially distributed with the constant failure rates of 0.055×10^{-3}, 0.10×10^{-3}, 0.70×10^{-3}, and 0.825×10^{-3} per hour, respectively. Evaluate the failure rate and MTTF of the system. Also calculate the reliability of each component of biogas power plant and the whole system at 400 hours.
10. Consider a four-component biogas energy system of which the components are exponentially distributed with the constant failure rates of 0.065×10^{-3}, 0.10×10^{-3}, 0.80×10^{-3}, and 0.925×10^{-3} per hour. Evaluate the failure rate and MTTF of the system. Also calculate the reliability of each component and the whole system at 300 hours.
11. Explain in detail biogas plant component failure and failure mode.
12. Explain the basic principle of fault tree analysis.
13. How we calculate reliability of biogas energy system through fault tree analysis.

Further reading

<http://web.iitd.ac.in/~arunku/files/CEL899_Y13/Failure_rate.pdf>.

<http://weibull.com/hotwire/issue7/relbasics7.htm>.

<http://www.engineeredsoftware.com/nasa/cumulati.htm>.

<http://www.engineeredsoftware.com/nasa/hazard.htm>.

<http://www.weibull.com/hotwire/issue94/relbasics94.htm>.

<https://courses.engr.illinois.edu/ece313/fa1999MW/Lectures/lec28.pdf>.

<https://warwick.ac.uk/fac/sci/wmg/ftmsc/modules/modulelist/peuss/slides/section_8b_peussdistributions_2_slides_compatibility_mode.pdf>.

<https://www.investopedia.com/terms/h/hazard-rate.asp>.

<https://www.techopedia.com/definition/8281/mean-time-to-failure-mttf>.

Balagurusamy, E., 2017. Reliability Engineering. McGraw Hill Education.

Beasley, M., 1991. Reliability for Engineers: An Introduction. Macmillan Education, UK.

Buth, B., Rabe, G., Seyfarth, T., 2009. Computer safety, reliability, and security. In: Proceedings of 28th International Conference, SAFECOMP 2009, Hamburg, Germany, September 15–18.

Ebeling, C., 2017. An Introduction to Reliability and Maintainability Engineering, 12th ed. McGraw Hill Education.

Kapur, K.C., Lamberson, L.R., 1977. Reliability in Engineering Design. Wiley, New York.

Myers, A., 2010. Complex System Reliability: Multichannel Systems With Imperfect Fault Coverage, second ed. Springer, edition.

Naikan, V.N.A., 2008. Reliability Engineering and Life Testing, first ed. Prentice-Hall of India Pvt. Ltd.

Pham, H., 2007. System Software Reliability. Springer, edition.

Thomas, M.U., 2006. Reliability and Warranties: Methods for Product Development and Quality Improvement. CRC, New York.

Verma, A.K., Ajit, S., Karanki, D.R., 2016. Reliability and Safety Engineering, second ed. Springer Nature.

Chapter 6

Biomass liquefaction

Chapter outline

Objectives

- To provide knowledge about the process of liquefaction
- To provide knowledge about the process of direct liquefaction
- To provide knowledge about the process of indirect liquefaction
- To provide knowledge about gasification

6.1 Introduction

In the electricity generation through renewable energy sources, biomass energy systems play an important role and biomass liquefaction is one of the processes or techniques of biomass energy systems. Biomass liquefaction is a primary level of thermal depolymerization framework, which is used to transform wet biomass into crude oil, and such type of crude oil, also known as bio-oil or biocrude oil, always worked at high-level pressure and at

Design and Optimization of Biogas Energy Systems. DOI: https://doi.org/10.1016/B978-0-12-822718-3.00006-X

variable temperature. The biocrude oil is a composition of 5%–25% weight of oxygen and different chemicals play an important role in the composition; such type of composition worked at higher density with a lower heating value of 34–37 MJ/kg. If we want to increase the quality of the composing product, then it is necessary to involve homogeneous and heterogeneous catalysts and they improve the performance indices of the composing product. For the example, biomass, peat, or low-positioned coals (lignite) are thermochemical materials and changed over into hydrophobic mixes with low consistency and high dissolvability. Fig. 6.1 shows the types of biomass liquefaction.

The main application of such type of fuel is in high-rating engines, including marine and locomotive, or upgraded to shipping fuels, such as diesel, gasoline, or jet fuels. Liquefaction of biomass is classified into two categories:

i. pyrolysis process in which liquefaction is done without any gasification medium; and
ii. methanol synthesis with gasification medium.

The working environment of biomass liquefaction is always preferred to be moderate temperature and moderate pressure with 260°C–560°C and 5–30 MPa, respectively, in which catalyst is moderate for 30–60 minutes at a maximum to minimum value of temperature. In wet biomass, some water is present and it is acting as a reactant, solvent, and catalyst to transform biomass into biocrude oil, and above the given moderate temperature and pressure, the water present in the biomass becomes either supercritical or subcritical.

The transformation of biomass substances into bio-oils has an accurate path, which is used to convert a carbonaceous solid material that is initially

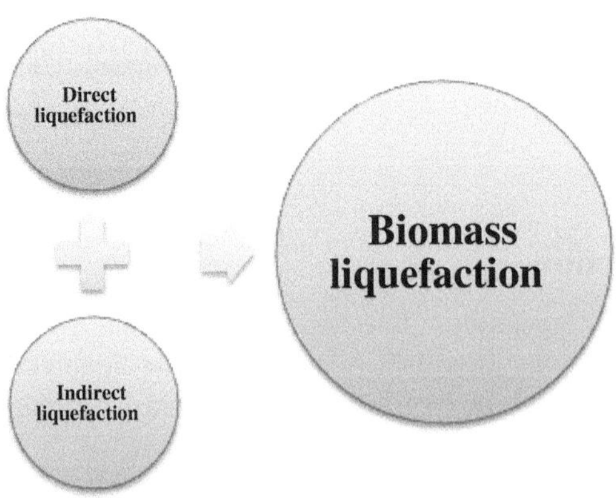

FIGURE 6.1 Types of biomass liquefaction.

tricky to grip, bulky, and of low-energy capability into pumpable oils having scientific characteristics, which allow storage and conversion through pumping systems and are then used either in direct combustion furnaces or as feedstocks for hydrotreatment leading to specific fuels and chemicals. The above definition can likewise be connected to coal, lignite, and peat. Since the essential item is "fluid oil," the term liquefaction has been all around acknowledged to speak to such a change course. Fig. 6.2 shows the elements of liquefaction.

The exact conversion of biomass into bio-oil is dependent on several variables or elements:

- composition of feedstock;
- temperature and heating value;
- pressure;
- solvent;
- residence time; and
- catalysts.

6.1.1 Feedstock

In the biomass energy system, feedstock is used to supply raw material or fuel to the machine or industrial processes. Hypothetically, any biomass can be changed over into bio-oil by utilizing aqueous liquefaction and little amount of water content, and try to develop different biomass products for the purpose of farming deposits, sewage slopes, sustenance process squanders and to increase nonnourishment biomass, for example, green growth.

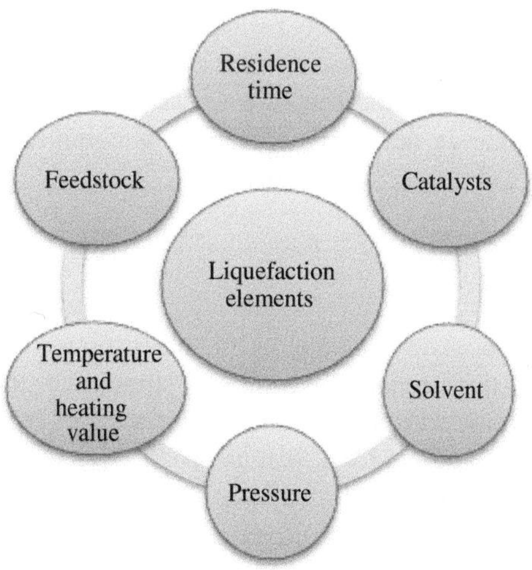

FIGURE 6.2 Liquefaction elements.

6.1.2 Temperature and heating rate

The temperature is always a key parameter when biomass is converted into bio-oil. Due to the moderate temperature, the process is categorized in two ways: depolymerization and repolymerization; in depolymerization, biomass is converted into bio-oil, and in repolymerization, biomass is converted into charcoal. The temperature of this conversion mainly depends on the type of feedstock, and if the temperature is higher than a critical value, then it increases the char and gas formation; on the other hand, if the temperature is lower than the critical value, then depolymerization is decreased. Optional responses become overwhelming in heating rates that are excessively low, prompting the development of burn. While high heating rates are required to frame fluid bio-oil, there is an edge heating rate and temperature where fluid generation is hindered and gas creation is supported in secondary reaction.

6.1.3 Pressure

Pressure (along with temperature) determines the super- or subcritical state of the solvents as well as the overall reaction kinetics and the energy inputs required to yield the desirable liquefaction products (oil, gas, chemicals, char, etc.).

6.1.4 Residence time

Residence time is the settling time of the product, because in biomass liquefaction, depolymerization is done properly if residence time is less. In the liquefaction process, residence times are measured in minutes, which is typically 15–60 minutes; however, the residence time is always dependent on the number of parameters such as reaction conditions, type of feedstock, solvent ratio, and temperature and pressure variation. As such, the optimization of the residence time is necessary to ensure a complete depolymerization without allowing further reactions to occur.

6.1.5 Catalysts

While water goes about as an impetus in the response, different impetuses can be added to the response vessel to streamline the transformation. Recently utilized impetuses incorporate water-solvent inorganic mixes and salts, including KOH and Na_2CO_3, just as progressive metal impetuses utilizing Ni, Pd, Pt, and Ru upheld on carbon, silica, or alumina. The expansion of these impetuses can prompt an oil yield increment of 20% or more prominent because of the impetuses changing over the protein, cellulose, and hemicelluloses into oil. This capacity of the impetuses to change over biomaterials other than fats and oils to bio-oil takes into consideration a more extensive scope of feedstock to be utilized.

6.2 Indirect liquefaction processes

Biomass liquefaction processes are always done in two ways: direct liquefaction and indirect liquefaction. Indirect liquefaction is the process in which oil is produced by synthesis processes, and this process is done in two steps:

- At the initial level or first step, the input material is converted at very high and moderate temperatures into a gas under the process of polymerization and hydrogenation, which contains a number of elements such as hydrogen, water, carbon monoxide, and carbon dioxide (CO_2), and the final product gas is in the form of synthetic gas.
- In the next step, the synthesis gas, which is the output of the first step, is converted into synthetic oil. The CO and H_2 atoms in the synthesis gas are then consolidated chemically to deliver intensifies that can be utilized as powers either hydrocarbon (HC) energizes, for example, synthetic gas or manufactured diesel, or oxygenated powers. The dispute of raising the H/C ratio is addressed by using the water–gas-shift reaction ($CO + H_2O \rightarrow H_2 + CO_2$) and removing the CO_2 thereby produced from the system. The properties of the oil just reflect the synthetic oil and it contains hydrogen, carbon, and oxygen, since it was obtained from union gas. The principal processes of indirect liquefaction processes are as follows:

6.2.1 Fischer–Tropsch synthesis

Fischer–Tropsch (F–T) is one of the processes of indirect liquefaction, where polymerization and hydrogenation of synthetic oil are done simultaneously, and in this process, synthetic oil always has a broad carbon number distribution. The key outcomes from the F–T synthesis are HCs, oxygenates, and water, which are used for making synthetic HCs in the place of normal HCs that can be assessed in a very simple manner by the following two catalytic reactions that build up large HC molecules from the small CO and H_2 molecules produced by gasification, with the oxygen in the CO feed being rejected in the steam:

$$CO + 2n\,H_2$$
$$H_2O + C_nH_{2n}(\text{olefins})$$
$$(2a)n\,CO + (2n + 1)$$
$$H_{2n} + H_2O + C_nH_{2n} + 2(\text{paraffins})$$

The final products always depend on the catalysts used and operating conditions of the reactor. Olefin-rich and paraffin-rich products are the two types of by-products, in which the first one in the range of 5–10 naphtha is used to develop synthetic gasoline at very high temperature and the second one in the range of 12–19 distillates is used to develop synthetic diesel and waxes at very low temperature so that F–T process is done at both levels of temperature: low level as well as higher level of temperature. Growth has emphasized producing synthetic diesel, because the raw distillate product is an excellent diesel fuel, whereas the raw naphtha product requires substantial subsequent refining to make an acceptable gasoline. F–T innovation is settled monetarily and is the focal point of the worldwide gas to liquid (GTL) endeavors to abuse minimal effort "stranded" gaseous petrol to make manufactured fluid transportation powers. Sasol in South Africa has broad development and working background with F–T innovation dependent on coal gasification and converts annually around 42 million t (Mt) of coal into 6 billion liters (GL) of engineered energizes and 2 GL of synthetic concoctions. Sasol F–T amalgamation innovation alongside a shell gasified will be utilized in a \$0.6 billion U.S. Department of Energy-supported exhibition venture in Gilbertown, Pennsylvania, which will be made from the coal squander materials 5000 b/day of the F–T fluids in addition to 41 MWe of power. As of date, a nitty gritty evaluation was done by the U.S. Department of Energy of the cogeneration of F–T fluids and power from coal by means of gasification everywhere scales. This examination, in the view of the E-gas gasified (presently claimed by PhillipsConoco) and slurry-stage reactors for F–T fluids' union, assessed that for a streamlined plant worked in the U.S. Midwest, the internal rate of return would be 10% if the power were sold for \$0.04/kWh and the F–T fluids for \$30 for each barrel; the proportional unrefined petroleum cost would be up to \$10 for each barrel not as much as this \$30 for each barrel cost (contingent upon processing plant arrangement and relative oil item requests), on the grounds that the F–T fluids would as of now be somewhat refined. Sulfur and sweet-smell-free F–T center distillates are as of now being utilized as mix stock with customary raw petroleum-determined diesel in California to give fuel that meets that state's stringent particulars for diesel.

6.2.2 Methanol synthesis

Methanol is the second step of indirect liquefaction and it is the process of partial CO hydrogenation; in this process, product is not really an oil, but methanol. Methanol is an entrenched chemical product, which is used throughout the world, and it can also be used as an indirect or direct way as a fuel. The primary reactions involved in making MeOH from synthetic gas are:

$$CO + H_2O = CO_2 + H_2 \text{(water, gas shift)}$$
$$CO + 2\,H_2 = CH_3OH \text{(methanol synthesis)}$$

The MeOH created can be additionally handled to make gas by the mobile procedure—a business innovation that can give gas at appealing expenses from ease stranded gaseous petrol by MeOH parchedness, or the MeOH can be utilized legitimately as fuel. In many parts of the world, MeOH is made by steam changing of gaseous petrol; however, in gas-poor areas, for example, China, it is made for the most part from coal-determined synthetic gas by means of gasification. Under the U.S. Department of Energy's Clean Coal Technology Program, Air Products and Chemicals, Inc. has conveyed to business status slurry-stage reactor innovation for MeOH creation.

6.3 Direct liquefaction processes

The main constituents of biomass are the biopolymers, cellulose, hemicellulose, and lignin. Furthermore, a number of inorganic compounds are present, representing the major source of ashes when the biomass is combusted. For thermochemical conversion processes leading to small molecules, not only these principal constituents but also the general structure of the plant are important. This general structure will influence the early process steps like the milling of the plant as well as the chemical or thermal decomposition of the raw material. The wall of the lingocellulotic plant cells typically consists of 30–35 wt.% cellulose, 15–35 wt.% hemicelluloses, and 20–35 wt.% lignin. Softwood has a relatively higher content of lignin than hardwood, while hardwood contains more hemicelluloses. Fresh biomass contains up to 60%–70% water; after drying, this value is reduced to about 10%–15%. Direct liquefaction of coal under expanded weight within the sight of hydrogen and an impetus is an around 100 years of age process examined in some detail in the principal half of the only remaining century for the most part by Bergius and Pier. The irreplaceable requirement for the hydrogen present in the process for coal as well as for biomass winds up clear when the transformation is considered in detail. Dry wood can be portrayed by the entirety recipe CH1.4O0.7, fluid powers by CH_2. Hydrogen has a twofold capacity.

The substance responses engaged with this procedure assemble a perplexing response organized. As ahead of schedule amid the primary oil emergency in the main portion of the seventies of the only remaining century, direct liquefaction of biomass came into center as one potential way far from unrefined petroleum as the reason for portability. Also the absence of the essential logical comprehension of the procedure prompted the disappointment of early quite huge show plants worked amid that period. Meanwhile, countless overall brought about a greatly improved comprehension of the procedure.

In Germany, a portion of the present specialized executions of direct liquefaction of biomass applies conditions fundamentally the same as those utilized with coal as a crude material, while others think of having the option to run the procedure without extra hydrogen and without increased pressure. A number of the current process implementations in Germany are based on the adventurous assumptions with respect to the chemical reactions involved. In some cases, the reaction equation is taken without sufficient reflection as the basis for the process design. An additional and very critical shortcoming of many of the processes is the lack of understanding of the chemical steps involved. Moreover, when aiming at producing liquid fuels for transportation purposes, the product has to comply with the respective national and international standards for these fuels, that is, with EN 228 for Otto fuels and EN 590 for diesel fuel.

6.3.1 Basic aspects of direct liquefaction of biomass

6.3.1.1 Structure of biomass

In the direct liquefaction of biomass, a proper structure of biomass is very necessary, and in that case, the main constituents of biomass are the biopolymers, cellulose, hemicelluloses, and lignin. On the other hand, a number of inorganic compounds are there, representing the key foundation of ashes when the biomass is composed. In direct liquefaction process thermochemical conversion process is the one of the key mechanism of biomass power plant. This common arrangement will influence the early process steps like the milling of the plant as well as the chemical or thermal decomposition of the raw material. The mass of the lingo cellulite plant cells commonly comprises of 30–35 wt.% cellulose, 15–35 wt.% hemicelluloses, and 20–35 wt.% lignin. Softwood has a moderately higher substance of lignin contrasted, while hardwood contains more hemicelluloses. New biomass contains up to 60%–70% water; in the wake of drying, this esteem is decreased to about 10%–15%.

6.3.1.2 Requirements of fuels

As pointed out, a key foundation for a suitable direct liquefaction process is the nature of the items. Here, the applicable European principles of the fluid fills must be obeyed concerning every single point.

6.3.1.3 Conceivable reaction pathways

The change of the woody biomass into fluid HCs contains the accompanying advances:

1. types and design of feedstock;
2. feedstock within a liquid carrier slurring;
3. heating the slurry;
4. addition of reducing gas at a given pressure;
5. reaction formation;

6. product division; and
7. solid–liquid division and recovery of solvent.

Normally, the catalyst of the different capacities is added to the slurry. These designing procedure ventures of liquefaction relate to the accompanying concoction steps:

- solvolysis of the biomass produced;
- depolymerization of the main components cellulose, hemicellulose, and lignin;
- chemical and thermal decompositions of monomers and smaller molecules, leading to new molecular rearrangements through bond ruptures, dehydration, and decarboxylation; and
- degradation of oxygen-containing functional groups in the presence of hydrogen.

As cellulose and hemicelluloses are polysaccharides containing a large amount of sugar and mash, paper businesses exist. Contingent upon substrate and temperature (hemicelluloses 120°C–180°C and cellulose >240°C), depolymerization in water media under the expansion of corrosives is trailed by the burst of ring securities or recourse of the action of the shaped monosaccharide. Trademark for the depolymerization of cellulose and hemicelluloses is the colossal number of various unwanted materials with a large number of oxygen-containing useful materials. From this useful material, just a couple is capable of further direct decrease of oxygen by the cleavage of CO_2. Along these lines, the nearness of hydrogen is the basis for the corruption of practical gatherings and to keep away from repolymerization. The depolymerization of lignin in a dissolvable prompts various distinctive substituted phenols. As in the previously mentioned case of celluloses and hemicelluloses, the nonattendance of hydrogen rapidly prompts the repolymerization of the items. Fig. 6.3 shows the thermochemical process of direct liquefaction. Fig. 6.4 shows different processes of thermochemical conversion.

Direct liquefaction happens in a nonoxidative climate, where the biomass is encouraged into a unit as water slurry at lower temperatures, with biorough in the fluid structure being the item. The essential focal point of these specific procedures is to deliver a fluid item that is a HC with a nuclear H:C proportion of ~2 and a bubbling scope of 170°C–280°C. A large number of procedures created depend on coal-to-fluid handling. The fundamental purposes of bringing coal and biomass into a fluid is to create fluids, to expel a portion of the less attractive parts (i.e., sulfur, oxygen, nitrogen, and minerals), and to make a higher energy density material that will flow.

One of the primary processes to convert coal into liquids directly is through a combination of thermal decomposition and hydrogenation under pressure. There are several single- and two-stage processes that have been developed, but have not been made commercial in the United States.

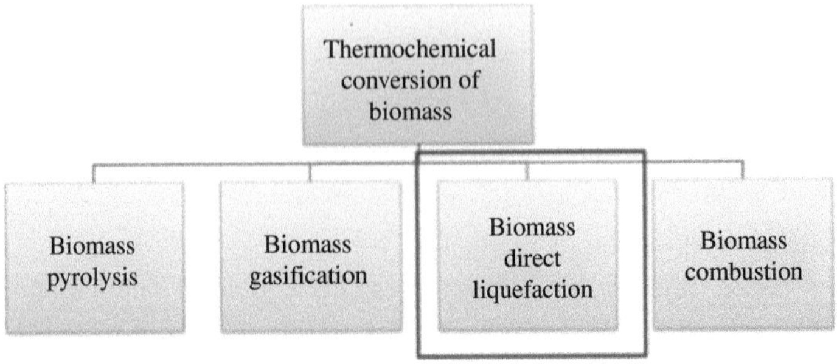

FIGURE 6.3 Thermochemical conversion process of direct liquefaction.

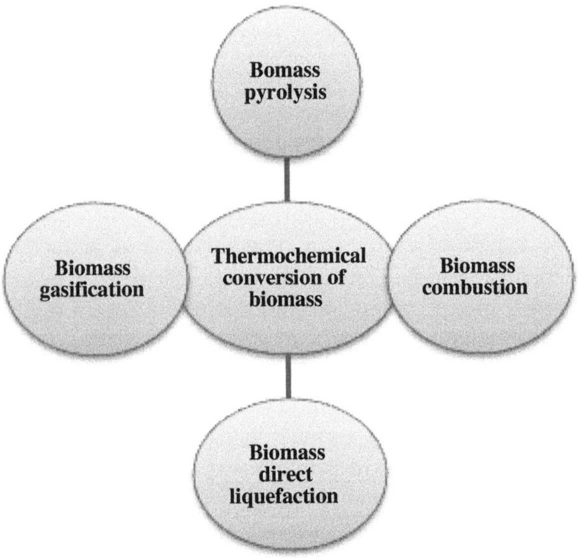

FIGURE 6.4 Thermochemical conversion of biomass.

However, China opened a commercial direct liquefaction plant partially based on U.S. designs in 2008.

Design considerations include:

1. moderate temperatures of ~450°C–550°C;
2. hydrogenation of carbon;
3. HC solvents;
4. aromatic in coal; and
5. sulfur and nitrogen.

Biomass can be set up thusly; nevertheless, biomass has out and out more oxygen and less fragrant blends, and separates interestingly rather than coal.

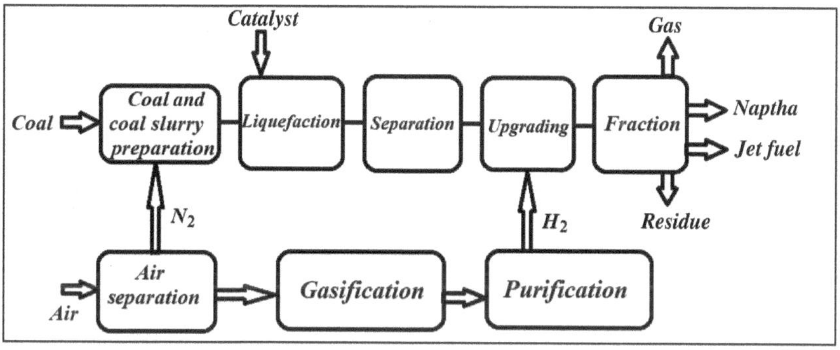

FIGURE 6.5 Direct coal liquefaction schematic from Shenhua plant in China.

Various systems have been delivered for biomass that appears to finish a prevalent activity of taking care of cellulose. Fig. 6.5 shows direct coal liquefaction schematic from the Shenhua plant in China.

Overall, what are the distinctions with direct liquefaction of biomass? Superficially, it looks practically equivalent to the procedure of coal liquefaction. It is a thermochemical change procedure of natural material into fluid biorough and coitems. Contingent upon the procedure, it is normally directed under moderate temperatures (300°C–400°C, lower than coal liquefaction) and weights (10–20 MPa, comparative or perhaps somewhat higher with principally hydrogen in coal to fluids) with included hydrogen or CO as a decreasing specialist. In contrast to coal, the biomass is "wet," or if nothing else wetter than coal, and can be handled as a watery slurry. At the point when handled as watery slurry, the procedure is alluded to in the writing as aqueous preparing and can be subcritical to supercritical for water. The basic reaction mechanisms can be described as:

1. depolymerization of biomass;
2. decomposition of biomass monomers by cleavage, dehydration, decarboxylation, and domination; and
3. recombination of reactive fragments.

Various sorts of biomasses respond distinctively relying on the biomass source. Sugars, for example, cellulose, hemicelluloses, and starch, can decay in aqueous water. The common item shaped under these conditions is glucose, and glucose would then be able to be aged to cause alcohols or further to distort in water to make glyceraldehydes and dihydroxyacetone. The substance made rely on the conditions: at temperatures ~180°C, substances are sugar monomers; however, at higher temperatures, 360°C–420°C, the aldehyde and $(CH_3)_2CO$ mixes are shaped. Lignin and unsaturated fats likewise decay in aqueous water, yet the items are altogether different on the grounds that the substrate is extraordinary. Bembenic and Clifford utilized aqueous

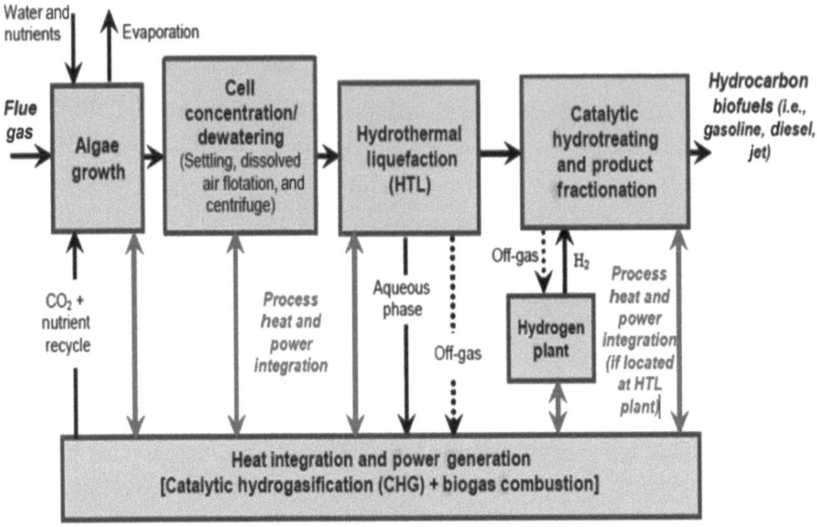

FIGURE 6.6 Schematic of an aqueous water procedure.

water at 365°C and ~13 MPa to shape methoxy phenols, utilizing various gases to change the item slate (hydrogen, carbon monoxide, CO_2, and nitrogen). For lipid or triglyceride (fats and oils) response in aqueous water at 330°C–340°C and 13.1 MPa, the primary items are the free unsaturated fats (HC–COOH) and glycerol ($C_3H_8O_3$). The free unsaturated fats would then be able to be responded to straight-chain HCs that can be utilized for diesel or stream fuel, in spite of the fact that the temperature for the most part should be somewhat higher (400°C) for this to happen. Fig. 6.6 demonstrates the schematic of an aqueous water procedure to change over green growth into fluid energizes, utilizing the heat from an incorporated heat and power framework. Pipe gas from a power age office is utilized to develop green growth. The green growth is then responded in an aqueous unit pursued by synergist hydrogenation to make the straight chain HC fluid energizes.

Numerous sorts of catalyst can be utilized, in spite of the fact that it relies on the procedure organized which catalyst are utilized and what feed material is utilized. In aqueous preparing, the more typical catalysts utilized are corrosive and base catalysts. Molecule estimate for biomass should be fine, with a size of <0.5 mm. Presentation of the feed into the reactor is additionally tested, as it is encouraged into a high-weight reactor. A few preferences of utilizing this procedure for biomass are as follows:

1. It is conceivable to process bolsters with high water content, as much as 90%.
2. It is conceivable to process a wide range of sorts of waste materials, including municipal solid waste (MSW), nourishment preparing waste, and creature compost.

3. The procedure serves the two ways of waste treatment and sustainable power source generation. Procedure parameters incorporate solids content, temperature, weight, settling time, and utilization of catalyst. Regularly synchronous responses are occurring, which makes the general comprehension of the responses confused. The sorts of responses occurring include solubilization, depolymerization, decarboxylation, hydrogenation, buildup, and hydrogenolysis. For one specific procedure, aqueous liquefaction requires the utilization of catalyst. One typical catalyst used is sodium carbonate combined with water and CO to produce sodium format:

$$Na_2CO_3 + H_2O + CO \rightarrow 2HCO_2Na + CO_2$$

This equation is not rendering properly due to an incompatible browser. See Technical Requirements in the Orientation for a list of compatible browsers.

This dehydrates the hydroxyl groups to carbonyl compounds and then reduces the carbonyl group to an alcohol:

$$HCO_2Na + C_6H_{10}O_5 \rightarrow C_6H_{10}H_4 + NaHCO_3H_2 + C_6H_{10}O_5 \rightarrow C_6H_{10}H_4 + H_2O$$

Other catalysts used that behave in a similar manner include K_2CO_3, KOH, NaOH, and other bases. For simultaneous decomposition and hydrogenation, nickel (Ni) catalysts are used. Similar to pyrolysis, the major product of this process is liquid biocrude, which is a viscous dark tar or asphalt material. Up to 70% of the carbon is converted into biocrude; lighter products are obtained when different catalysts are used. The liquid biofuel has a similar carbon-to-hydrogen ratio as in the original feedstock and is a complex mixture of aromatics, aromatic oligomers, and other HCs. In this process, the oxygen is reduced and is 10%–20% less than typical pyrolysis oils, with a heating value higher than pyrolysis oils, 35–40 MJ/kg on a dry basis. However, the U.S. Department of Agriculture (USDA) has developed a pyrolysis process using recycled gases that produce a fairly light HC with very little oxygen content. Table 6.1 shows a comparison of biocrude from various processes and feed materials. The quality of the biocrude shown from hydrothermal processing is for a heavy biocrude. Other processes will make a lighter material, but also produce more coproducts that must be utilized as well.

6.4 Other biomass liquefaction processes

Energy from biomass, bioenergy, is a promising source to supplant petroleum derivatives later on, as it is plenteous, clean, and CO_2 nonpartisan. Thermosynthetic liquefaction of biomass is generally examined as a promising technique to deliver one sort of fluid biofuel, specifically bio-oil. This

TABLE 6.1 Characteristics of different types of bio-oil.

Characteristics	Hydrothermal bio-oil	Fast pyrolysis bio-oil	Heavy-petroleum fuel	Oil oak
Water content (wt.%)	3.4	15–23	0.11	4.9
Insoluble solids (%)	1	0.5–0.75	0.01	NA
Heating value (HHV) (MJ/Kg)	30	17	40	34
Density (g/mL)	1.11	1.22	0.92	NA
Viscosity	3000–16,000	10–150	175	NA
Carbon (%)	75	39–56	85	80
Hydrogen (%)	8	7.5	11	6
Oxygen (%)	15	51	1	12
Nitrogen (%)	<0.1	<0.1	0.3	2.1
Sulfur (%)	<0.05	<0.05	2.3	NA

audit exhibits the ongoing exploration advance in the liquefaction of regular biomass from another point of view. Especially, this article abridges five parts of related work: first, the impact of the dissolvable kind of the liquefaction practices of biomass; second, the impact of biomass type of the liquefaction practices of biomass; third, the liquefaction of biomass in sub-/superbasic ethanol; fourth, the liquefaction of biomass in natural dissolvable water blended solvents; and fifth, the liquefaction of sewage muck.

Experiments on the properties of gases started early in the 17th century. By the middle of the 17th century, Robert Boyle had derived the inverse relationship between the pressure and volume of gases. About the same time, Guillaume Amontons started looking into temperature effects on gas. Various gas experiments continued for the next 200 years. During that time, there were efforts to liquefy gases. Many new facts on the nature of gases had been discovered. For example, early in the 19th century, Cagniard de la Tours had shown there was a temperature above which a gas could not be liquefied. There was a major push in the mid to late 19th century to liquefy all gases. A number of scientists including Michael Faraday, James Joule, and William Thomson (Lord Kelvin) did experiments in this area. In 1886 Karol Olszewski liquefied methane, the primary constituent of natural gas. By 1900, all gases had been liquefied except helium, which was liquefied in 1908.

The first large-scale liquefaction of natural gas in the United States was in 1918 when the U.S. government liquefied natural gas (LNG) as a way to extract helium, which is a small component of some natural gas. This helium was intended for use in British dirigibles for World War I. The LNG was not stored, but regasified and immediately put into the gas mains. The key patents having to do with natural gas liquefaction were in 1915 and the mid-1930s. In 1915 Godfrey Cabot patented a method for storing liquid gases at very low temperatures. It consisted of a Thermos bottle-type design that included a cold inner tank within an outer tank, the tanks being separated by insulation. In 1937 Lee Twomey received patents for a process of large-scale liquefaction of natural gas. The intention was to store natural gas as a liquid, so it could be used for shaving peak energy loads during cold snaps. Because of large volumes, it is not practical to store natural gas, as a gas, near atmospheric pressure. However, if it can be liquefied, it can be stored in a volume 600 times smaller. This is a practical way to store it, but the gas must be stored at −260°F. There are basically two processes for liquefying natural gas in large quantities. One is a cascade process in which the natural gas is cooled by another gas, which, in turn, has been cooled by still another gas, and hence a cascade. There are usually two cascade cycles prior to the liquid natural gas cycle. The other method is the Linde process. (A variation of the Linde process, called the Claude process, is sometimes used.) In this process, the gas is cooled regeneratively by continually passing it through an orifice until it is cooled to temperatures at which it liquefies. The cooling of gas by expanding it through an orifice was developed by James Joule and William Thomson and is known as the Joule–Thomson effect.

6.4.1 Liquefaction and carbonization

The processes are used for scientific, industrial, and commercial purposes. Many gases can be put into a liquid state at normal atmospheric pressure by simple cooling; a few, such as CO_2, require pressurization as well. Liquefaction is used for analyzing the fundamental properties of gas molecules (intermolecular forces), for the storage of gases, for example, LPG, and in refrigeration and air conditioning. There the gas is liquefied in the condenser, where the heat of vaporization is released and evaporated in the evaporator, where the heat of vaporization is absorbed. Ammonia was the first such refrigerant and is still in widespread use in industrial refrigeration, but it has largely been replaced by compounds derived from petroleum and halogens in residential and commercial applications. The importance of biomass materials as a renewable, inexpensive, and abundant source of energy is widely recognized. The decreasing reserves of fossil fuels and resultant negative effects of their use in terms of generation of greenhouse gases, acid rain, and changes in climate has also renewed the interest in utilization of biomass materials as fuel. Biomass materials hold a special promise for developing countries, which use biomass materials extensively or import fossil fuels to meet their energy demands. In countries like Australia, the use of biomass materials is also encouraged to meet the government's commitment to provide 2% of energy through renewable sources. Biogas is one such biomass material that is generated in large quantities during the processing of sugarcane in sugar mills. Biogas offers the advantage of being a cheap, plentiful, and low polluting fuel. In Australia, approximately 11.4 million tons of wet biogas is produced annually. On the average, biogas contains about 45%–50% of moisture, 43%–52% of fiber, and 2%–6% of soluble solids. The typical composition of biogas fiber is about 26.6%–54.3% of cellulose, 22.3%–29.7% of hemicelluloses, and 14.3%–24.45% of lignin. Biogas is characterized as a low-density fiber by its wide particle size distribution (less than 100 cm to greater than 10 cm). Biogas is used primarily in the sugar mills to meet their own internal energy needs. A part of the surplus is also used in paper, pulp, and board manufacture, compost, cattle feed, and ethanol manufacture. The burning of biogas in sugar mills for meeting internal energy needs, though justified on practical and economic grounds, is inefficient and wasteful. There has been a significant interest in converting this residue into higher energy dense products by means of pyrolysis, carbonization, liquefaction, gasification, and combustion. This study is focused on the carbonization of biogas. Carbonization is defined as the extensive thermal degradation of the carbonaceous component in the absence of air or in the presence of inert gas. During this process, the volatiles are removed in the form of oxides of carbon and hydrogen to enrich the carbon content in the solid char. Slow heating rates, coupled with a low final heating temperature and long gas and solid residence time, maximize the yield of char by secondary coking and repolymerization reactions. However, if the heating rate is maintained between 20°C/min and 100°C/min with a maximum temperature of 600°C, then the liquid and gas yields are markedly increased at the expense of char yield. There has been a considerable amount of interest in producing oil and liquid fuels from biomass materials recently, with studies being done to optimize the oil yield and quality. Char is a better fuel than the precursor biomass and finds application as a fuel, in the ferroalloy, aluminum, copper, tobacco, and cement industries, and for the production of chemicals and activated carbon. Though a considerable amount of studies have been conducted on combustion and on biological conversion, carbonization has received less interest. Most of the research studies in the past have focused on the theoretical rather than the practical aspects of carbonization. The thermal degradation behavior of biogas and its components has been studied under both an inert and an oxidizing atmosphere using thermogravimetric analysis at low heating rates in low-temperature regions and at high heating rates. This is an age-old pyrolysis process optimized for the production of charcoal. Traditional methods of charcoal production have centered on the use of earth mounds or covered pits into which the wood is piled. Control of the reaction conditions is often crude and relies heavily on experience. The conversion efficiency using these traditional techniques is believed to be very low; on a weight basis, Openshaw estimates that the wood-to-charcoal conversion rate for such techniques ranges from 6–12 tones of wood per tonne of charcoal.

Biomass is a heterogeneous, mainly organic, material that decomposes upon exposure to high temperatures. Decomposition of a compound, in the absence of oxygen, by the action of heat alone is known as pyrolysis. Pyrolysis products are divided into a volatile fraction, consisting of gases, vapors, and tar components, and a carbon-rich solid residue. The pyrolysis process consists of a very complex set of reactions involving the formation of radicals.

With reference to wood, pyrolysis processes are alternately referred to as carbonization, wood distillation, or destructive distillation processes. As generally accepted, carbonization refers to processes in which the char is the principal product of interest (wood distillation, the liquid, and destructive distillation, both char and liquid). At the usual carbonization temperature of about 675 K, char represents the largest component in wood decomposition products. Commercial char produced under these conditions contains about 80% of carbon, 1% ± 3% of ash, and 12% ± 15% of volatile materials. The elemental composition of char and its properties are dependent on the carbonization temperatures. Charcoal is a premium fuel that is widely used in many developing countries to meet household as well as a variety of other needs. Charcoal can be readily produced from wood with no capital investment in equipment through the use of charcoal piles, earth kilns, or pit kilns. As the names of these processes suggest, hardwood is carefully stacked in a mound or pit around a central air channel and then covered with dirt, humus, moss, clay, or sod. Utilization of different agricultural and forestry residues, such as hazelnut kernel husk, olive husk, hazelnut shell, spruce wood beech wood, and corncob, for producing bio coal, that is, briquetted charcoal, appears to be a novel way of simultaneously solving their disposal problems and partially substituting for biomass-derived charcoal. Not all residues, however, are equally attractive as raw material for biocoal. A ranking of the residues in terms of their suitability for biocoal production would aid proper planning of residue utilization and development of related hardware. The purpose of this study is to provide fundamental data on the conversion of biomass samples into useful gaseous, char, condensed liquid, and tar fuels by rapid carbonization.

The proximate and ultimate analysis and other characteristics of the biomass samples using elemental compositions of the samples were determined according to the ASTM D1103-80 standard methods. The higher heating values of the samples were determined by forming the samples into pellets weighing approximately 1 g each and burning the pellets in an adiabatic oxygen bomb calorimeter according to the ASTM D 2015 standard test method.

Any carbonization procedure requires an oxygen-free atmosphere as a basis for carbonaceous matter or semicoke production. This practice is required to prevent any combustion that may occur due to the production of volatile matter and combustible gases. It consists mainly of four separate units, the heating source unit, the sample handling unit, the heat exchanger units, and the gas collection unit. The reactor employed for the carbonization was 245-mm high and had an inner diameter of 21 mm. The temperature of the reactor was controlled by an electrical heating furnace. The heating source unit consists of a common purpose muffle and a furnace that can provide heat up to a temperature of 1600 K. The sample handling unit allows the system to be scaled up and organized depending on the heating source employed. The system consists primarily of a chamber that contains the sample. The pre-cooling heat exchanger (Fig. 6.5) is a tube and shell type. It is separated from the main heat exchanger by a manual control valve. The main heat exchanger is also a tube and shell type. The main exchanger is connected to a double-stage vacuum pump through a manual control valve, so the exiting gases can be collected for further analysis. The whole device is supported by a stand. The sample was preheated at 380 K for 1 hour to get rid of the extra moisture that affects further grinding. The sample was transferred to a Wiley mill, where it was ground to small grains and then screened to pass a 0.2-mm aperture. The ground meal was uniformly mixed before initiating gasification procedures. The carbonization treatments were performed from ambient temperature to 550, 650, 750, 850, 950, 1050, and 1150 K. The heating rate was held at 10 K/second. The carbonization system was depleted from air by means of the vacuum pump, after which the valves were both closed. The gasification chamber was then placed inside the muffle furnace, and the valve between the heat exchanger units was opened for enough time to allow the exiting gases to escape to the main heat exchanger. Then the valve was closed and the valve between the main heat exchanger and the vacuum pump was opened and the gases were sucked through the vacuum pump. This opening and closing cycle of valves was repeated every 3 minute for the whole period of carbonization. The outlet gases can be collected easily and separated for their constituents.

During carbonization, most of the volatile components of the wood are eliminated; this process is also called "dry wood distillation." Carbon accumulates mainly due to a reduction in the levels of hydrogen and oxygen in the wood. The wood undergoes a number of physic-chemical changes as the temperature rises. Between 100°C and 170°C, most of the water is evaporated; between 170°C and 270°C, gases develop containing condensable vapors, CO, and CO_2. These condensable vapors form pyrolysis oil, which can then be used for the production of chemicals or as a fuel after cooling and scrubbing. Between 270°C and 280°C, an exothermic reaction develops, which can be detected by the spontaneous generation of heat. The modernization of charcoal production has led to large increases in production efficiencies with large-scale industrial production in Brazil now achieving the efficiencies of over 30% (by weight).

There are three basic types of charcoal making:

- internally heated (by the controlled combustion of the raw material);
- externally heated (using fuel wood or fossil fuels); and
- hot circulating gas (retort or converter gas, used for the production of chemicals).

Internally heated charcoal kilns are the most common form of charcoal kiln. It is estimated that 10%–20% of the wood (by weight) is sacrificed; a further 60% (by weight) is lost through the conversion to, and release of, gases to the atmosphere from these kilns. Externally heated reactors allow oxygen to be completely excluded and, thus, provide better quality charcoal on a larger scale. They do, however, require the use of an external fuel source, which may be provided from the "producer gas" once pyrolysis is initiated. Recirculating heated gas systems offer the potential to generate large quantities of charcoal and associated by-products, but are presently limited by high investment costs for large-scale plant.

6.4.2 Cryogenic separation and liquefaction

The technical fields of cryogenic gas treatment comprise many different processes and have been used for the following three tasks for biogas:

- removal of trace contaminants, mainly in the landfill gas context;
- removal of main components such as CO_2 and N_2 (gas upgrading); and
- condensation of upgraded biomethane to bio-LNG [liquefied biogas (LBG)].

As a consequence, cryogenic processes cannot be treated as a homogeneous family of gas upgrading techniques. Furthermore, most processes currently available are under development, so little sound data are available. Cryogenic processes make use of low temperatures in order to achieve the wanted results. There is no fixed temperature level below which processes are considered "cryogenic." However, the processes described here operate at temperatures well below −50°C (approximately 220 K), that is, in an area where common gases become a liquid. Fig. 6.7 illustrates the condensation point of some common gases in the area of biogas.

6.4.2.1 Purification of landfill gas

Landfill gas is very similar to digester gas, being composed mainly of methane and CO_2. The main challenge in the use of landfill gas as a fuel is the varying and elevated content of trace contaminants such as siloxanes and halogen compounds, some of them being known and others not even being

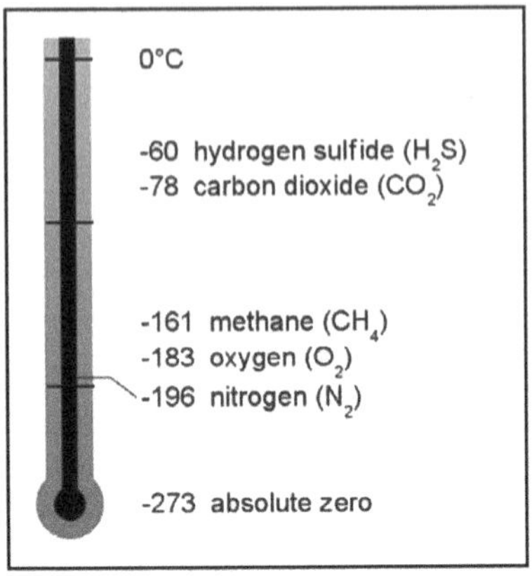

FIGURE 6.7 Boiling points at atmospheric pressure for some pure gases.

identified. Furthermore, landfill gas can contain elevated amounts of nitrogen, which cannot be removed by most conventional gas upgrading techniques. Techniques capable of removing nitrogen from biogas/landfill gas include pressure swing adsoprtion (PSA) and cryogenic upgrading to liquefied methane. One cryogenic technique specifically developed for the removal of trace contaminants is the CO_2 Wash process developed by U.S.-based Acrion Technologies. In this process, the raw gas stream is cleaned by liquid CO_2, even called "CO_2 Wash." The idea is that many impurities such as halogenated HCs or siloxanes have a much higher solubility in CO_2 than methane. The process is mostly interesting for the treatment of landfill gas with its high and varying contents of impurities. Usually, the CO_2 Wash process is combined with other gas treatment steps in order to produce compressed biogas (CBG) or LBG. After compression, H_2S removal, and drying, the gas stream enters a cryogenic column at the bottom and is then led upward, meeting a stream of liquid CO_2. Impurities are dissolved in the liquid CO_2 and bled off the system at the sump. The gas stream leaving the head of the column contains methane, a part of the CO_2 of the raw biogas as well as the oxygen and nitrogen from the raw gas. The rest of the CO_2 leaves the system in the upper half of the column as a liquid, food grade CO_2 stream.

The purified gas can be directly used in boilers and combined heat and power (CHP) engines; in this case, the CO_2 Wash system leads to less wearing of installation and a longer lifetime. Alternatively, the gas can be further upgraded to vehicle quality and liquefied by other upgrading techniques. The CO_2 Wash technology has previously been licensed separately for the production of compressed biomethane and LBG. Originally, the CBG license was held by firm green Inc. Today, Terracastus has the global rights to the CO_2 Wash technology for both CBG and LBG production.

6.4.2.2 *Upgrading of biogas*

Compared with other upgrading techniques, cryogenic approaches have the following potential or expected benefits:

- hope for low energy demand during upgrading;
- no contact between the gas and the chemicals;
- production of pure CO_2 as a subproduct;
- possibility to produce LBG; and
- possibility to remove nitrogen from the gas stream.

Since the condensation temperatures of the different compounds of biogas are quite different, it is principally easy to separate methane and CO_2 by gradually cooling down the mixture and, thereby, liquefying the CO_2. Other compounds with higher condensation temperature than methane such as water and hydrogen sulfide are removed simultaneously. It is also possible to lower the temperature in several steps, each of them removing certain compounds. This way, it is possible to minimize the irreversibilities of heat exchangers and optimize the refrigerant cycle. An example for this is the gross primary productivity (GPP) process of GtS. When cooling common digester gases, the first compounds to be removed are low-concentration impurities such as water, hydrogen sulfide, siloxanes, and halogens at a temperature down to approximately $-25°C$. These compounds can be removed mostly in their liquid state. When continuing to decrease the temperature, the next compound being condensed is CO_2. However, CO_2 will directly shift from the gaseous to the solid phase unless the cooling unit is operated at elevated pressure. Because of the high concentration, solid CO_2 can be a problem in the process by plugging pipes and devices such as heat exchangers. Therefore it can be an advantage to operate the process at a higher pressure where cold CO_2 is a liquid rather than a solid. However, since the solubility of methane in solid CO_2 is very low whereas more methane dissolves in liquid CO_2, the sublimation or freezing of CO_2 can be used for a more efficient separation of these two compounds in order to minimize methane losses. In this case, this step is designed as a batch process where operating conditions in the cooling step (normally a heat exchanger) are chosen such that the CO_2 is permitted to freeze or sublimate as a solid until the capacity of the heat exchanger is reached. Then, production is shifted to a parallel line, while the heat exchanger in the first line is defrosted. CO_2 is recovered in either gaseous or liquid state.

6.4.3 Liquefaction of upgraded biomethane

The third application of cryogenic techniques in the biogas context is the production of liquefied biomethane from conventionally upgraded gas streams, often called LBG or bio-LNG. LBG has a number of advantages

TABLE 6.2 Purity requirements for the liquefaction of biomethane.

Compound	Limit for liquefaction (ppm)
Water (H_2O)	0.5
Hydrogen sulfide (H_2S)	3.5
Carbon dioxide (CO_2)	50–125

compared with CBG, which are mostly related to the higher energy density. Gaseous methane of 1 Nm^3 is equal to 1.7 L of liquid methane. LBG has more than double the volumetric energy density of gaseous biomethane compressed to 250 bars(a):

- more efficient transport, bigger geographical marketing range;
- lower energy consumption and investment costs at filling stations; and
- opening up of new markets such as LNG backups, trucks, and ships.

Because of the low boiling point of methane, very low temperatures are needed to produce LBG. At these temperatures, common compounds such as water, CO_2, or hydrogen sulfide are in their solid state and have a very limited solubility in the liquid methane. In order to avoid plugging and freezing problems, some stringent purity requirements are valid for the gas entering a liquefaction step (Table 6.2)

Normally, the limiting purity requirement is the 50-ppm constraint for carbon dioxide, because it precipitates in the liquefaction unit at higher concentrations. This level is difficult to achieve with most upgrading technologies. At the moment, only optimized amine scrubbers and possibly cryogenic upgrading units are able to comply with the CO_2 constraint out of the box. Other upgrading plants such as water scrubbers, PSA units, and membrane separation plants must be completed with a polishing step to reduce the CO_2 concentration in the upgraded gas. Currently, the standard method used in this polishing step is a molecular sieve, working with temperature and pressure swing adsorption. Here, the difference in molecule size between methane and CO_2 is used to trap the CO_2 molecules, while methane molecules can more or less freely pass the columns. The purge gas from the polishing step typically contains between 60% and 70% methane (or 5%–10% of the total incoming methane) and can be returned to the raw gas inlet of the upgrading plant. The polishing step uses essentially the same technology as the PSA upgrading plants. However, due to the low amounts of CO_2 in the treated gas stream, the cycle time is much longer, typically several hours, so the valve wearing is not as critical as in the PSA upgrading plants. When the upgraded biogas is directly cooled down in an expansion valve, the stream leaves the valve in a two-phase regime where a part of the stream is in the liquid phase, whereas the other part still is gaseous. In this case, the cycle is constructed so that the gaseous fraction is recycled back to the compression step, while the liquid fraction is removed from the process as a product stream. As a side effect, the liquefaction process may be deployed to remove nitrogen impurities from the methane stream (nitrogen has a slightly lower boiling point than methane), which is especially interesting in the treatment of landfill gas. An alternative is to cool down the upgraded biogas in a heat exchanger with external cooling. In this case, the entire gas stream can be liquefied in one step.

Air Liquid Advanced Technologies has in 2012 commissioned their first plant for the liquefaction of biogas in Lidköping (Fig. 6.8), where it is part of a system for the production of liquid biogas. After a conventional fermentation and gas upgrading using a water scrubber plant owned by Swedish Biogas International, where gas with vehicle fuel quality is produced, the gas enters the liquefaction unit consisting of a temperature pressure swing adsorption unit-polishing step followed by the liquefaction process.

The technology for methane liquefaction is based on a reverse nitrogen Brayton cycle. This process extracts heat from methane, liquefying it in a plate-fin heat exchanger. The process can be divided into the following steps:

- Nitrogen is compressed inside a centrifugal compressor.
- Compressed nitrogen is then overpressurized in a turbobooster before entering the heat exchanger to be cooled down.
- The nitrogen stream is expanded, generating the cold power for the system at a temperature of 110 K (−163°C).
- The low-pressure stream flows back in the plate heat exchanger and cools down both methane and the high-pressure nitrogen stream.
- Bio methane is liquefied at 110 K and is released at a pressure of 3.5 bar. The outlet pressure can be adjusted depending on the customer requirements.

FIGURE 6.8 The Lidköping biogas plant.

6.4.4 Liquefaction process of biogas using Aspen HYSYS simulation

The liquefaction of biogas appears to be an appropriate solution in the cases where logistics constraints arise due to the absence of a transportation network. In the liquid form, biomethane is easy to transport to its point of use. Aspen HYSYS, computer-aided process design programs, widely used in chemical and thermodynamic process industries, has built-in process models that can effectively simulate the liquefaction process. Liquefaction process of biogas was simulated in Aspen HYSYS with the aim to assess the technology using cryogenic process to liquefy the biogas and obtain the liquefied biomethane and CO_2 as a by-product through the simulation results. The parameters such as biogas feed, temperature, and pressure are obtained from the case study from the Havys Biogas Power Plant owned by Cenergi Sdn. Bhd. Hence, with the working principle of the cryogenic process, a design of the process of liquefaction of biogas was simulated in the Aspen HYSYS simulation environment, where it consists of three stages: precooling stage, liquefaction stage, and subcooling stage. As a result, with a biogas feed of $1500\,m^3/h$ consisting of 60% of CH_4 and 40% of CO_2 at 200 kPa and 35°C, liquefied biomethane was obtained from the design of the simulation with a purity of 99% and liquid CO_2 was also obtained as a by-product of the liquefaction process. Currently, most of the studies focus on how to improve the efficiency of biogas upgrading. There has not been a comprehensive analysis from the perspective of the liquefaction of the biogas process. The design of the liquefaction process of biogas uses Aspen HYSYS simulator to obtain the liquefied biomethane and liquid CO_2. The liquefaction of biomethane appears to be an appropriate solution in cases where logistics constraints arise due to the absence of a transportation network. In liquid form, biomethane is easy to transport to its point of use. The advantages of the liquefaction of biomethane process are when the energy consumption is optimal, the process is reliable and efficient, and the technology is well understood and well managed. When biogas is condensed to a liquid by cooling it to about −162°C. This process reduces its volume by a factor of more than 600. Thus this will allow the liquefied biomethane to be shipped safely and efficiently.

6.4.5 Biogas liquefaction market

Rapid growth in the population, and increasing urbanization and modernization across the globe are increasing the demand for energy. Currently, major energy demand is supplied by fossil fuels such as coal, oil, and gas. Fast depleting fossil fuels, environmental concerns, and rising pollution have led to steady shift to more environment friendly and renewable fuel sources of clean energy. Biogas is a clean, renewable, and biological source of energy. It is produced when an organic material such as agricultural waste, manure, municipal waste, plant material, sewage, green waste, or food waste is decomposed under anaerobic conditions. It is also produced under controlled conditions in a digester chamber. Biogas primarily consists of methane, CO_2, hydrogen sulfide, moisture, and siloxanes. It is purified, processed, and liquefied to be used as a fuel. This process is called liquefaction of biogas. In a biogas liquefaction plant, purified and upgraded biogas is converted into LBG under specific pressure and temperature. The purest kind of liquid biogas consists of approximately 97% methane. LBG is approximately 600 times more space competent than that in gaseous state; it is much easier to transport LBG. LBG is produced in two main ways such as cryogenic technology and conventional upgrading technology. Certain procedures such as biogas production, pretreatment, liquefaction, storage, and transport are carried out in a biogas liquefaction plant.

The biogas liquefaction industry is expected to register significant growth in the near future. Countries such as Sweden and Netherlands have started the use of LBG for running vehicles. Certain trains and trucks also use LBG as fuel. Public transport buses in Sweden and other regions now use LBG as fuel. Cargo ships have also started the use of LBG in certain regions. Production of biogas also leads to proper waste management. LBG burns more cleanly and emits less CO_2, which helps in controlling the global warming and environmental pollution. As petrol and diesel are polluting and impermanent fuels, many vehicles have switched to LBG. To support this rising demand for LBG, many biogas liquefaction plants and units have been installed worldwide. The global biogas liquefaction market can be segmented based on the technology and geography. The two main technologies for biogas liquefaction are cryogenic technology and conventional upgrading technology connected with a small-scale liquefaction plant. Europe and North America are comparatively mature markets for biogas liquefaction. The United States, Sweden, Switzerland, and Germany are the major drivers in these regions. Asia Pacific and the Middle East are in the nascent stage but are projected to register higher growth rate in the near future. In recent years, China has made huge investments in biogas plants and other alternate sources of energy. Australia has built many big- and small-sized biogas liquefaction plants in recent years. Rapid shift to LBG as a primary source of fuel in vehicles would spur the growth of biogas liquefaction plants.

An effort from South Korea, Indonesia, Malaysia, India, and South Africa to promote the use of biogas is expected to enhance the growth in these markets. High biogas demand could directly or indirectly fuel the demand for biogas liquefaction plants across the globe. As per the recent regulations in Sweden, the country imports waste to produce biogas, which may fuel additional growth in the biogas liquefaction market. As many as 11 new biogas liquefaction plants have been built in Sweden. France, Sweden, and United Kingdom have committed to produce LBG with a goal to meet 20% of their energy demand through biogas. Comprehensive shift to renewable energy from fossil fuels may increase biogas demand. However, improper waste management could confine the market growth. However, recent global shift to renewable and clean energy along with technological developments in biogas processing and applications around the globe points toward a positive outlook for the biogas liquefaction industry.

Some of the key players in the global biogas liquefaction market include Ameresco Inc., APROVIS Energy Systems GmbH, Awite Bioenergie GmbH, BioConstruct GmbH, Eisenmann SE, Energy & Waste S.L., Enspar GmbH, EnviTec Biogas AG, ETW Energietechnik GmbH, Groth Corporation, Himark BioGas, Nova Analytical Systems Inc., and Schmack Biogas GmbH.

The report offers a comprehensive evaluation of the market. It does so via in-depth qualitative insights, historical data, and verifiable projections about the market size. The projections featured in the report have been derived using proven research methodologies and assumptions. By doing so, the research report serves as a repository of analysis and information for every facet of the market, including but not limited to regional markets, technology, types, and applications.

The study is a source of reliable data on the following:

- market segments and subsegments;
- market trends and dynamics;
- supply and demand;
- market size;
- current trends/opportunities/challenges;
- competitive landscape;
- technological breakthroughs; and
- value chain and stakeholder analysis.

The regional analysis covers the following:

- North America (United States and Canada);
- Latin America (Mexico, Brazil, Peru, Chile, and others);
- Western Europe (Germany, United Kingdom, France, Spain, Italy, Nordic countries, Belgium, Netherlands, and Luxembourg);
- Eastern Europe (Poland and Russia);
- Asia Pacific (China, India, Japan, ASEAN, Australia, and New Zealand); and
- Middle East and Africa (Gasification combined cycle (GCC), South Africa, and North Africa).

The report has been compiled through extensive primary research (through interviews, surveys, and observations of seasoned analysts) and secondary research (which entails reputable paid sources, trade journals, and industry body databases). The report also features a complete qualitative and quantitative assessment by analyzing data gathered from industry analysts and market participants across key points in the industry's value chain.

A separate analysis of prevailing trends in the parent market, macro- and microeconomic indicators, and regulations and mandates is included under the purview of the study. By doing so, the report projects the attractiveness of each major segment over the forecast period.

6.5 Process of gasification

Gasification is a process that converts organic or fossil fuel-based carbonaceous materials into carbon monoxide, hydrogen, and CO_2. This is achieved by reacting the material at high temperatures ($>700°C$), without combustion, with a controlled amount of oxygen and/or steam. The resulting gas mixture is called syngas (from synthesis gas) or producer gas and is itself a fuel. The power derived from gasification and combustion of the resultant gas is considered a source of renewable energy if the gasified compounds were obtained from biomass. The advantage of gasification is that using the syngas (synthesis gas H_2/CO) is potentially more efficient than direct combustion of the original fuel, because it can be combusted at higher temperatures or even in fuel cells, so that the thermodynamic upper limit to the efficiency defined by Carnot's rule is higher or (in the case of fuel cells) not applicable. Syngas may be burned directly in gas engines, used to produce methanol and hydrogen, or converted via the F-T process into synthetic fuel. Gasification can also begin with the material that would otherwise have been disposed of such as biodegradable waste. In addition, the high-temperature process refines out corrosive ash elements such as chloride and potassium, allowing clean gas production from otherwise problematic fuels. Gasification of fossil fuels is currently widely used on industrial scales to generate electricity.

6.5.1 How gasification works?

Gasification is most simply thought of as choked combustion or incomplete combustion. It is burning solid fuels like wood or coal without enough air to complete combustion, so the output gas still has combustion potential. The unburned gas is then piped away to burn elsewhere as needed.

Gas produced by this method goes by a variety of names: *wood gas, syngas, producer gas, town gas, generator gas*, and others. It is sometimes also called *biogas*, though biogas more typically refers to the gas produced via microbes in anaerobic digestion. In the context of biomass gasification using air-aspirated gasifiers, the term *producer gas* is the term we will be using, since the other terms have implications that do not necessarily apply to the gas produced by our gasifiers.

6.5.2 Five process of gasification

True gasification is a bit more than just the choked combustion summary presented above. It is more accurately understood as *staged combustion*. It is a series of distinct thermal events put together to convert solid organic matter into specific HC gases as output. Simple incomplete combustion is a dirty mess. The goal in gasification is to take control of the discrete thermal processes usually mixed together in combustion, and reorganize them toward desired end products. In digital terms, "Gasification is the operating system of fire." Once you understand its underlying code, you can pull fire apart and reassemble it to your will, as well as a stunning variety of end products and processes. Fig. 6.9 shows the process of gasification.

Gasification is made up of five discrete thermal processes: *drying*, *pyrolysis*, *combustion*, *cracking*, and *reduction*. All these processes are naturally present in the flame you see burning a match, though they mix in a manner that renders them invisible to eyes not yet initiated into the mysteries of gasification. Gasification is merely the technology to pull apart and isolate these separate processes, so that we might interrupt the "fire" and pipe the resulting gases elsewhere. Fig. 6.10 shows the process of gasification.

Three of these processes tend to confuse all newcomers to gasification. Once you understand these three processes, all the other pieces fall in place quickly. These three nonobvious processes are pyrolysis, cracking, and reduction. Here is the quick cheat sheet.

6.5.3 Pyrolysis

Pyrolysis is the application of heat to raw biomass, in the absence of air, so as to break it down into charcoal and various tar gasses and liquids. It is essentially the process of charring. Biomass begins to decompose rapidly with heat once its temperature rises above around 240°C. The biomass breaks down into a combination of solids, liquids, and gasses. The solids that remain we commonly call *charcoal*. The gasses and liquids that are released

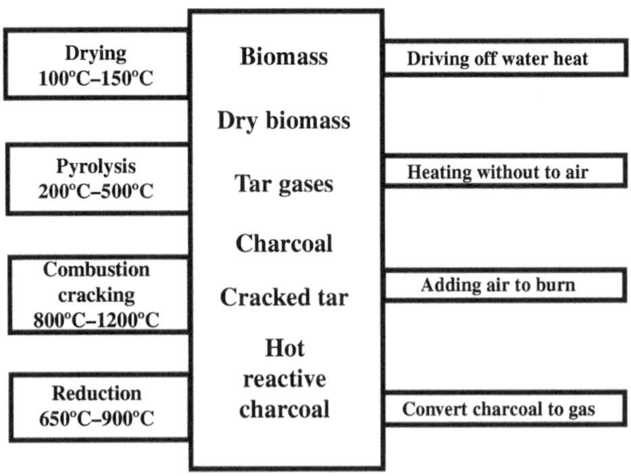

FIGURE 6.9 Five steps of gasification.

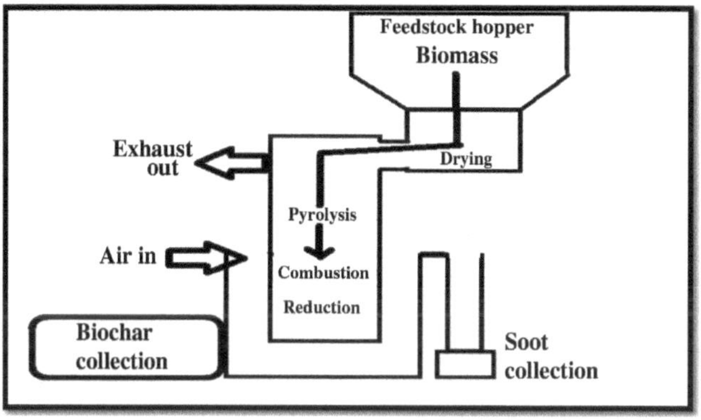

FIGURE 6.10 Process of gasification.

we collectively call *tars*. The gasses and liquids produced during lower temperature pyrolysis are simply fragments of the original biomass, which break off with heat. These fragments are the more complicated H, C, and O molecules in the biomass that we collectively refer to as volatiles. As the name suggests, volatiles are reactive. Or more accurately, they are less strongly bonded in the biomass than the fixed carbon, which are the direct C–C bonds. The input to gasification is some form of solid carbonaceous material, typically biomass or coal. All organic carbonaceous material is made up of carbon (C), hydrogen (H), and oxygen (O) atoms though in a dizzying variety of molecular forms. The goal in gasification is to break down this wide variety of forms into simple fuel gasses of H_2 and CO. Both hydrogen and carbon monoxide are burnable fuel gasses. We do not usually think of carbon monoxide as a fuel gas, but it actually has very good combustion characteristics (despite its poor characteristics when interacting with human hemoglobin). Carbon monoxide and hydrogen have about the same energy density by volume. Both are very clean, as they only need to take on one oxygen atom during burning, in one simple step, to arrive at the proper end states of combustion, CO_2 and H_2O. This is why an engine run on producer gas can have such clean emissions. The engine becomes the "afterburner" for the dirtier and difficult early stages of combustion that now are handled in the gasifier. Thus in review, pyrolysis is the application of heat to biomass in the absence of air/oxygen. The volatiles in the biomass are evaporated off as tar gases, and the fixed carbon-to-carbon chains are what otherwise known as charcoal remains.

6.5.3.1 *Cracking*

Cracking is the process of breaking down large complex molecules such as tar into lighter gases by exposure to heat. This process is crucial for the production of clean gas that is compatible with an internal combustion engine, because tar gases condense into sticky tar that will rapidly foul the valves of an engine. Cracking is also necessary to ensure proper combustion, because complete combustion only occurs when combustible gases thoroughly mix with oxygen. In the course of combustion, the high temperatures produced decompose the large tar molecules that pass through the combustion zone.

6.5.3.2 *Reduction*

Reduction is the process of stripping oxygen atoms off the combustion products of HC molecules, to return the molecules to forms that can burn again. Reduction is the direct reverse process of combustion. Combustion is the combination of combustible gases with oxygen to release heat, producing water vapor and CO_2 as waste products. Reduction is the removal of oxygen from these waste products at high temperature to produce combustible gases. Combustion and reduction are equal and opposite reactions. In fact, in most burning environments, they are both operating simultaneously, in some form of dynamic equilibrium, with repeated movements back and forth between the two processes. Reduction in a gasifier is accomplished by passing CO_2 or water vapor (H_2O) across a bed of red hot charcoal (C). The carbon in the hot charcoal is highly reactive with oxygen; it has such a high oxygen affinity that it strips the oxygen off the water vapor and CO_2, and redistributes it to as many single bond sites as possible. The oxygen is more attracted to the bond site on the C than to itself; thus no free oxygen can survive in its usual diatomic O_2 form. All available oxygen will bond to available C sites as individual O until all the oxygen is gone. When all the available oxygen is redistributed as single atoms, reduction stops. Through this process, CO_2 is reduced by carbon to produce two CO molecules, and H_2O is reduced by carbon to produce H_2 and CO. Both H_2 and CO are combustible fuel gases, and those fuel gasses can then be piped off to do desired work elsewhere.

6.5.3.3 *Combustion and drying*

These are the most easily understood of the five processes of gasification. They do what we think by common understanding, though now they do it in the service of pyrolysis and reduction. Combustion is the only net exothermic process of the five processes of gasification; ultimately, all of the heat that drives drying, pyrolysis, and reduction either comes directly from combustion or is recovered indirectly from combustion by the heat exchange processes in a gasifier. Combustion can be fueled by either the tar gasses or the char from pyrolysis. Different reactor types use one or the other, or both. In a downdraft gasifier, we are trying to burn the tar gasses from pyrolysis to generate heat to run reduction, as well as the CO_2 and H_2O to reduce in reduction. The goal in combustion in a downdraft is to get good mixing and high temperatures so that all the tars are either burned or cracked, and thus will not be present in the outgoing gas. The char bed and reduction contribute a relatively little to the conversion of messy tars to useful fuel gasses. Solving the tar problem is mostly an issue of tar cracking in the combustion zone. Drying is what removes the moisture in the biomass before it enters pyrolysis. All the moisture needs to be (or will be) removed from the fuel before any above 100°C process happen. All of the water in the biomass will be vaporized out of the fuel at some point in the higher temperature processes. Where and how this happens is one of the major issues that has to be solved for successful gasification. High moisture content fuel, and/or poor handling of the moisture internally, is one of the most common reasons for failure to produce clean gas. More simply, you might just think of gasification as burning a match, but interrupting the process by piping off the clear gas you see right above the match, not letting it mix with oxygen, and complete combustion. Or you might think of it as running your car engine extremely rich, creating enough heat to break apart the raw fuel, but without enough oxygen to complete combustion, thus sending burnable gasses out the exhaust. This is how a hot rod gets flames out the exhaust pipes.

6.5.3.4 *Types of gasifier*

Gasifier equipment is generally classified as upward draft, downward draft, and cross-draught gasifier, based on the direction of air/oxygen flow in the equipment. It should be noted that there are types of gasifier equipment, which are different from the types of gasification processes. Gasification processes can be categorized into three groups: entrained flow, fluidized bed, and moving bed (sometimes called, somewhat erroneously, fixed bed).

6.5.3.4.1 Updraught or countercurrent gasifier

The oldest and simplest type of gasifier is the countercurrent or updraught gasifier shown schematically in Fig. 6.11. The air intake is at the bottom and the gas leaves at the top. Near the grate at the bottom, the combustion reactions occur, which are followed by reduction reactions somewhat higher up in the gasifier. In the upper part of the gasifier, heating and pyrolysis of the feedstock occur as a result of heat transfer by forced convection and radiation from the lower zones. The tars and volatiles produced during this process will be carried in the gas stream. Ashes are removed from the bottom of the gasifier.

The major advantages of this type of gasifier are its simplicity, high charcoal burnout, and internal heat exchange, leading to low gas exit temperatures and high equipment efficiency, as well as the possibility of operation with many types of feedstock (sawdust, cereal hulls, etc.).

Major drawbacks result from the possibility of "channeling" in the equipment, which can lead to oxygen breakthrough and dangerous, explosive situations and the necessity to install automatic moving grates, as well as from the problems associated with the disposal of the tar-containing

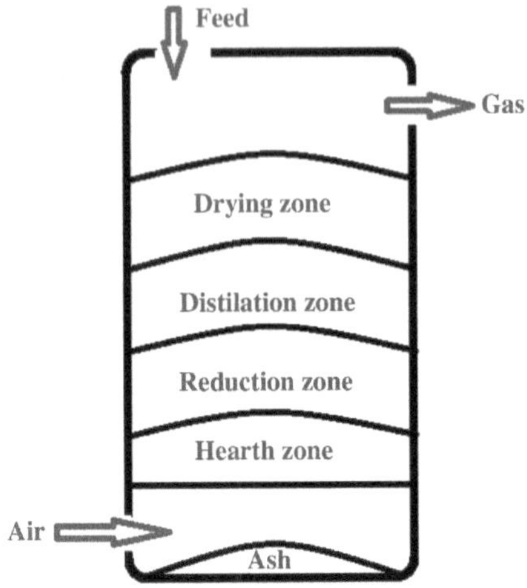

FIGURE 6.11 Updraught gasifier.

condensates that result from the gas cleaning operations. The latter is of minor importance if the gas is used for direct heat applications, in which case the tars are simply burnt. Fig. 6.11 shows the schematic diagram of the updraught gasifier.

6.5.3.4.2 Downdraught or cocurrent gasifier

A solution to the problem of tar entrainment in the gas stream has been found by designing the cocurrent or downdraught gasifiers, in which primary gasification air is introduced at or above the oxidation zone in the gasifier. The producer gas is removed at the bottom of the apparatus, so that fuel and gas move in the same direction, as schematically shown in Fig. 6.12.

On their way down the acid and tarry distillation products from the fuel must pass through a glowing bed of charcoal and, therefore, are converted into permanent gases hydrogen, CO_2, carbon monoxide, and methane. Depending on the temperature of the hot zone and the residence time of the tarry vapors, a more or less complete breakdown of the tars is achieved. The main advantage of the downdraught gasifiers lies in the possibility of producing a tar-free gas suitable for engine applications. In practice, however, a tar-free gas is seldom if ever achieved over the whole operating range of the equipment: tar-free operating turndown ratios of a factor 3 are considered standard; a factor 5–6 is considered excellent. Because of the lower level of organic components in the condensate, downdraught gasifiers suffer less from environmental objections than the updraught gasifiers.

A major drawback of the downdraught equipment lies in its inability to operate on a number of unprocessed fuels. In particular, fluffy, low-density materials give rise to flow problems and excessive pressure drop, and the

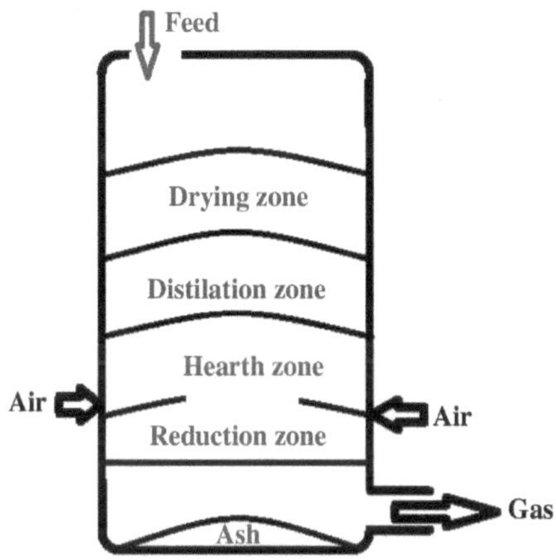

FIGURE 6.12 Downdraughts or cocurrent gasifier.

solid fuel must be pelletized or briquetted before use. Downdraught gasifiers also suffer from the problems associated with high ash content fuels (slagging) to a larger extent than the updraught gasifiers. Minor drawbacks of the downdraught system, as compared with the updraught, are somewhat lower efficiency resulting from the lack of internal heat exchange as well as the lower heating value of the gas. Besides this, the necessity to maintain uniform high temperatures over a given cross-sectional area makes impractical the use of downdraught gasifiers in a power range above about 350 kW (shaft power).

6.5.3.4.3 Cross-draught gasifier

Cross-draught gasifiers are an adaptation for the use of charcoal. Charcoal gasification results in very high temperatures (1500°C and higher) in the oxidation zone, which can lead to material problems. In cross-draught gasifiers, insulation against these high temperatures is provided by the fuel (charcoal) itself. Advantages of the system lie in the very small scale at which it can be operated. Installations below 10 kW (shaft power) can under certain conditions be economically feasible. The reason is the very simple gas-cleaning train (only a cyclone and a hot filter), which can be employed when using this type of gasifier in conjunction with small engines. A disadvantage of cross-draught gasifiers is their minimal tar-converting capabilities and the consequent need for high-quality (low-volatile content) charcoal. It is because of the uncertainty of charcoal quality that a number of charcoal gasifiers employ the downdraught principle, in order to maintain at least a minimal tar-cracking capability.

6.5.3.4.4 Fluidized bed gasifier

The operation of both up- and downdraught gasifiers is influenced by the morphological, physical, and chemical properties of the fuel. Problems commonly encountered are the lack of bunkerflow, slagging, and extreme pressure drop over the gasifier Air is blown through a bed of solid particles at a sufficient velocity to keep these in a state of suspension. The bed is originally externally heated and the feedstock is introduced as soon as a sufficiently high temperature is reached. The fuel particles are introduced at the bottom of the reactor, very quickly mixed with the bed material and almost instantaneously heated up to the bed temperature. As a result of this treatment, the fuel is pyrolyzed very fast, resulting in a component mix with a relatively large amount of gaseous materials. Further gasification and tar-conversion reactions occur in the gas phase. Most systems are equipped with an internal cyclone in order to minimize char blowout as much as possible. Ash particles are also carried over the top of the reactor and have to be removed from the gas stream if the gas is used in engine applications. The major advantages of the fluidized bed gasifiers, as reported by Van der Aarsen and others, stem from their feedstock flexibility resulting from easy control of temperature, which can be kept below the melting or fusion point of the ash (rice husks), and their ability to deal with fluffy and fine grained materials (sawdust, etc.) without the need for preprocessing. Problems with feeding, instability of the bed, and fly-ash sintering in the gas channels can occur with some biomass fuels. Other drawbacks of the fluidized bed gasifier lie in the rather high tar content of the product gas (up to 500-mg/m^3 gas), the incomplete carbon burnout, and poor response to load changes. Particularly, because the control equipment needed to cater for the latter difficulty, very small fluidized bed gasifiers are not foreseen and the application range must be tentatively set at above 500 kW (shaft power). Fluidized bed gasifiers are currently available on a semicommercial basis from several manufacturers in Europe and United States.

In principle, gasification can proceed from just about any organic material, including biomass and plastic waste. The resulting syngas can be combusted. Alternatively, if the syngas is clean enough, it may be used for power production in gas engines, gas turbines or even fuel cells, or converted efficiently to dimethyl ether by methanol dehydration, methane via the Sabatier reaction, or diesel-like synthetic fuel via the F-T process. In many gasification processes most of the inorganic components of the input material, such as metals and minerals, are retained in the ash. In some gasification processes (slagging gasification) this ash has the form of a glassy solid with low leaching properties, but the net power production in slagging gasification is low (sometimes negative) and costs are higher.

Regardless of the final fuel form, gasification itself and subsequent processing neither directly emit nor trap greenhouse gases such as CO_2. Power consumption in the gasification and syngas conversion processes may be significant though and may indirectly cause CO_2 emissions; in slagging and plasma gasification, the electricity consumption may even exceed any power production from the syngas.

Combustion of syngas or derived fuels emits exactly the same amount of CO_2 as would have been emitted from direct combustion of the initial fuel. Biomass gasification and combustion could play a significant role in a renewable energy economy, because biomass production removes the same amount of CO_2 from the atmosphere as is emitted from gasification and combustion. While other biofuel technologies such as biogas and biodiesel are carbon-neutral, gasification in principle may run on a wider variety of input materials and can be used to produce a wider variety of output fuels.

There are at present a few industrial-scale biomass gasification plants. Since 2008, in Svenljunga, Sweden, a biomass gasification plant has been generating up to 14 MW_{th}, supplying industries and citizens of Svenljunga with process steam and district heating, respectively. The gasifier uses biomass fuels such as Chronium copper arsenate (CCA) or creosote-impregnated waste wood and other kinds of recycled wood to produces syngas that is combusted on site. In 2011 a similar gasifier, using the same kinds of fuels, is being installed at Munkfors Energy's CHP plant. The CHP plant will generate 2 MW_e (electricity) and 8 MW_{th} (district heating).

Examples of demonstration projects include:

- The 32-MW dual fluidized bed gasification of the GoBiGas project in Gothenburg, Sweden, produced around 20 MW of substitute natural gas from forest residues and fed it into the natural gas grid since December 2014. The plant was permanently closed due to technical and economic problems in April 2018. Göteborg Energi had invested €175 million in the plant and intensive attempts to sell the plant to new investors had failed for a year.
- Those of the Renewable Energy Network Austria, including a plant using dual fluidized bed gasification that has supplied the town of Güssing with 2 MW of electricity, produced utilizing reciprocating gas engines and 4 MW of heat, generated from wood chips, since 2001.
- Go Green Gas' pilot plant in Swindon, United Kingdom, has demonstrated methane production from waste feedstocks at 50 kW. The project has prompted the construction of a £25 million commercial facility that aims to generate 22 GWh/annum of grid-quality natural gas from waste wood- and refuse-derived fuel.

Exercise

1. What is liquefaction?
2. What are the classifications of liquefaction?

3. What is direct liquefaction?
4. What is indirect liquefaction?
5. What are the different elements of liquefaction?
6. Explain the process of Fisher-Tropsch synthesis.
7. Explain the process of methanol synthesis.
8. Write the basic aspects of direct liquefaction of biomass.
9. Explain the process of thermochemical conversion of biomass.
10. What is carbonization?
11. Explain the five steps process of gasification.
12. Explain the working of a gasifier.

Chapter 7

Advances in biogas power plant

Chapter outline

Design and Optimization of Biogas Energy Systems. DOI: https://doi.org/10.1016/B978-0-12-822718-3.00007-1

Objectives

- To provide knowledge about environmental assessment of biogas plant
- To provide knowledge about economic assessment of biogas plant
- To provide knowledge about assessment of biogas plant by game theory
- To provide knowledge about assessment of biogas plant by big data

7.1 Environmental assessment of biogas power plant

The environmental benefits of biogas technology are often highlighted, as a valid and sustainable alternative to fossil fuels. Together with the reduction of greenhouse gas (GHG) emissions, biogas can enhance energy security thanks to its high energetic potential. As a renewable energy source, it allows exploiting agricultural and zoo technical byproducts and municipal wastes, with a lower impact on air quality than combustion-based strategies for these biomasses. Furthermore, while ashes from combustion find scarce agronomic applications, the by-product of anaerobic digestion, that is, digestate, looks as a reliable material for agricultural uses. Another important advantage of biogas technology is its easy scalability, allowing exploiting the energetic potential of decentralized biomass sources. Finally, biogas can be upgraded to biomethane, suitably used as a vehicle fuel, or injected into national natural gas grids. A main objective of biogas industry is the reduction of fossil fuel consumption, with the final goal of mitigating global warming. However, anaerobic digestion is associated with the production of several GHGs, namely carbon dioxide, methane (CH_4) and nitrous oxide (N_2O). As a consequence, dedicated measures should be taken in order to reduce these emissions. According to Hijazi, the main measures to improve the global warming reduction potential of biogas plants are to use a flare avoiding CH_4 discharge, to cover tanks, to enhance the efficiency of combined heat and power (CHP) units, to improve the electric power utilization strategy, to exploit as much thermal energy as possible, and to avoid leakages. Similar conclusions were obtained by Buratti and coworkers for the specific case study of cereal crops in Umbria, Italy. Biomethane chain exceeds the minimum value of GHG saving (35%) mainly due to the open storage of digestate; usual practices to improve GHG reduction (up to 68.9%) include using heat and electricity produced by the biogas CHP plant, covering digestate storage tanks. The impact induced by biogas plants on global warming needs to be studied case by case. Bachmaier and coworkers calculated the GHG impact of 10 agricultural biogas plants. GHG emissions coming from electricity production in the investigated biogas plants ranged from −85- to 251-g CO_2-eq/kWh_{el}, and the GHG saving was 2.31−3.16 kWh_{fossil}/kWh_{el}. The results obtained also highlighted that reliable estimates of GHG emissions in the case of electricity production from biogas can be only made on the basis of individual monitoring data, for instance, reduction of direct CH_4 emission and leakage, exploiting of heat obtained from cogeneration, amount and nature of input material, N_2O emission (e.g., from energy crop cultivation), and digestate management. Battini and coworkers, in a case study of an intensive dairy farm situated in the Po valley (Italy), calculated a GHG emission reduction due to anaerobic digestion ranging between −23.7% and −36.5%, depending on digestate management. In a Finnish case study, the GHG release reduction was estimated to be equal to 177.0, 87.7, and 125.6 Mg of CO_2 eq./year for dairy cows, sow, and pig farms, respectively. Optimizing all process parameters looks important with regard to the final environmental impact, for instance, a specific case study on wastewater treatment showed that the process optimization could result into the emission abatement equal to 1103-kg CO_2 eq./day for N_2O, 256 kg eq./day for CO_2, and 87-kg CO_2 eq./day for CH_4.

7.1.1 Carbon dioxide emissions

Harmful compounds and air contaminants are introduced into the environment during biogas production and use through both combustion processes and diffusive emissions. Considering carbon dioxide, combustion of biogas leads to efficient CH_4 oxidation and conversion into CO_2, with a rate of 83.6 kg per GJ (based on a biogas with 65% CH_4 and 35% CO_2). Other releases of this contaminant are related to transport and storage of biomass, as well as digestate use. In the case of both biogas combustion and biomass/digestate emission, CO_2 is considered as biogenic and calculated neutral with regard to the impact on climate. Taking into account the reduction of fossil fuel, it can be demonstrated that biogas production leads globally to mitigation of anthropogenic greenhouse impact of the environment. Poeschl and coworkers have investigated the CO_2 emissions associated with biogas production from several feedstocks, and the relative contribution of feedstock supply, biogas plant operation and infrastructure, biogas utilization, and digestate management.

7.1.1.1 Methane emissions

CH_4 released by biogas processes is not considered relevant for health issues: though exposure to hydrocarbon mixtures can have some adverse effects on humans, no evidence exists of relevant interactions between CH_4 and biologic systems. However, CH_4 is a GHG whose global warming power is estimated to be 28−36 times higher than CO_2 over 100 years: as such, it is the second major component among anthropogenic greenhouse chemicals. Hence, in evaluating the impact of biogas industry on climate change, CH_4 emissions are a point of primary importance. CH_4 can be released during biogas incomplete combustion; however, a strong contribution to this contaminant comes out from diffusive emission related to biomass storage and digestate management. On the other hand, other biomass management strategies must be taken into account to abate emissions related to biogenic CH_4. In the above-mentioned study of Poeschl and coworkers, CH_4 emissions were also discussed; in all investigated cases, the emission rates were below 5 g/kg. Considering cattle manure, important reductions in CH_4 emission are related to digestate processing and handling, since this kind of biomass is characterized by high CH_4 emission rate when spread in the field without any pretreatment.

7.1.2 Impact of feedstock and digestate storage and treatment

In the biogas combustion management, feedstock and digestate storage and treatments can be the most important processes to achieve the global warming benefits of biogas production processes. Indeed, the impact of a biogas plant on GHG emission is heavily influenced by feedstock storage: most of N_2O can be abated when a closed storage is used for manure and codigestion feeding. Emissions from uncovered biomass storage have also been identified as the main ammonia source along the whole biogas production chain, and closed storage is strongly advised.

In a specific French case study of anaerobic digestion and composting plant for municipal solid waste, Beylot and coworkers have identified the following four conditions for process operation, which highly influence the impact of the whole plant:

1. the features of degradation of the fermentable fraction;
2. the collection efficiency of gas streams released by biological operations;
3. the abatement effectiveness of collected pollutants; and
4. NO_x emission rate from biogas combustion.

The importance of digestate storage step has been highlighted by Battani and coworkers, in the above-mentioned case study of intensive dairy farm situated in the Po valley (Italy): GHG emission reduction due to AD, calculated as equal to −23.7%, can reach 36.5% when a gas-tight tank is used for digest storage. A proper design and management of feedstock and digestate storage units looks also important in order to mitigate the odor impact of the plant. Indeed, the two major sources of the olfactory annoyance are biomass storage, and production of biogas and digestate composting units; closed-operated hydrothermal hydrolysis has positive effects on overall fugitive odor control in plants; on the other hand, eventual fugitive emissions during high temperature and seemingly open pretreatments can be the principal source of odors. In conclusion, gas-tight storage should always be advised, since the corresponding GHG and ammonia fugitive emissions are even more important for those coming from fertilizers. As mentioned above, avoiding leakages and using closed tanks are among the most important ways to reduce the global warming impact of biogas plants.

7.2 Economic assessment of biogas power plant by optimization technique

In simple terms, economics can be defined as a discipline that studies the behavior patterns of human being. The main aim of economics is to analyze how individuals, households, organizations, and nations use their scarce resource to achieve maximum profit.

7.2.1 Basic economics problems

Every economy has some root problems related to the proper allocations and utilizations of resources. In the case of biogas energy system, economy of the overall system mainly depends on the perfect utilization of waste material with effective solid concentration, pH value, etc. Some problems are emerging because the world's prosperous, developed, and rich economic resources available are scarce and limited in nature. Fig. 7.1 shows the basic economic problem of biogas power plant.

- *What to produce*: This is the first problem of the economic condition of biogas energy system and an economy can resolve it by determining the type and quantity of electricity generated by the biogas power plant at minimum losses. However, from the biogas industry point of view, both consumer and producers act together to decide the perfect utilization of waste material for generation of electricity.
- *How to produce*: This question is related to the selection of biogas technology of generation of electricity to be used. It also determines different factor of prefeasibility analysis of biogas energy system and the allocations of various factors of electricity generation, labor, capital, and land cost.

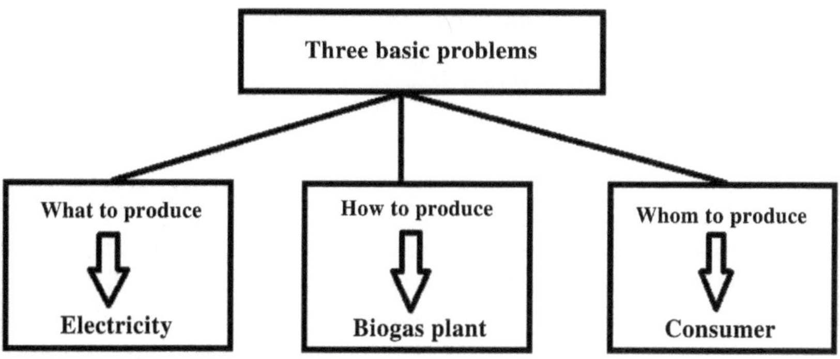

FIGURE 7.1 Three basic economic problems.

- *For whom to produce*: This problem is also called the "problem of distribution" and it is related to the distribution of electricity for consumer and in rural areas. In other words, it determines the share of electricity, when we utilize the hybrid biogas energy system and services that an individual receives from an economy with their given level of income.

7.2.2 Supply function of biogas power plant

The electricity supply function is the mathematical expression of the law of supply of electricity, which is a part of economics. In other words, supply function quantifies the relationship between electricity supplied and the price of the electricity for the consumer, while keeping the other factors constant. The supply function can be expressed as:

$$Q_{\mathrm{SE}} = f(P_{\mathrm{AE}}, P_{\mathrm{AB}}, P_{\mathrm{CE}}, T, G) \tag{7.1}$$

where Q_{SE} is the electricity supply; P_{AE} is the price of electricity supply; P_{AB} is the price of other factors of biogas energy system; P_{CE} is the price of factor input of biogas energy system; T is the technology used in biogas energy system; G is the government policy toward biogas energy system.

7.2.3 Total utility and marginal utility

Total electricity utility is defined as the sum of the utility derived from an electric consumer who takes electricity from different units of biogas power plant. Assume that an individual consumer takes electricity from five units of biogas power plant X at a given period of time and derives utility out of the consumptions of each unit as $U1$, $U2$, $U3$, $U4$, and $U5$. The total utility is measured as follows:

$$\mathrm{TU} = U1 + U2 + U3 + U4 + U5 \tag{7.2}$$

Apart from total utility, the concept of marginal utility is equally important for utility analysis. It can be defined as the addition to the total utility of a commodity resulting from the consumption of an additional unit.

$$\text{Marginal Utility} = \frac{\text{Change of Total Utility}}{\text{Change in Quantity}} \tag{7.3}$$

7.2.4 Elasticity of electricity demand from biogas power plant

Income elasticity of representation of sensitivity of the quantity demanded of a certain product to a change in real income of buyers who buy this product, keeping all other things to remain constant. The formula for calculating the income elasticity of demand is the percent change in quantity demanded divided by the percent change in income.

$$\text{Income Elasticity of Demand} = \frac{\%\ \text{Change in Quantity Demanded}}{\%\ \text{Change in Consumer Income}} \tag{7.4}$$

$$\text{Income Elasticity of Demand} = \frac{\frac{Q_F - Q_I}{0.5(Q_F + Q_I)}}{\frac{I_F - I_I}{0.5(I_F + I_I)}} \tag{7.5}$$

Q.7.1: Suppose the monthly income from biogas power plants is increased from \$10,000 to \$15,000, which increases his demand for electricity from 20 to 25 kWh. Calculate the income elasticity of demand and interpret the result.

Solution:

Here, Q_F and Q_I is the final and the initial quantity, respectively, and I_F and I_I final and initial income of the consumer.

Here, the value of $Q_F = 25$, $Q_I = 20$, $I_F = 15{,}000$, and $I_I = 10{,}000$.

$$\%\ \text{Change in quantity of clothes} = \frac{Q_F - Q_I}{0.5(Q_F + Q_I)} = \frac{25 - 20}{0.5(25 + 20)} = 0.22$$

$$= 0.22 \times 100 = 22\%$$

$$\%\ \text{Change in income level} = \frac{I_F - I_I}{0.5(I_F + I_I)} = \frac{15{,}000 - 10{,}000}{0.5(15{,}000 + 10{,}000)} = 0.4$$

$$= 0.4 \times 100 = 40\%$$

Now put the value in Eq. (7.5)

Income Elasticity of Demand $= \frac{0.22}{0.4} = 0.55$.

Since electricity units have the positive income elasticity of demand, they are normal goods and also called superior goods.

7.2.5 Cross elasticity

Cross elasticity of demand is the ratio of the percentage change in quantity demanded of a product of percentage change in price of a related product.

Cross elasticity of demand can be calculated using the following formula:

$$\text{Cross Elasticity Demand } E_{A,B} = \frac{\%\ \text{Increase in Quantity Demanded of } A}{\%\ \text{Increase in Price of Product } B} \tag{7.6}$$

$$\text{Income Elasticity of Demand} = \frac{\frac{Q_F - Q_I}{0.5(Q_F + Q_I)}}{\frac{P_F - P_I}{0.5(P_F + P_I)}} \tag{7.7}$$

TABLE 7.1 Types of price elasticity of demand.

Numerical value	Types of price elasticity of demand	Condition
= Infinity	Perfectly elastic demand	Greater change in electricity demand for biogas energy system for smaller change in price of electricity.
= 0	Perfectly inelastic demand	No change in electricity demand for biogas energy system for smaller change in price of electricity.
>1	Relatively elastic demand	A change in electricity demand for biogas energy system greater than the price of electricity.
<1	Relatively inelastic demand	A change in electricity demand for biogas energy system less than the price of electricity.
= 1	Unitary elastic demand	A change in electricity demand for biogas energy system equivalent to the price of electricity.

where Q_F and Q_I are the final and initial quantities demanded of product A, respectively, and P_F and P_I are the final and initial prices of product B. Table 7.1 shows the types of prices elasticity of the system.

Q.7.2: The electricity quantity demanded from biogas power plant to run a school in rural areas has increased from 100 to 160 units with an increase in the price of the electricity from \$40 to\$ 50. Calculate the cross elasticity of demand between electricity and price.

Solution:

Here, the value of $Q_F = 25$, $Q_I = 20$, $I_F = 15{,}000$, and $I_I = 10{,}000$

$$\% \text{ Increase in electricity unit} = \frac{Q_F - Q_I}{0.5(Q_F + Q_I)} = \frac{160 - 100}{0.5(160 + 100)} = 0.46$$

$$= 0.46 \times 100 = 46\%$$

$$\% \text{ Increase in price} = \frac{P_F - P_I}{0.5(P_F + P_I)} = \frac{50 - 40}{0.5(50 + 40)} = 0.22$$

$$0.22 \times 100 = 22\%$$

Put the value in Eq. (7.7).

Cross elasticity of demand $= \frac{0.46}{0.22} = 2.09$.

7.2.6 Demand forecasting of electricity

The moving average method is used to calculate the mean of averages of the given data over a period of time. In demand forecasting, this method is used to calculate the overall trend of demand for a specific period of time and it also determines the demand for the upcoming period. Under this method, demand values of the "*N*" period data series are used to forecast the demand for "*N* + 1th" or a next period. The moving averages for the given period are calculated as follows:

Data Series	*N*1	*N*2	*N*3	*N*4	*N*5	*N*6

$$\text{Moving average(for } N \text{ period)} = \text{Aggregate value for } N \text{ period}/N$$

where Aggregate value for N years $= N1 + N2 + N3 + N4 + N5 + N6$

$N =$ Total number of periods

Q.7.3: Calculate two-period moving averages from the waste collection department in biogas power plant and forecast the demand for the year 2017 (Table 7.2).

Solution:

Two-period moving averages will be calculated as follows:

$$\text{Moving average for year } ''N'' = \frac{[(\text{Value in Year } ''n'') + (\text{Value for year } ''n-1''\text{)}]}{2}$$

$$\text{Moving average for year 2008} = 130 \text{ tones}$$

$$\text{Moving average for year 2009} = \frac{170 + 130}{2} = 150 \text{ tones}$$

$$\text{Moving average for year 2010} = \frac{190 + 170}{2} = 180 \text{ tones}$$

$$\text{Moving average for year 2011} = \frac{230 + 190}{2} = 210 \text{ tones}$$

$$\text{Moving average for year 2012} = \frac{240 + 230}{2} = 235 \text{ tones}$$

$$\text{Moving average for year 2013} = \frac{260 + 240}{2} = 250 \text{ tones}$$

$$\text{Moving average for year 2014} = \frac{290 + 260}{2} = 275 \text{ tones}$$

$$\text{Moving average for year 2015} = \frac{310 + 290}{2} = 300 \text{ tones}$$

TABLE 7.2 Yearwise waste material demand.

Year	Waste material demand (in tones)
2008	130
2009	170
2010	190
2011	230
2012	240
2013	260
2014	290
2015	310
2016	325
2017	?

$$\text{Moving average for year 2016} = \frac{325 + 310}{2} = 317.5 \text{ tones}$$

Forecast for the year 2017 will be equal to the moving averages for the year 2017 = 317.5 tones.

7.2.6.1 Optimization technique

Each genuine issue from financial to logical and designing fields is at last gone up against with a typical errand viz., enhancement. A streamlining issue can be characterized by indicating the arrangement of every achievable hopeful and a measure for assessing their value. The objective is to locate the best solution(s). In the plan of airfoils, for example, the parameters that characterize the geometry of the air foil are advanced to accomplish the coveted surface weight conveyance. In the plan of a satellite receiving wire, the radio wire design is enhanced to boost the fundamental shaft pick up, while limiting the side projection pick up. In robot direction, arranging, the position, introduction, speed, and quickening that indicates the robot direction are enhanced for possible hindrance free movement. Fig. 7.2 shows the flowchart of optimization technique.

Three main types of problems can be handled by the classical optimization techniques:

- single variable functions where overall cost of the biogas power plant; and
- multivariable functions with no constraints, in biogas power plant economic evaluation in terms of fixed cost, variable cost, and marginal cost, are the part of multivariable functions.

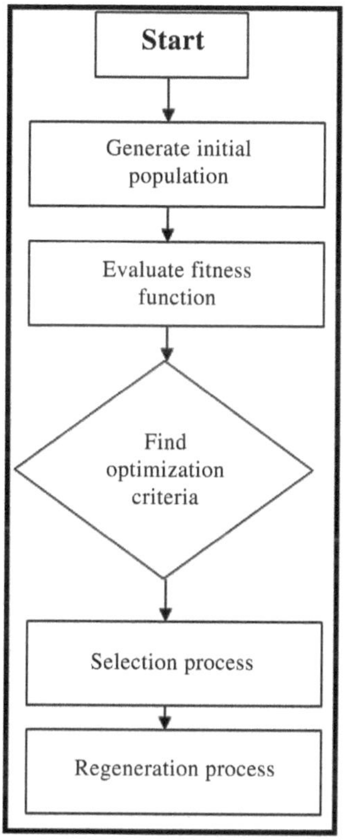

FIGURE 7.2 Flowchart of optimization technique.

Optimization technique programming is subdivided into the following category:

- *Linear programming*: This studies the case in which the objective function *f* of the cost functions of biogas power plant is linear and set A is specified using only linear equalities and inequalities (*A* is the design variable space).
- *Integer programming*: It studies linear programs in which some or all cost-optimized variables of biogas power plant are constrained to take on integer values.
- *Quadratic programming*: It allows the cost objective function of biogas power plant to have quadratic terms, while set A must be specified with linear equalities and inequalities.
- *Nonlinear programming*: This studies the general case which has the cost objective function of biogas power plants or the constraints or both contain nonlinear parts.

- *Stochastic programming*: It studies the case in which some of the constraints depend on random variables.
- *Dynamic programming*: This studies the case in which the optimal strategy is based on splitting the problem into smaller subproblems. In the case of biogas power plant, if we assess economic theory of overall plant, it is necessary for economic evaluation of individual component of biogas power plant.
- *Combinatorial optimization*: This is concerned with problems where the set of feasible solutions is discrete or can be reduced to a discrete one.

The objective function of biogas power plant is cost minimization function, which is optimized with different optimization techniques. At present, a large number of countries produce electricity by conventional or nonrenewable energy system. These systems produce large amount of atmospheric pollution. This problem is largely overcome by intensive use of alternative or renewable energy system. The cost minimization function of biogas power plant is given by following equation:

$$\begin{aligned}&(C^{i}_{\text{capital, Biogas Digester}} + C^{i}_{O\&M,\text{ Biogas Digester}} + C^{i}_{\text{Replacement, Biogas Digester}})\\&+(C^{i}_{\text{capital, mixing tank}} + C^{i}_{O\&M,\text{ mixing tank}} + C^{i}_{\text{Replacement, mixing tank}})\\&+\sum_{j=1}^{N_{\text{WT}}} N^{j}_{\text{Battery}} \times \left(C^{j}_{\text{capital, Battery}} + C^{i}_{O\&M,\text{ WT}} + C^{i}_{\text{Replacement, Battery}}\right)\\&+C_{\text{capital cost}_{\text{Biogas}}\text{ Generator}}\\&+C_{O\&M\text{ cost Biogas Generator}} + C_{\text{fuel cost}_{\text{Bio}}\text{ Generator}}\end{aligned} \tag{7.8}$$

Following are some techniques that are utilized for solved out the cost optimization function of biogas power plant.

7.2.6.2 Particle swarm optimization

Particle swarm optimization (PSO) is a developmental calculation, streamlining, procedure (a hunt strategy in view of a characteristic framework) created by Kennedy and Eberhart. The framework at first has a populace of arbitrary specific arrangements. PSO strategy can produce amazing arrangements inside shorter figuring time and have more steady merging qualities than other stochastic strategies. PSO is a met heuristic as it makes few or no suspicions about the issue being streamlined and can look extensive spaces of applicant arrangement. The decision of PSO parameter cans largely affect streamlining execution. In connection to PSO, the word unions normally means one of the two things, despite the fact that it is regularly not cleared up, which definition is implied and now and then they are erroneously thought to be indistinguishable. Initial step is to locate a potential arrangement. Every potential arrangement is known as a molecule. Every molecule is given an arbitrary speed and is flown through the issue space. The particles have memory and every molecule monitors its past best position (called as Pbest) and its relating wellness. There exist various Pbest for the separate particles in the swarm, and the molecule with the most prominent wellness is known as the worldwide best (gbest) of the swarm.

Swarm intelligence principle: Swarm intelligence can be described by considering five fundamental principles:

1. *Proximate principle*: The population should be able to carry out simple space and time computation.
2. *Quality principle*: The population should be able to respond to quality factor in the environment.
3. *Diverse response principle*: The population should not commit its activity along excessively narrow channels.

TABLE 7.3 Cost assumption of different ocean energy parameters.

Parameter		Minimum cost ($)	Maximum cost ($)
$C^i_{\text{capital, Biogas Digester}}$	Capital cost of biogas digester	160,000	190,000
$C^i_{\text{O\&M, Biogas Digester}}$	Operation and maintenance cost of biogas digester	50,000	54,000
$C^i_{\text{Replacement, Biogas Digester}}$	Replacement cost of biogas digester	7400	7800
$C^i_{\text{capital, mixing tank}}$	Capital cost of mixing tank	730,000	790,000
$C^i_{\text{O\&M, mixing tank}}$	Operation and maintenance cost of mixing tank	250,000	280,000
$C^i_{\text{Replacement, mixing tank}}$	Replacement cost of mixing tank	150,000	155,000
$C^j_{\text{capital, Battery}}$	Capital cost of battery	125,500	126,200
$C^i_{\text{O\&M, Battery}}$	Operation and maintenance cost of battery	12,050	12,200
$C^i_{\text{Replacement, Battery}}$	Replacement cost of battery	55,250	55,400
$C_{\text{capital cost\_Generator}}$	Capital cost of generator	248,000	253,000
$C_{\text{O\&M cost\_Generator}}$	Operation and maintenance cost of generator	290,000	297,000
$C_{\text{fuel cost\_Generator}}$	Replacement cost of generator	11,200	11,650

4. *Stability principle*: The population should not change its mode of behavior every time with the environment change.
5. *Adaptability change*: The population should be able to change its behavior made when it is worth the computational price.

We assume that the cost of different variables is then optimized by MATLAB code of PSO, and the big bang-big crunch (BB-BC) optimization technique for 25-MW ocean energy power plant. Table 7.3 shows the assumed cost data of the ocean energy parameter. Here, the cost of ocean energy system is classified into capital, operational and maintenance, and replacement cost of the device or equipment.

7.2.7 MATLAB code of particle swarm optimization for above 12 variables

```
clc
clear all
d =12; %number of variable
n=10; % number of popuation
x1min=160000;x1max=190000;x2min=50000;x2max=54000;x3min=7400;x3max=7800;x4min
=730000;x4max=790000;x5min=250000;x5max=280000;x6min=150000;x6max=155000;x7mi
n
=125500;
x7max=126200;x8min=12050;x8max=12200;x9min=55250;x9max=55400;x10min=248000;x1
0
max=253000;x11min=290000;x11max=297000;x12min=11200;x12max=11650;
wmin=0.6;wmax=0.9;
c1=2.01;c2=2.01;
for iter= 1:itermax
w = wmin+((wmax-wmin)/itermax).*iter;
end
x1 = x1min+rand(n,1,1).*(x1max-x1min);x2 = x2min+rand(n,1,1).*(x2max-x2min);x3 =
x3min+rand(n,1,1).*(x3max-x3min);x4 = x4min+rand(n,1,1).*(x4max-x4min);
x5 = x5min+rand(n,1,1).*(x5max-x5min);x6 = x6min+rand(n,1,1).*(x6max-x6min);x7 =
x7min+rand(n,1,1).*(x7max-x7min);x8 = x8min+rand(n,1,1).*(x8max-x8min);
```

```
x9 = x9min+rand(n,1,1).*(x9max-x9min);x10 = x10min+rand(n,1,1).*(x10max-x10min);x11 =
x11min+rand(n,1,1).*(x11max-x11min);
x12 = x12min+rand(n,1,1).*(x12max-x12min);
v = 0.4.*rand(n,d,1);
x = [x1,x2,x3,x4,x5,x6,x7,x8,x9,x10,x11,x12];
tic
for i = 1:n
f(i,:,1)=
x(i,1,1)+x(i,2,1)+x(i,3,1)+x(i,4,1)+x(i,5,1)+x(i,6,1)+x(i,7,1)+x(i,8,1)+x(i,9,1)+x(i,10,1)+x(i,11,
1
)+x(i,12,1);
end
[c,I]= min(f(:,:,1));
gbest(1,:,1) = x(i,:,1);
for i=1:n
for r =1:d
G(i,r,1)=gbest(1,r,1);
end
end
pbest = x;
v(i,:,2)= w.*v(i,:,1)+c1.*rand.*(pbest-x)+ c2.*rand.*(gbest-x);
x(i,:,2)= x(i,:,1)+v(i,:,2);
155
ofb(1,1,1)=
gbest(1,1,1)+gbest(1,2,1)+gbest(1,3,1)+gbest(1,4,1)+gbest(1,5,1)+gbest(1,6,1)+gbest(1,7,1)+g
be
st(1,8,1)+gbest(1,9,1)+gbest(1,10,1)+gbest(1,11,1)+gbest(1,12,1);
for j= 2:100
for i =1:n
f(i,:,j)=
x(i,1,1)+x(i,2,1)+x(i,3,1)+x(i,4,1)+x(i,5,1)+x(i,6,1)+x(i,7,1)+x(i,8,1)+x(i,9,1)+x(i,10,1)+x(i,11,
1
)+x(i,12,1);
end
[D1,I3] = min(abs(f(:,:,j)));
gbest(1,:,j)= x(I3,:,j);
ofb(1,:,j)=
gbest(1,1,1)+gbest(1,2,1)+gbest(1,3,1)+gbest(1,4,1)+gbest(1,5,1)+gbest(1,6,1)+gbest(1,7,1)+g
be
st(1,8,1)+gbest(1,9,1)+gbest(1,10,1)+gbest(1,11,1)+gbest(1,12,1);
[D,I] = min(abs(ofb(1,:,:)));
if ofb(1,:,j)<=D
gbest(1,:,j) = gbest(1,:,j);
else
gbest(1,:,j) = gbest(1,:,I);
end
```

```
for p =1:n
for r = 1:d
G(p,r,j) = gbest(1,r,j);
end
end
ofbest(1,:,j) =
gbest(1,1,1)+gbest(1,2,1)+gbest(1,3,1)+gbest(1,4,1)+gbest(1,5,1)+gbest(1,6,1)+gbest(1,7,1)+g
be
st(1,8,1)+gbest(1,9,1)+gbest(1,10,1)+gbest(1,11,1)+gbest(1,12,1);
for i = 1:n
[D,I]= min(f(i,:,:));
if f(i,:,j) <=D
pbest(i,:,j) = x(i,:,j);
else
pbest(i,:,j) = x(i,:,j);
end
end
v(:,:,j+1) = w.*v(:,:,j)+ c1.*rand(pbest(:,:,j)-x(:,:,j))+ c2.*rand(G(:,:,j)-x(:,:,j));
x(:,:,j+1) = x(:,:,j)+ v(:,:,j+1);
for i1= 1:n
for q1=1:d
if (x(i1,q1,j+1)<17)||(x(i1,q1,j+1)>22000)
x(i1,q1,j+1) = x(i1,q1,j);
else
x(i1,q1,j+1) = x(i1,q1,j+1);
end
end
156
end
end
```

After the optimization of 12 variables of the cost function through PSO, Table 7.4 shows the result of that algorithm and result shows that cost is reduced by the PSO algorithm.

- *Big bang-big crunch optimization*:

The BB-BC algorithm by two-phase's performances in the search space allowed navigates solution vector for minimizing the objective function. Global search is done in the first phase by this algorithm to find that the optimum solution exists in this region, and then the second phase is begun around the best answer obtained from the first phase.

The BB-BC approach takes the following steps:

1. Determine the center of mass that has global best fitness using equation. The candidates are arranged in the ascending order of their fitness (fitness) and the first candidate will be the candidate with the best fitness (minimum loss).
2. Generate new candidates around the center of mass by adding/subtracting a normal random number according to equation ().
3. Calculate the fitness function values of all the candidate solution.

TABLE 7.4 Cost assessment by particle swarm optimization.

Parameter		Cost ($)
$C^i_{\text{capital, Biogas Digester}}$	Capital cost of biogas digester	180,000
$C^i_{\text{O\&M, Biogas Digester}}$	Operation and maintenance cost of biogas digester	53,000
$C^i_{\text{Replacement, Biogas Digester}}$	Replacement cost of biogas digester	7500
$C^i_{\text{capital, mixing tank}}$	Capital cost of mixing tank	750,000
$C^i_{\text{O\&M, mixing tank}}$	Operation and maintenance cost of mixing tank	270,000
$C^i_{\text{Replacement, mixing tank}}$	Replacement cost of mixing tank	153,000
$C^j_{\text{capital, Battery}}$	Capital cost of battery	126,000
$C^i_{\text{O\&M, Battery}}$	Operation and maintenance cost of battery	12,100
$C^i_{\text{Replacement, Battery}}$	Replacement cost of battery	55,300
$C_{\text{capital cost\_Generator}}$	Capital cost of generator	25,000
$C_{\text{O\&M cost\_Generator}}$	Operation and maintenance cost of generator	295,000
$C_{\text{fuel cost\_Generator}}$	Replacement cost of generator	11,600

Find the center of mass according to the following equation:

$$x^{\rightarrow c} = \frac{\sum_{i=1}^{N} \frac{1}{f^i} x^{\rightarrow i}}{\sum_{i=1}^{N} \frac{1}{f^i}} \tag{7.9}$$

where x_c = center of mass, x_i = is a point within an n-dimensional search space generated, f_i = is a fitness function value, and N = population size in the big bang phase.

$$x^{\text{new}} = \frac{x^c + l.r}{k} \tag{7.10}$$

where l is the upper limit of the parameter, r is the normal random number, and k is the iteration step.

7.2.8 MATLAB code of big bang-big crunch optimization for 12 variables

```
clc
clear all
d1 =12; %number of variable
n=10; % number of popuation
x1min=160000;x1max=190000;x2min=50000;x2max=54000;x3min=7400;x3max=7800;x4min
=730000;x4max=790000;x5min=250000;x5max=280000;x6min=150000;x6max=155000;x7mi
n
=125500;
x7max=126200;x8min=12050;x8max=12200;x9min=55250;x9max=55400;x10min=248000;x1
0
max=253000;x11min=290000;x11max=297000;x12min=11200;x12max=11650;
c1=2.01;c2=2.01;
x1 = x1min+rand(n,1,1).*(x1max-x1min);x2 = x2min+rand(n,1,1).*(x2max-x2min);x3 =
x3min+rand(n,1,1).*(x3max-x3min);x4 = x4min+rand(n,1,1).*(x4max-x4min);

x5 = x5min+rand(n,1,1).*(x5max-x5min);x6 = x6min+rand(n,1,1).*(x6max-x6min);x7 =
x7min+rand(n,1,1).*(x7max-x7min);x8 = x8min+rand(n,1,1).*(x8max-x8min);
x9 = x9min+rand(n,1,1).*(x9max-x9min);x10 = x10min+rand(n,1,1).*(x10max-x10min);x11 =
x11min+rand(n,1,1).*(x11max-x11min);
160
x12 = x12min+rand(n,1,1).*(x12max-x12min);
x = [x1,x2,x3,x4,x5,x6,x7,x8,x9,x10,x11,x12];
kmin=[x1min,x2min,x3in,x4min,x5min,x6min,x7min,x8min,x9min,x10min,x11min,x12min];
kmax=[x1max,x2max,x3max,x4max,x5max,x6max,x7max,x8max,x9max,x10max,x11max,x12
max];
tic
for i=1:n
f(i,:,1)=
x(i,1,1)+x(i,2,1)+x(i,3,1)+x(i,4,1)+x(i,5,1)+x(i,6,1)+x(i,7,1)+x(i,8,1)+x(i,9,1)+x(i,10,1)+x(i,11,
1
)+x(i,12,1);
end
C,I]=min(abs(f(:,:,1)));
xbest=x(I,:,1);
for i=1:n
for l=1:d1
z(i,l,1)=x(i,l,1)./f(i,1,1);
z1(i,1,1)=1./f(i,1,1);
end
end
B=0.2;
den1=sum(z1);
for i=1:d1
num1(i)=sum(z(:,i,1));
xc(i,1,1)=num1(i)./den1;
end
for j=2:100
for i1=1:n
for i=1:d1
x(i1,i,j)=0.1.*xc(i,1,1)+((1-0.1).*xbest(1,i,j-1))+((rand.*0.3.*(kmax(1,i,1)-kmin(1,i,1)))./j);
end
f(i1,1,j) =
x(i1,1,1)+x(i1,2,1)+x(i1,3,1)+x(i1,4,1)+x(i1,5,1)+x(i1,6,1)+x(i1,7,1)+x(i1,8,1)+x(i1,9,1)+x(i1,1
0
,1)+x(i1,11,1)+x(i1,12,1);
end
for i=1:n
[C,I]=min(f(i,1,:));
if f(i,1,j)<=C
x(i,:,j)=x(i,:,j);
f(i,1,j)=f(i,1,j);
```

```
else
x(i,:,j)=x(i,:,I);
f(i,:,j)=f(i,:,I);
end
end
[C,I]=min(abs(f(:,:,1)));
xbest(1,:,j)=x(I,:,j);
161
for i=1:n
for l=1:d1
z(i,l,1)=x(i,l,j).*f(i,1,j);
z1(i,1,1)=f(i,1,j);
end
end
B=0.2;
den1=sum(z1);
for i=1:d1
num1(i)=sum(z(:,i,1));
xc(i,1,1)=num1(i)./den1;
end
f
end
toc
```

Table 7.5 shows the cost assessment of biogas power plant by BB-BC optimization technique.

At present, ocean energy-based system is not cost-competitive against conventional fossil fuel-based stand alone or grid-interfaced power sources. In the present investigation, the cost assessment is done by the PSO and BB-BC algorithm for cost minimization analysis. Table 7.6 shows the comparative cost assessment through different optimization techniques. Simulation results show that the algorithm performs better than the simulation software and can be efficiently employed to solve the cost minimization function. PSO is providing good solution, but the tremendous potential solution is found by chaotic PSO. The BB-BC optimization has several advantages over other evolutionary methods: most significantly, a numerically simple algorithm and heuristic methods with relatively few control parameters and the ability to solve problems that depend on a large number of variables.

7.3 Assessment of biogas power plant by game theory concept

The well-defined method for a formal diversion theoretic investigation is the investigation of a duopoly of Antoine Cournot in 1838. The mathematician Emile Borel recommended a formal hypothesis of diversions in 1921, which was advanced by the mathematician John Von Neumann in 1928 out of a hypothesis of parlor recreations. Diversion hypothesis was set up as a field in its own particular directly after the 1944 distribution of the amazing volume hypothesis of amusements and financial conduct by Von Neumann and the business analyst Oskar Morgenstern. In 1950 John Nash showed that limited amusements have dependably had a balance point, at which all players pick activities that are best for them given their rival's decisions.

In the 1960s, amusement hypothesis was widened hypothetical and connected to issues of war and governmental issues. Since the 1970s, it has driven an unrest in financial hypothesis. Furthermore, it has discovered applications in humanism and brain science and set up joins with development and science. Amusement hypothesis got exceptional consideration in 1944 with the granting of the Nobel Prize in financial matters to Nash, John Harsanyi, and Reinhard Selten. Toward the finish of the 1990s, a prominent use of diversion hypothesis has been planned of closeouts. Fig. 7.3 demonstrates the order of amusement hypothesis idea.

As a scientific device for the leader, the quality of game theory is the approach, and it accommodates organizing and breaking down issues of key decisions. The procedure of formally displaying a circumstance as an amusement requires the leader to count unequivocally the players and their methodology alternatives and to think about their inclinations and responses. Game theory hypotheses are acquainted with this part with displaying the arranging of cost investigation of the tidal sustainable power source framework.

7.3.1 Basic concept of game theory

Game theory is the branch of science worried about the examination of methodologies for managing focused circumstances where the result of a member's decision of activity depends basically on the activities of alternate members. Game theory is an investigation of key basic leadership. In particular, it is the investigation of numerical models of contention and participation between insightful reasonable leaders. Game theory is the way toward demonstrating the vital cooperation between at least two players in a circumstance containing set guidelines and results. Theory based on decision-making is the formal investigation of basic leadership, where a few players must settle on decisions that conceivably influence the premiums of alternate players.

Let us start out by defining a few terms commonly used in the study of game theory:

1. *GAME*: It is any arrangement of conditions that has an outcome subject to the activities of at least two chiefs.
2. *Players*: This is a key leader inside the setting of the amusement.
3. *Strategy*: This describes a total arrangement of move a player will make given the arrangement of conditions that may emerge inside the diversion.
4. *Payoff*: This is the result a player gets from landing at a specific result. The result can be in any quantifiable from dollars or rupees in utility.

TABLE 7.5 Cost assessment by big bang-big crunch.

Parameter		Cost ($)
$C^i_{\text{capital, Biogas Digester}}$	Capital cost of biogas digester	163,000
$C^i_{\text{O\&M, Biogas Digester}}$	Operation and maintenance cost of biogas digester	51,500
$C^i_{\text{Replacement, Biogas Digester}}$	Replacement cost of biogas digester	7330
$C^i_{\text{capital, mixing tank}}$	Capital cost of mixing tank	733,000
$C^i_{\text{O\&M, mixing tank}}$	Operation and maintenance cost of mixing tank	259,000
$C^i_{\text{Replacement, mixing tank}}$	Replacement cost of mixing tank	147,000
$C^i_{\text{capital, Battery}}$	Capital cost of battery	123,000
$C^i_{\text{O\&M, Battery}}$	Operation and maintenance cost of battery	12,025
$C^i_{\text{Replacement, Battery}}$	Replacement cost of battery	55,110
$C_{\text{capital cost\_Generator}}$	Capital cost of generator	24,600
$C_{\text{O\&M cost\_Generator}}$	Operation and maintenance cost of generator	278,000
$C_{\text{fuel cost\_Generator}}$	Replacement cost of generator	11,230

5. *Information set*: These are the data accessible at a given point in the diversion. The term data set is most normally connected when the diversion has a consecutive part.
6. *Equilibrium*: It is the point in a division where the two players have settled on their choice and a result is come to.
7. *Nash equilibrium*: It is additionally called key harmony that is a rundown of procedures for every player, which has the property that no player can singularly change his technique and show signs of improved result.

Game theory has a wide range of applications in the field of economics, but in the field of computer science, especially in the optimization algorithm is seldom used. Table 7.7 shows the comparison between the biogas energy system and the game theory concept.

7.3.2 Solution concepts

A Game is only a model of collaboration, and it does not determine how to play and what moves should the players make, with a specific end goal to accomplish their targets (boosting result for our situation). The focal inquiry

TABLE 7.6 Comparative cost analysis.

Parameter		PSO	BB-BC
		Cost ($)	Cost ($)
$C^{i}_{\text{capital, Biogas Digester}}$	Capital cost of biogas digester	180,000	163,000
$C^{i}_{O\&M,\text{ Biogas Digester}}$	Operation and maintenance cost of biogas digester	53,000	51,500
$C^{i}_{\text{Replacement, Biogas Digester}}$	Replacement cost of biogas digester	7500	7330
$C^{i}_{\text{capital, mixing tank}}$	Capital cost of mixing tank	750,000	733,000
$C^{i}_{O\&M,\text{ mixing tank}}$	Operation and maintenance cost of mixing tank	270,000	259,000
$C^{i}_{\text{Replacement, mixing tank}}$	Replacement cost of mixing tank	153,000	147,000
$C^{j}_{\text{capital, Battery}}$	Capital cost of battery	126,000	123,000
$C^{i}_{O\&M,\text{ Battery}}$	Operation and maintenance cost of battery	12,100	12,025
$C^{i}_{\text{Replacement, Battery}}$	Replacement cost of battery	55,300	55,110
$C_{\text{capital cost\_Generator}}$	Capital cost of generator	25,000	24,600
$C_{O\&M\text{ cost\_Generator}}$	Operation and maintenance cost of generator	295,000	278,000
$C_{\text{fuel cost\_Generator}}$	Replacement cost of generator	11,600	11,230

that game theory is endeavoring to answer is the manner by which players pick their techniques of play.

There are two main points of view:

- *Prescriptive*—analyze and specify how players should play (recommendation); and
- *Descriptive*—analyze and predict how players will play (explanations and prediction).

These two methodologies are connected; we need to configure amusements utilizing the prescriptive approach and break down them utilizing the portrayed approach. At last, we need the clear investigation to foresee that players would play in the way we proposed when we outlined the diversion. We center around the prescriptive approach. This is on the grounds that our fundamental objective is to configuration recreations, which have decent

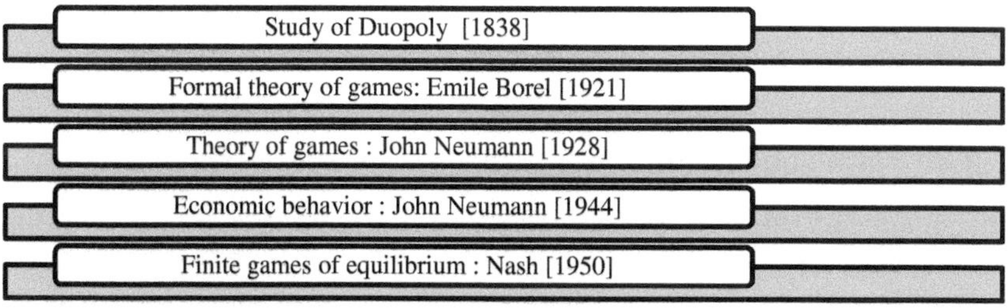

FIGURE 7.3 Chronology of game theory.

TABLE 7.7 At a glance comparison between biogas and game theory.

Biogas energy system	GAME theory
Component	Player
Study area	Research space
Prefeasibility	Strategies
Cost minimization function	Payoff function

equilibrium(s), a great result for all players. An answer idea offers proposals on what moves to make.

7.3.3 Assumptions

1. The amusement is given infrequently it is composed by us. Everyone knows the tenets (every accessible activity). It is very much characterized and players cannot transform it.
2. *Awareness*: All utilities of the greater part of the players are known, and a player knows all activities accessible to him.
3. *Perfect information*: A player knows all the actions of other players. For strategic form games, this is the same as 2. For extensive form games, this is not necessary.
4. No computational restrictions (here cryptography will come into play). Is finding a strategy feasible? Is playing a strategy feasible?
5. Players are rational: they try to maximize the payoff.
6. All actions of all players are taken simultaneously.

Game theory is invigorating in light of the way that regardless of the way that the measures are fundamental and the applications are clearing. Apparently, the most entrancing beguilements incorporate correspondence, since such immense quantities of layers of framework are possible. Delight speculation can be used to plot sound obligations, threats, or ensures, or to assess proposals and clarifications offered by others.

7.3.4 Nash equilibrium

In game theory, the Nash balance is an answer idea of an agreeable diversion including at least two players, in which every player is accepted to know the balance procedures of alternate players, and no player has anything to pick up by changing just their own technique. Let (s,f) be a game with N players, where S_k is the strategy set for player k, $S = S_1 \times S_2 \times \ldots\ldots \times S_N$ is the set of strategy profiles, and $f(x) = (f_1(x), \ldots\ldots f_N(x))$ is its payoff function evaluated at $x \in S$. Let x_k be a strategy profile of player k and x_{-k} be a strategy profile of all players except for player k when each player $k \in (1\ldots N)$ chooses strategy x_k resulting in strategy profile $x = \{x_1 \ldots\ldots\ldots x_N\}$ and then player k obtain payoff $f_k(x)$.

7.3.5 Game theory in biogas renewable energy system

Game theory is the branch of decision theory concerned with independent decision. A game is given by $G = (N, S, U)$ in which the set of player is N, the set of strategies is S, and the set of the payoff function (utility function) is U. A payoff is a number, also called utility, which reflects the desirability of an outcome for a player, for whatever reason. When the outcome is random, the payoffs are usually weighted by their probabilities. If the set of player N is equal to 3 = (Biogas energy, Anaerobic Digester, battery is the set of player), biogas energy, battery, and generator worked as player 1, player 2, and player 3, respectively. Fig. 7.4 shows the key step of game theory in biogas energy system.

A. Set the strategies: *S*

- The biogas energy system is sufficient to fulfill the entire requirement and the battery bank is in recharge mode.
- The biogas energy system are providing sufficient amount of energy and the battery is set to supply power to the load.

If power demand is higher than the power available from biogas energy sources and battery current is higher than the maximum discharging current, the hybrid controller automatically turns on to the diesel generator to provide the extra current.

According to the GAME theory, concept of fairness is applicable in biogas energy system in the following way.

- *Parity*: Equal treatment of equals, which means an equal contribution of biogas and grid energy system in the generation of electricity.
- *Proportionality*: There are several proportionality conditions that are applied in biogas energy system, such as the following.

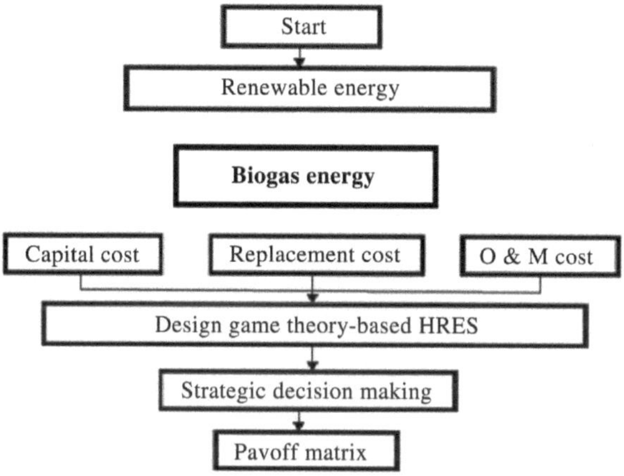

FIGURE 7.4 Key step of game theory in biogas energy system.

- The biogas energy system output is directly proportional to the amount of waste material.
- Biogas power is directly proportional to the solid concentration.

- *Normal form*:

The *strategic* and normal form of game is generally displayed by a lattice that demonstrates the players, techniques, and settlements. All the more by and large, it can be spoken to by any capacity that partners a result for every player with each conceivable blend of activities. In the going with mixture framework, there are two individual frameworks: one picks the line and alternate picks the section. At the point when a diversion is displayed in ordinary shape, it is assumed in this technique that the measure of the biogas energy and biogas generators is being ascertained in the most troublesome month.

7.3.6 Mixed strategy Nash equilibrium

A mixed strategy is the one in which a player plays his accessible unadulterated methodologies with specific probabilities. Blended systems are best comprehended with regard to rehashed recreations, where every player's point is to keep alternate player(s) speculating. In the event that every player in an n-player diversion has a limited number of unadulterated techniques, at that point, there exists no less than one balance in blended systems. In the event that there is no unadulterated procedure balance, there must be special blended system balance. Table 7.8 presents the blended technique of biogas sustainable power source framework. Let P be the probability of sufficient amount of waste material that gives biogas energy system worked properly, so that $1 - P$ is the probability when the amount of waste material is less and condition of retention time is better.

TABLE 7.8 Mixed strategy of biogas energy system.

	Waste material			
Retention time		Waste material	Retention time	Q-Mix
	Waste material	50,50	80,20	50Q + 80 (1 − Q) 50Q + 20 (1 − Q)
	Retention time	90,10	20,80	90Q + 20 (1 − Q) 10Q + 80 (1 − Q)
	P-mix	50P + 90 (1 − P) 50P + 10 (1 − P)	80P + 20 (1 − P) 20P + 80 (1 − P)	

- Let Q be the probability that the condition of retention time is better and biogas energy system worked properly, so that $1 - Q$ is the probability when retention time is not perfect and condition of waste material is better.
- To find mixed strategies, we add the P-mix and Q-mix strategies to the payoff matrix.

7.3.7 Algebraically

$$50P + 10(1 - P) = 20P + 80(1 - P)$$

$$50P + 10 - 10P = 20P + 80 - 80P$$

$$40P + 10 = 80 - 60P$$

$$100P = 70, \mathrm{SO}$$

$$P = \frac{70}{100} = 0.7$$

If the quantity of waste material is perfect, then the probability of success of biogas energy system is 70% and success rate of generator is 30%.

$$50Q + 80(1 - Q) = 90Q + 20(1 - Q)$$

$$50Q + 80 - 80Q = 90Q + 20 - 20Q$$

$$80 - 30Q = 70Q + 20$$

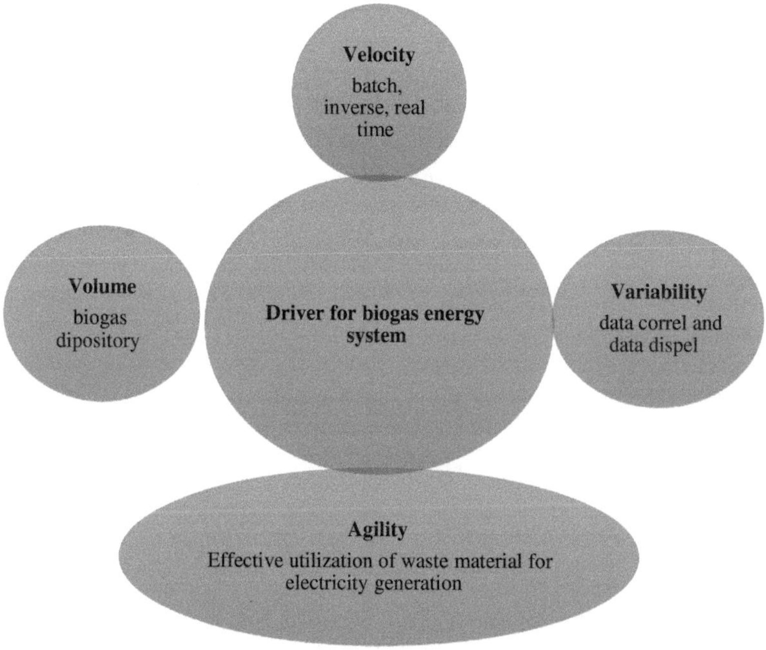

FIGURE 7.5 Big data driver for biogas energy system.

$$60 = 100Q, \text{SO}$$

$$Q = \frac{60}{100} = 0.60$$

If retention time is perfect, then the probability of success of tidal energy system is 60% and success rate of generator is 40%.

7.4 Assessment of biogas power plant by big data analysis

Big data refer to the massive data set that is collected from a variety of data sources for implementing biogas energy system at a particular place and to reveal new insights for better decision-making. Based on different data analyses of the any study area, it is observed worldwide and a lot of places are available where lots of waste material are available in abundance. The accessibility of nonconventional energy resources at any place is a significant feature to develop the biogas energy system for different purposes. Big data generate values related to biogas energy system for the storage and processing of very large quantities of information that cannot be analyzed with traditional computing techniques. Big data are categorized into three parts: volume, velocity, and the variety and assess the prefeasibility assessment with the help of these three features, which is shown in Fig. 7.5.

With phenomenal development in the field of electricity generation through renewable energy system, biogas energy data sources have risen sharply. Exhaustive use of waste material big data can provide an effective way for safe operation of high-quality power supply of a biogas energy system. The effective goal of big data research in the field of biogas energy system is to "become aware of unawareness," and objectively, there are well-known facts and unknown facts that affect the working of biogas energy system. Big data help us in learning what we do not know and are done in two ways: from small to big and from big to small. It is necessary to develop a proper insight view of big data in the field of biogas energy system and model out the biogas system based on big data. Most big data applications in biogas energy system consist of two parts: data correl and data dispel. In a data correl, lot of information about the biogas energy system project is collected and stored in biogas cargo space, which is used to store the data and is utilized for further suitable application.

7.4.1 Biogas energy system by Hadoop environment

Hadoop ecosystem is a framework of various types of complex and evolving tools and components. Some of these elements may be very different from each other in terms of their architecture. If we apply the concept of Hadoop ecosystem in the field of biogas energy system, then it is categorized into four types: data management, data access, data processing, and data storage. Fig. 7.6 shows that Hadoop biogas energy ecosystem is also categorized into three types of prefeasibility assessment in terms of loading rate, solid concentration, retention time data assessment, and load demand data assessment. In Hadoop system, data management is done by Oozie, Chukwa, Flume, and Zookeeper, where Oozie is an open-source Apache Hadoop service used to manage and process submitted tasks. Hadoop works by the divide and conquer approach. Once a problem is divided, it is approached and processed by using distributed and parallel processing technique across Hadoop cluster.

Data management of biogas energy system through the Hadoop system is always part of resource characterization and site assessment. In biogas energy system, site analysis is usually carried out aforementioned to consumption to set up the comprehensive corporeal surrounding for a particular renewable energy project, with the following objectives:

- to assess the biogas energy production throughout the life of the project;
- to describe low and high solid concentration of waste material and high and low retention periods; and
- to describe the temperature of the substrate and toxic substances of biogas material.

Objective and nature of resource assessment, biogas energy conversion characteristics, site condition, constraints, and physical boundary of assessment are the part of the biogas project description, which is the first step of resource allocation. If the project is in its initial stages and consists mainly of site screening, the resource assessment should be qualified as regional if the area of study is very large and incorporates many potential sites, whole country, or a large portion thereof.

7.4.2 Energy aware cluster node management of biogas energy system

Cluster node management is one of the parts of big data analysis, and a cluster manager usually is a backend graphical user interface or command line software that runs on one or all cluster nodes. The cluster manager works together with a cluster management agent. A cluster is the process of making a group of abstract objects into classes of similar objects. The main advantage of clustering over classification is that it is adaptable to changes and helps single out useful features that distinguish different groups.

7.4.3 Clustering method in biogas energy system

- *Partitioning method*:

  Suppose that we are given a database of prefeasibility assessment of biogas energy system, which is represented by "n" database and the partition method construct "k" partition of data. Then, a prefeasibility assessment of the data is partially in the form of project description, estimation of current speed, result presentation, data analysis, and available and extractable energy data, so that "n" database is partitioned into $k=5$ partitioned.
- *Hierarchical methods*:

In the hierarchical method of clustering, we identify step-by-step process of data gathering. In this method, data collection and decomposition are done in two ways: first one is agglomerative and another one is divisive approach. In the agglomerative process, merging the object and group that are close to one another and in the case of biogas energy system data are collected in the following manner:

1. collection of prefeasibility data of biogas energy system;
2. collection of data for modeling of biogas energy system;
3. collection of data for controlling of biogas energy system; and
4. collection of data for reliability assessment of biogas energy system.

In divisive approaches, we start with all of the objects in the same cluster, and in the continuous iteration, a cluster is split into smaller clusters. According to the divisive approach, agglomerative data are distributed and divided into following manner.

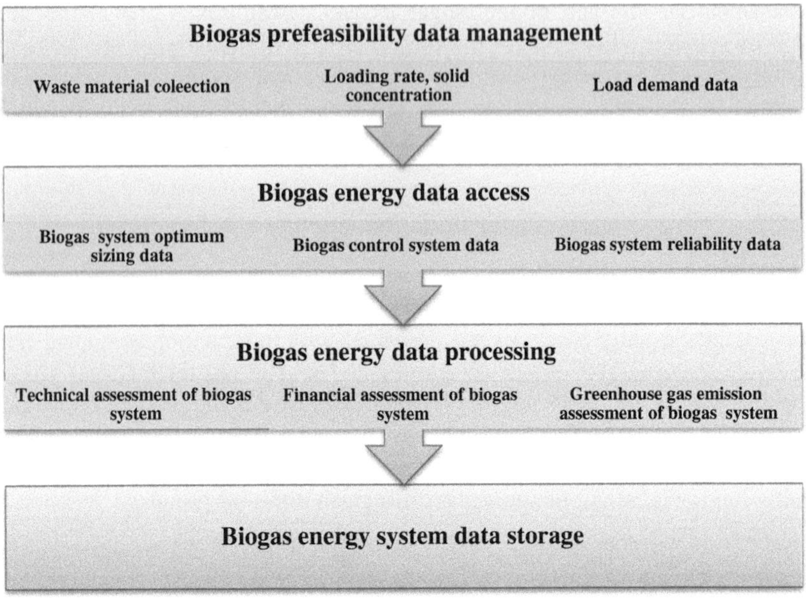

FIGURE 7.6 Hadoop ecosystem for biogas energy system.

1. Collection of prefeasibility data of solar and wind energy system:
   a. location of the site;
   b. geographical condition of the site;
   c. data of waste material;
   d. data of design of anaerobic digester;
   e. data of temperature;
   f. data of retention period; and
   g. data on consumer demand.
2. Collection of data for modeling of biogas energy system:
   a. data of electricity required;
   b. data of specification of biogas energy;
   c. data of specification of direct current (DC) generator and alternating current (AC) generator; and
   d. data of specification of battery.
3. Collection of data for controlling of biogas energy system:
   a. data of different types of errors;
   b. data of different control strategies of biogas energy system; and
   c. data for stability analysis of biogas power plant.
4. Collection of data for reliability assessment of biogas energy system:
   a. data of failure distribution model of biogas energy system;
   b. data of time-dependent failure model of a biogas energy system; and
   c. data of constant failure rate model.

7.4.4 Basic big data measures for biogas energy data, text retrieval

In the text retrieval, we need to check the accuracy of the data, because it is very necessary part in the development of biogas power plant. Let the set of biogas energy system documents relevant to a query be denoted as (relevant biogas data) and the set of retrieved documents as (retrieved biogas data). The set of biogas documents that are relevant and retrieved can be denoted as:

(Relevant biogas data) ∩ (Retrieved biogas data).

This can be shown in Fig. 7.7 in the form of a Venn diagram as follows:

There are three fundamental measures for assessing the quality of biogas energy system retrieval:

- precision;
- recall; and
- F-score.

Precision: Precision is the percentage of biogas energy system retrieved documents that are in fact relevant to the consumer query. Precision of biogas power plant can be defined as:

$$\text{Precision of biogas energy system data} = \frac{(\text{Relevant biogas data}) \cap (\text{Retrieved biogas data})}{(\text{Retrieved biogas data})}$$

Recall: Recall is the percentage of biogas energy system documents that are relevant to the consumer query and were in fact retrieved. Recall is defined as:

$$\text{Recall of biogas energy system data} = \frac{(\text{Relevant biogas data}) \cap (\text{Retrieved biogas data})}{(\text{Relevant biogas data})}$$

F-score: F-score is the commonly used trade-off. The information retrieved system often needs to trade off for precision or vice versa. F-score is defined as the harmonic mean of recall or precision as follows:

$$F\text{-score} = \frac{\text{Recall} \times \text{Precision}}{0.5 \times (\text{Recall} + \text{Precision})}$$

7.4.5 Application of map reduces in solar and wind energy system

The MapReduce algorithm contains two important tasks namely, map and reduce.

- The map task is done by means of Mapper Class.
- The reduce task is done by means of Reducer Class.

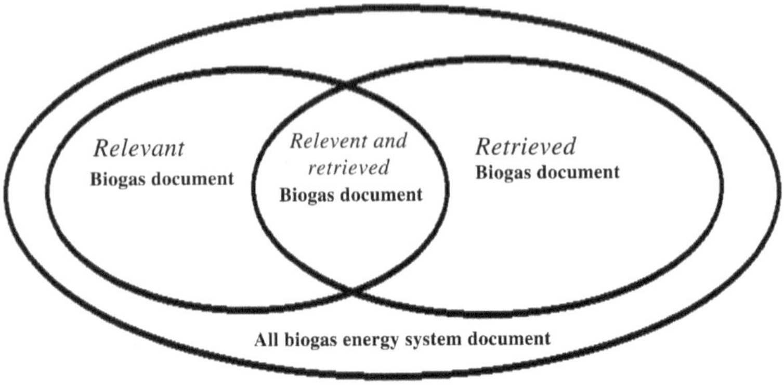

FIGURE 7.7 Biogas text retrieval.

The Mapper class takes the input, tokenizes it, maps, and sorts it. The output of Mapper class is used as input by Reducer class, which in turn searches matching pairs and reduces them.

Map Reduce implements various mathematical algorithms to divide a task into small parts and assign them to multiple systems. In technical terms, Map Reduce algorithm helps in sending the Map and Reduce tasks to appropriate servers in a cluster.

7.4.5.1 Sorting

Sorting is one of the basic Map Reduce algorithms to process and analyze data. Map Reduce implements sorting algorithm to sort automatically the output key-value pairs from the mapper by their keys.

- Sorting methods are implemented in the mapper class itself.
- In the Shuffle and Sort phases, after tokenizing the values in the mapper class, the context class (user-defined class) collects the matching valued keys as a collection.
- To collect similar key-value pairs (intermediate keys), the Mapper class takes the help of raw-comparator class to sort the key-value pairs.
- The set of intermediate key-value pairs for a given Reducer is automatically sorted by Hadoop to form key-values (K2, {V2, V2, ...}) before they are presented to the Reducer.

7.4.5.2 Searching

Searching plays an important role in the Map Reduce algorithm. It helps in the combiner phase (optional) and in the Reduce phase. Let us try to understand how Searching works with the help of an example.

- Let us assume we have waste collection data in four different files—W, X, Y, and Z. Let us also assume there are duplicate waste collection records in all four files because of importing the waste collected data from all database tables repeatedly. See the following illustration and keep in mind value of waste collection in tonnes both are lying in 0–1.

Countrywise site area, avg. waste collection in tonnes	**Countrywise site area, avg. waste collection in tonnes**	**Countrywise site area, avg. waste collection in tonnes**	**Countrywise site area, avg. waste collection in tonnes**
India, 5.2	China, 6	India, 5.2	India, 5.2
Japan, 5	Japan, 5	United States, 9	Japan, 5
Germany, 7	Germany, 7	Germany, 7	New Zealand, 8
Austria, 5	Austria, 5	Austria, 5	Austria, 5

- *The Map phase* processes each input file and provides the biogas energy system data in key-value pairs (<k, v>: <Site area, waste collection>). See the following illustration.

Countrywise site area, avg. waste collection in tonnes	**Countrywise site area, avg. waste collection in tonnes**	**Countrywise site area, avg. waste collection in tonnes**	**Countrywise site area, avg. waste collection in tonnes**
<India, 5.2>	<China, 6>	<India, 5.2>	<India, 5.2>
<Japan, 5>	<Japan, 5>	<United States, 9>	<Japan, 5>
<Germany, 7>	<Germany, 7>	<Germany, 7>	<New Zealand, 8>
<Austria, 5>	<Austria, 5>	<Austria, 5>	<Austria, 5>

- *The combiner phase* (searching technique) will accept the input from the Map phase as a key-value pair with site area and waste material. Using searching technique, the combiner will check all the countrywise site area to find the highest waste collection availability in each file. See the following snippet.

```
<k: country wise site area, v- waste collection>
Max = the value of waste collection of a first country. Treated as max waste
collection
if(v(second country).waste_collection > Max)
{
Max = v(waste_collection);
}
else
{
Continue checking;
}
```

The expected result is as follows:

<United States, 9>	<New Zealand, 8>	<Germany,7>	<China, 5>

- *Reducer phase*—Form each file, you will find the highest solar radiation or wind velocity. To avoid redundancy, check all the <k, v> pairs and eliminate duplicate entries, if any. The same algorithm is used in between the four <k, v> pairs, which are coming from four input files. The final output should be as follows:

```
United States <7>
```

7.4.5.3 *Indexing*

Normally indexing is used to point to a particular data and its address. It performs batch indexing on the input files for a particular Mapper. The indexing technique that is normally used in Map Reduce is known as inverted index. Search engines like Google and Bing use inverted indexing technique. Let us try to understand how Indexing works with the help of a simple example. The following text is the input for inverted indexing. Here, *X* [0], *X* [1], and *X* [2] are the file names and their waste material data are in double quotes.

```
X[0] = "5.2, 7.2, 6.3, 9.4"
X[1] = "9.4, 7.2, 8.1"
X[2] = "7.2, 6.3, 8.8, 4.5"
```

After applying the Indexing algorithm, we get the following output of waste material:

```
"7.2": {0,1,2}
"5.2": {0}
"6.3": {0,2}
"4.5": {2}
```

Here, "4.5": {2} implies that the term "a" appears in the *X*[2] file. Similarly, "7.2": {0, 1, 2} implies that the term "is" appears in the files *X*[0], *X*[1], and *X*[2].

7.4.6 Market basket model in biogas energy system

It shows that many relationships between two concepts “items” and “Baskets” and each basket consist of a number of items. In the context of biogas energy system, market basket model is considered in three ways: prefeasibility assessment, modeling, and reliability assessment, and it contains a lot of items that is related to the biogas energy system and shown in Fig. 7.8. Basket of prefeasibility assessment contains a lot of information, because prior to installation and operation, the prefeasibility study of biogas energy system should be done. In biogas energy projects, an initial study is undertaken to determine whether it is worthwhile to continue to the feasibility study stage. A precise feasibility study should provide a chronological background of the projects. In addition to climate condition of the application site, availability of biogas energy sources, the potential of biogas energy sources, and load demand of application sites are included to find out the best location to develop a biogas renewable energy system. Generally, feasibility precedes technical development and project implementation. It must, therefore, be conducted with a balanced approach to provide information upon which decisions can be based. Modeling basket contains different techniques that are used to model the solar and wind energy system such as through Hybrid optimization and modelling of renewable energy (HOMER), fuzzy logic, and analytical technique. Modeling is the first step to design a system according to the different parameter and constraint. Modeling of biogas energy system is based on annual cost, battery autonomy function, sizing criteria, and ecological statistical factor. Step-by-step optimization practice is used to find out the efficient result of the solar system model. Third and most important basket is reliability basket, which is very important in the recent scenario, because it evaluates the failure rate of individual component and overall biogas energy system.

7.4.7 Market basket model based frequent item set mining for biogas energy system

Let $K = \{k_1 \ldots\ldots\ldots\ldots k_n\}$ be a set of parameters. Let “E” be the task relevant data and be a set of database parameter where each parameter T is a set of parameters such that $E \subseteq K$. Each parameter is associated with an identifier, called TKE. Biogas energy system parameter is referred to as a parameter set. We can call a parameter K a “frequent item set” only if its support count is sufficiently large. We prescribe a minimum support “s” and any K, which has support greater than or equal to “s,” is a frequent parameter set.

If we consider parameters of biogas energy system = {Loading rate (1), solid concentration (2), Retention time (3), Temperature of the substrate (4), Toxic substances (5)}, Minimum support $s = 3$.

Transactions

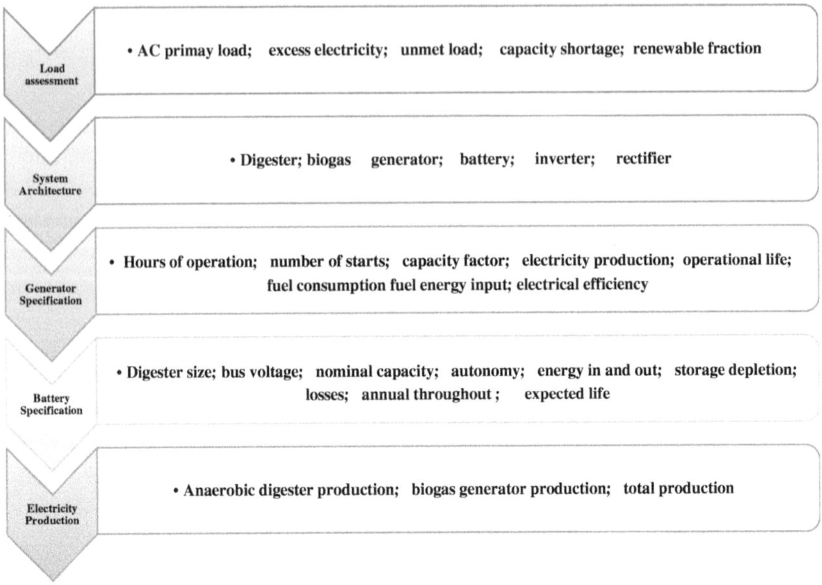

FIGURE 7.8 Market basket model for optimum sizing and modeling of biogas energy system.

1. $T1 = \{1,2,4\}$;
2. $T2 = \{1,3,5\}$;
3. $T3 = \{1,4\}$;
4. $T4 = \{2,5\}$;
5. $T5 = \{1,3,4\}$;
6. $T6 = \{1,2,4,5\}$;
7. $T7 = \{2,4,5\}$; and
8. $T8 = \{4,2\}$.

Frequent parameters: {1}, {2}, {4}, {5}, {1, 4}, {2, 4}, {5, 2}.

7.4.8 Associate rule mining in biogas energy system

The main purpose to discovering frequent parameters from a large data set is to discover a set of "if-then" rules called Association rules. The form of association rules is *K-j*, where *K* is a set of parameters of solar or wind power plant. Let $K = \{K1, K2...Kn\}$ be a set of *n* distinct attributes of biogas energy system, which is also called literals of renewable power plants. Let "*E*" be a database of biogas energy system, where each records supply and demand sides, and *T* has a unique identifier and contains set of parameters from the set *K*. An association rules is an implication of the form $X \rightarrow Y$, where X, $Y \subseteq K$ are item sets and $X \cup Y$ is a frequent item set. Thus the frequency of occurrence in $X \cup Y$ is at least equal to the minimum support *s*.

TABLE 7.9 Transaction of biogas energy system database.

Transaction ID	Parameters
T1	Digester, mixing tank, biogas generators
T2	Mixing tank, biogas generators, consumer demand
T3	Mixing tank
T4	Digester, mixing tank

To test the reliability of biogas energy system, we try to define "confidence" of the rule, which presents performance parameter of biogas power plant. Let $X \rightarrow Y$ be an association rule. The confidence of the rule that assess the performance of biogas power plant is defined as the fraction of the biogas power plant parameters that support the rule among those that support the antecedent:

$$\text{Confidence } (X \rightarrow Y) := P\left(\frac{Y}{X}\right) = \frac{\text{support } (X \cup Y)}{\text{support}(X)}$$

The confidence of the rule indicates the degree of correlation between certain parameters of biogas energy system, such that value of retention time also depends on the quantity of waste material, so it is necessary to determine correlation between the waste material and the retention time.

Consider a small database with four parameters of the combined biogas energy system $K = \{$Digester Specification (DS), Mixing Tank Specification (MT), Biogas Generator Specification (BG), Consumer demand in Kw (CD)$\}$ and four transactions of these parameters shown in Table 7.9. Table 7.10 shows all the parameters for K. Suppose that the minimum support and minimum confidence of an association rule are 40% and 60%, respectively, which is shown in Table 7.11.

7.4.9 Framework for frequent parameter mining

The market baskets are also organized in memory. Generally, market basket data are stored in a file basket by basket. Generally, market basket data of biogas energy system are stored in a memory basket by basket.

7.4.10 Monotonicity and Apriori algorithm property of biogas energy parameters

Given a database of transaction "**E**" over "**K**" and two sets X, $Y \subseteq K$. Then

$$X,\ \ Y \subseteq K \Rightarrow \text{support}(Y) \leq \text{support}(X).$$

TABLE 7.10 **Support for parameters in table and large parameters with a support of 40%.**

Parameters	Support "s" (%)	High/low
Digester	50	High
Mixing tank	100	High
Biogas generators	50	High
Consumer demand	25	Low
Digester, mixing tank	50	High
Digester, biogas generator	25	Low
Digester, consumer demand	0	Low
Mixing tank, biogas generator	50	High
Mixing tank, consumer demand	25	Low
Biogas generator, consumer demand	25	Low
Digester, mixing tank, biogas generators	25	Low
Digester, mixing tank, consumer demand	0	Low
Digester, biogas generators, consumer demand	0	Low
Mixing tank, biogas generators, consumer demand	**25**	**Low**
Digester, mixing tank, biogas generators, consumer demand	**0**	**Low**

TABLE 7.11 **Confidence of some association rule where confidence interval = 60%.**

Rule	Confidence (%)	Rule-hold
Digester ⇒ mixing tank	**100**	**Yes**
Mixing tank ⇒ digester	**50**	**No**
Mixing tank ⇒ biogas generators	**50**	**Yes**
Biogas generators ⇒ mixing tank	**100**	**No**

The monotonicity property of support also allows us to compact the information about frequent biogas energy parameters, as shown in Table 7.12. First, some definition is given below:

1. Biogas energy system parameters are closed if none of its immediate parameter has the same count as the parameter.

TABLE 7.12 Indicating frequent, closed, and maximal parameters.

Parameter	Count	Frequent	Closed	Closed frequency	Maximum frequency
Capital cost	4	Yes	No	No	No
Replacement cost	5	Yes	Yes	Yes	No
Operation and maintenance cost	3	Yes	No	No	No
Capital cost, replacement cost	4	Yes	Yes	Yes	Yes
Capital cost, Operation and maintenance cost	2	No	No	No	No
Replacement cost, operation and maintenance cost	3	Yes	Yes	Yes	Yes
Capital cost, replacement cost, peration and maintenance cost	2	No	Yes	No	No

2. Biogas energy system parameter is closed frequently if it is frequently closed.
3. Biogas parameter is maximal frequent if it is frequent and none of its immediate superset is frequent.

Financial analysis of biogas energy system consists of three parameters = {Capital cost, Replacement cost, Operation & Maintenance cost} and the following baskets:

1. {Capital cost, Replacement cost};
2. {Capital cost, Replacement cost};
3. {Replacement cost, Operation & Maintenance cost};
4. {Capital cost, Replacement cost, Operation & Maintenance cost}; and
5. {Capital cost, Replacement cost, Operation & Maintenance cost}.

Assume the support threshold $s = 3$.

Monotonosity of biogas energy system is also done by Apriori algorithm, and in this algorithm, let us be the minimum support required. Let "n" be the number of items. In the first pass, we read the baskets and count and performance in main memory the occurrence of each parameter. In the second pass, we assess the basket again and count in main memory only those pairs where both parameters are frequent parameters.

This algorithm is also used in biogas energy system for finding the most utilized parameter without counting all the necessary parameters that can be extended to find larger frequent parameters without an exhaustive count of all data sets of biogas energy system. In the Apriori algorithm, one step of each parameter of biogas energy system is taken for each set size K. The pattern of moving from one size K to the next size $K+1$ can be summarized as follows. For each size K, there are two sets of parameters:

1. D_K is the set of parameter of size K, the biogas energy parameter that we must assess in order to determine whether they are in fact frequent.
2. MK is the set of truly frequent parameters of size K.

The pattern of moving from one set to the next and one size to the next is depicted:

Assume that we have assessed reliability of biogas energy system and consider some parameters = {Fault Rate (FR), Levelized Cost of Energy (LCE), Loss of Power Supply Probability (LPSP), Level of Autonomy (LA), Minimum Time to Failure (MTF)} in the reliability basket.

1. {FR, LCE};
2. {FR, LCE, LPSP};
3. {FR, LCE, LA};
4. {LCE, LPSP, LA};
5. {FR, LCE, LPSP, LA}; and
6. {FR, LCE, LA, MTF}.

Let the support threshold $s = 3$. The Apriori algorithm is as follows:

1.

(a) Construct D1 = {{FR},{LCE},{LPSP},{LA},{MTF}}.

(b) Assess the support of parameters of reliability assessment of biogas energy system in D1.

(c) Remove infrequent parameter to get M1 = {{FR},{LCE},{LPSP},{LA}}.

2.

(a) Construct D2 = {{FR, LCE},{LCE, LPSP},{FR, LA},{LCE, LPSP},{LCE, LA},{LPSP, LA}}.

(b) Assess the support of parameters of reliability assessment of biogas energy system in D2.

(c) Remove infrequent parameter to get M2 = {{FR, LCE},{LCE, LPSP},{FR, LA},{LCE, LPSP}}.

3.

(a) Construct D3 = {{FR, LCE, LPSP},{ FR, LCE LA},{ LCE, LPSP, LA }}.

(b) Assess the support of parameters of reliability assessment of biogas energy system in D3.

(c) Remove infrequent parameter to get M3 = {{FR},{LCE},{LA}}.

4.

Construct D4 = {Empty set}

The above assessment through market basket model shows that frequent item set plays an essential role in many data mining tasks of biogas energy system. The identification of different parameters, characteristics, and often together technical and financial assessments of biogas energy system is a one of the basic tasks of the market basket model.

7.5 Role of biogas power plant in clean development mechanism

The clean development mechanism (CDM) is one of the adaptability components characterized in the Kyoto Protocol that accommodates emanation decrease ventures, which create Certified Emission Reduction (CER) units, which might be exchange outflow exchanging plans.

The CDM is viewed as a standout among the most essential globally executed market-based systems to lessen carbon outflows. Power age through biogas energy framework is one of the strategies from the sustainable power source framework, which decrease the GHGs amid the age of power then again made under the Kyoto Protocol, and the CDM was intended to enable created countries to meet household ozone depleting substance (GHG) diminishment duties by putting resources into ease outflow lessening ventures in creating nations. The CDM has rapidly developed to finance a large number of tasks worldwide and achieve a few billion-euro advertise esteem. The CDM is defined in Article 12 of the Protocol, and is intended to meet two objectives:

1. to help parties excluded in Annex I in accomplishing maintainable improvement and in adding to a definitive target of the United Nations Framework Convention on Climate Change (UNFCCC), which is to avert hazardous environmental change.
2. to help parties incorporated into Annex I in accomplishing consistence with their evaluated outflow constraint and lessening responsibilities [ozone depleting substance (GHG) discharge tops]. "Add I" parties are those nations that are recorded in Annex I of the bargain and are the industrialized nations. Non-Annex I parties are creating nations.

The CDM tends to the second goal by permitting the Annex I nations to meet piece of their emanation decrease duties under the Kyoto Protocol by purchasing CER units from CDM outflow lessening ventures in creating nations. The activities and the issue of CERs are liable to endorse to guarantee that these outflow decreases are genuine and "extra." The CDM is regulated by the CDM Executive Board and is under the direction of the Conference of the Parties (COP/MOP) of the UNFCCC.

The flexible mechanisms consist of:

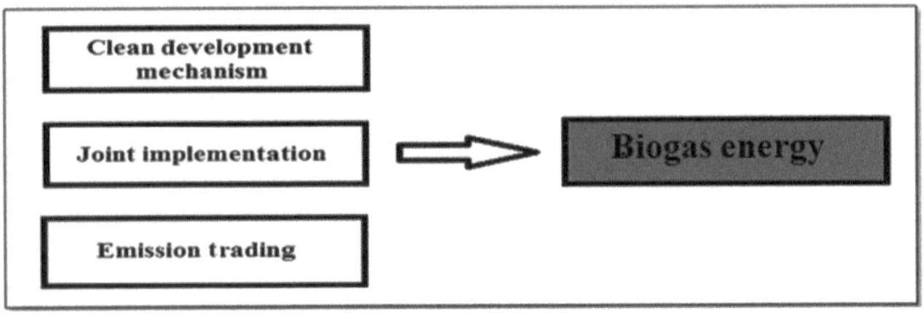

FIGURE 7.9 Clean environment toward biogas energy system.

- CDM, a project-based mechanism for international trading with GHGs (article 12);
- joint implementation, a project-based mechanism described in article 6 of the protocol, which applies only to Annex B countries; and
- emission trading, a way to trade with emission rights without being tied to a project.

The CDM is a piece of the Kyoto convention natural understanding and the CDM ventures are intended to fortify economic improvement through individuals focused and participatory formative action to lessen GHG outflows. The historical backdrop of CDM starts from the mid-1990s with the expanding acknowledgment and activities by the researchers and others because of increment in GHG outflows from the human exercises like substantial businesses and enormous horticulture hones with higher thought processes of benefits and aimless creations, which have included the climatic change and ascend in temperature not in the Indian circumstance, but rather likewise at worldwide level (Fig. 7.9).

An extensive piece of the world has concurred that the world's GHG emanation needs to back off. This was showed in the Kyoto Protocol in 1997.The Kyoto Protocol was embraced at the third COP. COP is held by the UNFCCC as an approach to accumulate the nations of the world and examine environmental change and measures. The understanding in the Kyoto Protocol was that the Annex I countries should bring down their outflow on a normal of 5.2% from 1990 years level; the decrease ought to be accomplished by the time of year 2008–12 (the primary responsibility time frame). The non-Annex I2 nations were not given any headings other than to gauge the discharge and to report the outcome.

7.5.1 Clean development mechanism's role in technology transfer

While historically developed countries have been the main contributors of GHG emissions in the atmosphere, more recently emissions from developing countries, primarily the large emerging economies, have been on the rise. An

International Energy Agency study concluded that 56% of the growth in emissions between now and 2030 will be from China and India. By incentivizing the private sector to seek out low-cost emission reduction opportunities in developing countries, international offset programs bring human resources, information, and capital to these countries while also transferring mitigation technologies.

The UNFCCC and Kyoto Protocol both require participating countries to promote and cooperate in the development, application, and diffusion (transfer) of technologies. Although the CDM does not have an explicit technology transfer mandate, it does appear to contribute to the transfer of technology to developing countries.

In each CDM PDD, project developers include a description of whether or not the technologies proposed are environmentally safe and if there will be a transfer in technological "know-how" to the developing country host. It may be inferred from the information in PDDs that project developers almost universally interpret technology transfer to mean the acquisition of equipment and/or knowledge not previously available in the host country. Seres and Haites (2008) analyzed the claims made in 3296 PDDs and found that technology transfer was involved in 36% of the reviewed projects accounting for 59% of the annual emission reductions. They also concluded that technology transfer varied widely across project types, but occurred more often in larger projects. It was also more common for projects that have foreign participants, possibly because those projects tend to be larger. Five countries were the sources of over 70% of the transfer of equipment or knowledge under the CDM: Japan, Germany, the United States, France, and Great Britain (16% of the high tech equipment and 11% of the hired expertise used in CDM projects came from the United States). Greater participation by the United States in the CDM would likely increase the technology transfer of U.S. technologies to developing countries, thereby leading to greater exports and business opportunities for U.S. businesses.

7.5.2 Clean development mechanism and sustainable development

The notion of the sustainable development within the CDM can also be seen from different perspectives and the indicators will necessarily have to relate to the perspective taken. The following are some possible perspectives that can be taken for regarding sustainable development in the CDM:

- *From the global investment perspective*:

  This would include the perspective of both the UNFCCC as well as global investment funders. Their objective would be to maximize both the climate change reduction potential and the sustainable development of a portfolio of CDM investments across the globe. It would, therefore, encompass elements that enable some geographic distribution across the developing countries (to ensure some level of equitable distribution of projects) as well as across technologies and sectors. Experience so far has shown that it is likely that a small number of developing countries (e.g., China, India, South Africa, Brazil, and few Latin American countries) could effectively account for almost all CDM projects if there is no concerted effort to enable other (smaller and poorer) developing countries to access the CDM market. This is explicitly recognized in the UNFCCC negotiating text for the Kyoto Protocol where it recommends an equitable geographical distribution of CDM projects.

- *From the individual project's perspective*:

  Each potential CDM project needs to be assessed on its own merits with respect to sustainable development indicators that will need to apply in a very specific and local context. This includes carrying out appropriate Environmental Impact Assessments and Social Impacts Assessments as required by the national laws of each host country (or by the international funding agencies). One important consideration is the need for transparency and openness in making decision including the involvement of stakeholders and communities in decision-making.

  For specific CDM projects, countries (and project developers) have approached the issue of defining sustainable development criteria in different ways. In a number of cases, attempts have also been made to develop specific sustainability criteria and indicators for CDM projects by different groups (see Annex 1 for an example). Examples of assessing sustainable development benefits of potential CDM projects have been made for India, China, Zimbabwe Columbia, Kenya, Nepal, Sri Lanka, Bangladesh, and Indonesia. These studies have shown that CDM projects indeed can have substantial potential for enhancing sustainable development locally as well as nationally if they are designed with the sustainable development goals as part of their criteria.

- *From a development agency's perspective*:

One of the criteria agreed regarding the CDM is that it should be for projects that represent additional resources to those provided by Organisation for Economic Co-operation and Development (OECD) countries under their official development assistance. The question has, therefore, been raised as to whether development aid should be used to finance CDM projects. This is a difficult question to answer definitively. However, as one of the objectives of the CDM is to promote sustainable development in developing countries, it may be possible for development agencies to fund certain activities related to CDM. These could include capacity building (specially for the LDCs) who may not be able to attract projects from the private sector market on their own, also for identifying potential CDM projects in the LDCs. The World Bank's "CDM Assist" project in Africa is a good example of such development support. In addition, it is possible that a part of the CDM market (albeit only a niche) may put a higher value on the sustainable development elements of certain projects then only on their CERs and price of CERs. Thus it is conceivable that, in some cases, the sustainable development goals of projects may be the *primary* goal for the investor and the CERs the *secondary* goal. There is the evidence already from the experience of the World Bank's Prototype Carbon Fund that investors value projects with clear sustainable development benefits higher than others.

7.5.3 Clean development mechanism and biogas energy promotion

The CDM increases the revenues for biogas energy generation. Most of the biogas energy projects in the CDM pipeline involve electricity-generating technologies. The basic economic barrier is the relatively higher electricity generation costs for renewable energy technology (RET), although the scale of the difference varies from technology to technology and from country to country. For example, in Thailand, costs for electricity from biogas are almost competitive with the average electricity tariff, while wind electricity is about double the average price. The revenues attained from selling CERs from a CDM project can help compensate for this price difference to an extent.

Nevertheless, biogas energy projects do not get as much out of the CDM as other project types. Among the 1700 projects currently at an advanced stage of the CDM project cycle, biomass projects make up the largest share, accounting for 21%, followed by hydropower projects (including large hydro) at 19% and wind energy at 12%. In total, renewable energy projects constitute 59% of the project portfolio.

In terms of CERs, the market is dominated by projects to reduce hydrofluorocarbons (HFCs), N_2O, and CH_4, which in total account for about two-thirds of all expected CERs. This is due to the high global warming potential of these gases, which in the case of HFC-23 is 11,700 times that of CO_2. In fact, a mere 41 HFC, PFC, and N_2O reduction projects account for 40% of all expected CERs from the more than 1700 projects in the project cycle.

Biogas energy projects, thus, receive a disproportionately small financial benefit from the CDM. At current CER prices, the increase in the internal rate of return from the sale of CERs from a CO_2-based renewable energy project is estimated at about 2%. This additional CER revenue can be enough to lift projects across the threshold of being economically viable. However, in some countries, energy subsidies tilt the energy market against RET such that CER revenues are not enough for a single one of the RET CDM projects currently in the pipeline to become profitable. Instead, these projects must rely on further support from official development assistance. The *carbon intensity* of a country's electricity mix is also an important factor. For example, countries like Thailand or Egypt have an average electricity carbon intensity of approximately 500-kg CO_2/MWh. This is much lower than the likes of China (916-kg CO_2/MWh) and India (896-kg CO_2/MWh). The economic outcome under the logic of the CDM is that high-carbon-intensity countries benefit almost double from CERs for each conventional kWh substituted by renewable energy.

7.6 Exercise

1. Explain the environmental assessment of biogas power plant.
2. Explain the economic assessment of biogas power plant.
3. Write the significance of particle swarm optimization to economic assessment of biogas energy system.
4. Write the significance of big bang-big crunch to economic assessment of biogas energy system.
5. Explain the basic concept of game theory in the field of decision-making.
6. Explain basic concept of big data analysis.
7. Explain the impact of feedstock and digestate storage and treatment system with the concern of environment.
8. Explain basic economic problem of biogas energy system.
9. Explain different supply function of biogas power plant.
10. Explain the concept of elasticity of electricity demand from biogas power plant.
11. Suppose the monthly income from biogas power plants is increased from \$ 20,000 to \$ 25,000, which increases his demand for electricity from 20 to 25 kWh, calculate the income elasticity of demand and interpret the result.
12. The electricity quantity demanded from biogas power plant to run a school in rural areas has increased from 120 to 180 units with an increase in the price of the electricity from \$ 40 to \$ 50. Calculate the cross elasticity of demand between electricity and price.

Further reading

Bersa, R.J., 2018. Future trends for big data application in power system. Big Data Application in Power System. Elsevier, Cambridge, MA, pp. 223–242.

Eckroth, J., 2018. A course on big data analysis. J. Parallel Distrib. Comput. 118, 166–176.

Guan, L., Zhang J., 2017. Enhancing security and resilence of bulk power systems by multi-source big data learning. In: IEEE Power and Energy Society General Meeting, Chicago, IL, pp. 1–5.

Hangxun, T., Hongang, W., 2018. Measuring system of power quality by big data analytics. In: IEEE International Conference on Cloud Computing and Big Data Analytics, Chengdu, China, pp. 248–252.

Hejazi, H.A., Rad, H.M., 2018. Power system big data analytics: an assessment of paradigm shift barriers and prospectus. Energy Rep. 4, 91–100.

Huang, J., Niu, L., Zhan, J., 2014. Technical aspects and case study of big data based conditioning monitoring of power apparatuses. In: IEEE PES Asia Pacific Power and Energy Engineering Conference, Hong Kong, China, pp. 1–4.

Junaidi, N., Shaaban, M., 2018. Big data application in electric energy system. In: IEEE International Conference on Computational Approaches in Smart System Design and Applications, pp. 1–4.

Percuku, A., Minkovska, D., Stoyanova, L., 2018. Big data and time series use in short term forecasting in power transmission system. Procedia Comput. Sci. 141, 167–174.

Shyam, R., Bharati, G.H.B., Kumar, S., 2015. Apache spark based big data analytics plate form for smart grid. Procedia Technol. 21, 171–178.

Torrecilla, J.L., Romo, J., 2018. Data learning from big data. Stat. Probab. Lett. 136, 15–19.

Wanxing, S., Keyan, L., 2016. The anomalous data identification study of reactive power optimization system based on big data. In: International Conference on Probabilistic Method Applied to Power System, Beijing, China, pp. 1–5.

Yang, Z., Gao, C., Zhao, M., 2019. Utilizing big data to explore the running opportunity of power to gas in energy system. Energy Procedia. 158, 2341–2347.

Yao, H.W., Wang, X.W., Wu, L.S., 2018. Prediction method for smart meter life based on big data. Procedia Eng. 211, 1111–1114.

Zhan, J., Huang, J., 2014. Study of the key technologies of electric power big data and its application prospects in smart grid. In: IEEE PES Asia Pacific Power and Energy Engineering Conference, Hong Kong, China, pp. 5–8.

Index

Note: Page numbers followed by "*f*" and "*t*" refer to figures and tables, respectively.

C

D

Q

R

S

T

U

V

W

FSC
www.fsc.org
MIX
Papier | Fördert
gute Waldnutzung
FSC® C083411

Kolibri 360 GmbH
Ferdinand-Jühlke-Straße 7
99095 Erfurt, Deutschland
produktsicherheit@kolibri360.de